PARISH of
Glenfarquhar
Strath Fenella Hill
Garrow Wood
FORDON
Pittarrow
Redmyre
Windsford
Laurence Fair
PARISH
of
Halkerton
C O N V E T H P A R I S H
Dirva
Bruff Park
Whitefield
Laurencekirk
Thornton
of
Garvock
Garvock Kirk
G A R V O C K
Johnston
Cairn of Johnston
MARYKIRK
Newton
Hill
Craig
PARISH of BENHOLM
Benholm
Glebe
John haven
Common
Hill of Woodston
Kirktonhall
Glebe
Criggie
Sundown
Canterland
Woodston
Johnshaven
Harbour
Milton of Mathers
PARISH of St CYRUS
Morphie
Kirkside
Logie K.
North Esk River
Newbridge
P A R

FLITTING THE FLAKES

STONEHAVEN the County Town
Harbour
Dunnotar Castle
DUNNOTAR
Cratown Ness
Cratown Harbour
Hilton
Cave of Hilton
Todhead Point
A North View of Dunnotar Castle
Scale of British Miles
Scale of Scotch Chains
Scale of English Chains
Represents the Boundaries of Parishes
The Boundaries of Estates
The Boundaries of Parishes & Estates
A Map of KINCARDINE SHIRE
Drawn from a Survey taken ANNO
MDCCLXXIV

Flitting the Flakes

The Diary of J. Badenach a Stonehaven Farmer 1789 - 1797

EDITED BY

Mowbray Pearson

ABERDEEN UNIVERSITY PRESS
and
NATIONAL MUSEUMS
OF SCOTLAND
Edinburgh
1992

First published 1992
Aberdeen University Press

In Our Own Words Series: A Publication of the European Ethnological Research Centre, National Museums of Scotland, Queen Street, Edinburgh EH2 1JD

British Library Cataloguing in Publication Data

Flitting the Flakes: Analysis and Transcription of the Diary of James Badenach, Esq. of Whiteriggs, 1789 to 1797. – (In Our Own Words Series)
I. Pearson, Mowbray II. Series
630.92

ISBN 1 85752 008 4

Typeset from author generated discs by
Hewer Text Composition Services
Printed by BPCC–Wheatons Ltd, Exeter

CONTENTS

PREFACE

Flitting the Flakes, a more attractive phrase than *Moving the Hurdles,* was chosen as the title for this analysis and transcription of *The Diary of James Badenach Esq of Whiteriggs,* because that recurring chore on the farm seemed to epitomise the very hard, labour-intensive work that was so characteristic of farming in the eighteenth century.

The analysis, by no means comprehensive or exhaustive, is intended to highlight certain aspects of themes that can be recognised within the diary and to examine them on their own merits. However, it must be emphasised that it is only by reading the diary itself that the full flavour of the outdoor life on an eighteenth century farm can really be appreciated. It is then possible to have some sympathy with the author for the hardships, the frustrations, the anxieties, the disappointments, and the occasional successes.

The original form of the writing of the diary has been preserved as far as is possible in print and the spelling, which varies enormously, has been retained. For example the word *women* is spelt *wewmen, wowmen* and, towards the end of the diary, *women.* For the sake of clarity *weather* has been written out in full, although throughout the diary the word is abbreviated to *w*^{r}. There are occasional gaps where figures were not given by James Badenach and where words could not be deciphered. Care has been taken to ensure, as far as possible, that apparent errors are genuine and not typographical.

At the time when the diary was written the estate of Whiteriggs, situated in the Parish of Fordoun, Kincardineshire, comprised *parts of the Barony of Garvock ARTHURS comp Thornylee, Suttie Wells* (Register of Sasines) in addition to the land of Whiteriggs itself as shown on the map of Kincardineshire by William Garden (1776), republished in 1797. On 27th February 1792 James Badenach bought the properties of Easterlair and Westerlair from Mr Smart of Polburn. Thereafter the names of the farms were combined as Waterlair. Parts of both Thornylee and Waterlair were farmed by James Badenach and parts were leased to tenants, as detailed in the accounts in Appendix 1 to the diary. Arthurhouse and Sooty (Sutty, Suttie) Wells were leased.

Before it was acquired by James Badenach in 1788, Whiteriggs was owned and improved by James Leith whose descendants bought the property back in 1807. In 1848 Leithfield House was erected by Mr Arnott of Leithfield, W.S., who succeeded the Leith family, and changed the name of the estate from Whiteriggs to Leithfield. The *mansion house* of Whiteriggs that features in the diary is now incorporated within the buildings of the Mains of Leithfield, shown on some maps as Whiteriggs. The estate is now owned and farmed by Mr George Anderson of Kair House to whom, with his family, I am most grateful for allowing access to the estate and for pointing out features of the farmland.

The diary was amongst the archives of Glenbervie House, Kincardineshire, and I should like to express my thanks to the present owner of that estate, Mrs Jill Macphie, for kindly releasing the document for transcription, analysis and publica tion. I am grateful to Professor Alexander Fenton of the School of Scottish Studies and the European Ethnological Research Centre for his encouragement and advice. I should also like to thank Ray Harris and Norma Henderson of the Department of Geography, University of Edinburgh for the preparation for publication of diagrams and photographic material.

M.G.Pearson
Juniper Green 1991

Part One

Thematic Analysis of the Diary

of – August 1792 –

Friday 17	Almost constant Rain very heavy from the N.E. — One Plough at Haugh Turnips and people at Do thinning — Brought Back from 16 B Oats Kiln Meal 15 B 9 P Meal
Saturday 18th –	Fair Wr from the North but very little drought. Flited the cattle Stakes morning and at Horse & Hand Hoeing of Turnips nearly done for the first dressing — after Breakfast all people at turning the Hay Ross Hillock & after Dinner at Turnips again indifferent weather.
Monday 20th –	Yesterday Coldish dry Wr from the Northward White frost in the night, but day fair and some drought from N.W. — finished first dressing of Turnips Haugh, and after 16 Hands at Coling of Hay all day of which did about 60 Coles in tolerable Order, Some of the Oats & Barley lying.
Tuesday 21st –	a Shower early in the morning — day fair and not unpleasant except little drought — two Carts spread some lime on the the two ridges west side of Thornybed 32 B per acre, second dressing Turnips at Bridge Shade — Sheep Market fair very cheap —
Wedy 22d –	Warm Wr and after Breakfast turned out drought — about thirteen people at the Hay Turning & after Coling — which finished by 6 Pm in good Order — Appearance of Rain —
Thursy 23d –	dry fair Wr most of the day with some Showers — at Laurence fair a Great many Cattle but not much Sale except amongst the best Cattle. Observe none lying Corns to the Market. from that to Boynton Lord Inverury's Meeting —
Friday 24th	Bleak dry Wr from the N.W. till 4 Pm — rain moderate, 2 Carts to the Hill for Peats, and others at 2d dressing of Turnips Bridge Shade
Saturd 25th –	Last night and all this day almost constant rain but not very heavy from S.E. — Sent two Carts to Corbett Hill for Peats, and one Cart to Laurencekirk for 30 feet of lining Stone for new Kitchen, the Oats much lodged and the rankest of Barley, very Bad weather
Monday 27th –	Yesterday Moderate Wr — this morning came on a Heavy rain from the Northward which lasted till past 10 AM — people flited the Stakes before dinner and Pm at Turnips and at Sundries — the Highland reedy Cows
Tuesdy 28th –	Frosty dew in the night & morning, Pulled the Lint tolerable Crop, and after at Sundries — at Lord Inverurie's Shooting Quarters a kind of Fête Champêtre — plenty of Drink — the Crop in that part not behind ours —

DR JAMES BADENACH

In an earlier diary James Badenach, the author of the diary under review, described himself as a *Doctor of Medicine and Member of the Philosophical Society of Edinburgh.* He took possession of the Mains of Balmakewan (Balmaqueen), Kincardineshire, in November 1776 and farmed it until September 1788 when he purchased Whiteriggs, also in Kincardineshire, from Alexander Leith. Alexander's father, James Leith, Sheriff Substitute of Kincardine, had done a great deal to improve Whiteriggs from about 1760 onwards. Balmakewan itself had been improved by a distinguished improver, William Graham of Morphie (Murphy) from 1754. James Badenach's wife was a daughter of William Graham and she might well have influenced him in his farming activities, both at Balmakewan and at Whiteriggs.

Little was said in the diary about James Badenach's family. It transpired that his father lived at Johnston, a neighbouring estate, which he farmed successfully, judging by remarks made by his son about the sale of his cattle at high prices. His occupancy must have been as tenant because, according to the record of sasines at Register House, the property was owned by members of the Garden family at that time.

In a marginal note on 16th June 1790 came the first probable mention of James Badenach's sons when he wrote: *Inoculated two of my Boys at Snaton 10 Miles distance.* No further details were given. On 4th June 1792 his eldest son James required two horses to go *mostly to Aberdeen* for an undisclosed purpose. James was mentioned again on 22nd November 1793 when there was *a Cart to Aberdeen with my Oldest son's Apparel for the East Indies.*

His second son, William, was provided with two horses to take him to Aberdeen *for the Navy* on 25th July 1795.

When Whiteriggs was sold back to the Leith family in 1806 the names of the children involved in the transaction, in addition to Ann Graham, his widow, were Walter, Robert and Catherine. None of them was named in the diary.

James Badenach took a keen interest in his neighbours' crops as well as observing, as he travelled on horseback, how farmers were faring further afield and what was the state of the countryside.

On Monday, 29th August 1791 he wrote of a journey that he had made on the previous day to see *Lord Adam Gordon's new Erecting Seat at the Gannochy Bridge, N Esk,* namely The Burn. He found there *fine Thriving Plantations of wood.* On his way home he dined with *Sir A.Ramsay* at Fasque.

On 18th June 1792, a day of very heavy rain, he was *Compleatly Drenched* during a visit to Stonehaven *on a Corn Jury Bill.* On 28th August of the same year he was at Bogton, *Lord Inverurie's Shooting Quarters* where there was *a kind of Fete Champetre – plenty of Drink* and his handwriting at that point was less regular than usual. At St Cyrus on 29th December 1792 he attended the burial of a Mr Graham, presumably his father-in-law.

In April 1793 he was at Aberdeen on Assize business and in the same month he attended the annual meeting of the County when it was agreed to *Carry on the Business of a Bill for Turnpikes without delay.* Those meetings were held annually on 30th April and at the meeting in 1794 there was *an Appearance of Turnpikes being brought to Effect,* but James Badenach suspected it would not *be in a Hurry owing to the Money Matters.* At the meeting in 1796 it was *an Uncommon large Meeting of the County about the New Turnpike.* Apparently some progress was made, because, on 22nd June, he *Rode some Miles on the New formed Turnpike road benorth Stonehaven* which he thought would be *pleasant Instead of the Dreary present One.* On 23rd June 1797 he rode on the *New Turnpike road from Muchalls to Stonehaven* and two miles on the Whiteriggs side of it – *fine road if finished.* James Badenach was never away from home for more than two or three days at a time and such expeditions were often to Aberdeen. On 24th January 1796 he was able to ride from Aberdeen to Whiteriggs *in four Hours with no difficulty.* At Aberdeen on 22nd June 1797 he painted a dismal picture when he *Visited the Camp of about 1500 Men, beEast Don Mouth, Cold Quarters for Men & Horses, some latter dead.*

Hares caused considerable damage on the estate and James Badenach took part in hunting them as reported in January 1790.

Salmon netting on the river Bervie was mentioned in August 1793. It was new to him and, after a little practice, he achieved some success.

Reference to operations on animals on the farm was evidence that his medical training was put to good use.

On 7th November 1797 a note in the margin stated simply *self Indisposed.* The last entry in the diary was on Saturday, 25th November 1797.

He must have died shortly afterwards because his son took over the estate on 30th January 1801. It was sad to discover that the farm was neglected during the years between James Badenach's death and November 1807 when it was sold back to the Leith family. George Robertson in his *General View of the Agriculture of Kincardineshire* published in 1813 described how Whiteriggs had been for a time out of the hands of Mr Leith's family. After it had been repurchased and was under the management of the Misses Leith, in the absence of their brother, *it was quickly returning to the same productive state in which it had been in Mr Leith's time. Luxuriant crops of wheat were grown where windlestraws hardly were produced.*

When George Robertson made his review he had apparently found no trace of the good work that James Badenach had done at Whiteriggs during his nine years in residence. The only mention of a Badenach in his book was of James Badenach Gordon, Esq. of Arthurhouse, Thornylee and Waterlair. Those 800 acres, which had formed part of Whiteriggs during James Badenach's possession of the estate, were described as having a *northern exposure and steep declination as the lands of Davo* and were *highly susceptible of improvement.* He was not aware that James Badenach had been responsible for *the very thriving plantation of from 20 to 30 acres of wood* on the estate, as he obviously had never heard of him. In appropriate sections of this book are to be found details of the planting of that woodland and the improvements which James Badenach made to that part of his property, as well as to the estate in general.

WEATHER

Details of the weather were given every day throughout the Diary. Actual temperatures in degrees Fahrenheit appeared occasionally, but not consistently enough to build up any quantitative analysis. However, in this review of the weather, some significant temperatures have been quoted. James Badenach realised the importance of siting the thermometer in the shade, but he seldom said where it was placed. He usually specified at what time of day the reading was made.

In June 1793 a *Pluviometer* was acquired and monthly records of rainfall were kept. Amounts of snow were estimated by their depth on the ground and James Badenach seemed to be well aware of the difficulties of making such assessments with accuracy. The most recent average annual rainfall figures available for the area are 30 inches (762 mm.), much the same as those that can be derived from James Badenach's figures.

Relative humidity was not measured, but there was a great awareness of the amount of *drought* within the atmosphere and constant reference was made to its presence or absence.

Occasional readings were made from a mercury barometer. There was seldom a day on which the direction of the wind was not mentioned and frequently changes during the day were noted. In addition there was usually some comment about the general nature of the weather.

The principal features of the seasons throughout the nine years from June 1789 to November 1797 were studied with particular reference to their effects on the activities on the farm. Outstanding features included the rainfall during the harvest of 1792, what were perceived by James Badenach to be such high temperatures during the summer of 1794, the spectacular snowfalls and heavy rain in 1795, and the very heavy rainfall in every month from July to November 1797.

1789

The first summer (1789) was pleasant with a mixture of dry and wet weather with temperatures in both June and August recorded in the 60s. On 18th June, 68° at 1 p.m. was described as *very warm;* between 15th and 18th August with temperatures of 66° or 67° it was *uncommonly fine weather – very warm.* At the end of August there was enough rain to cause some flooding *injuring some of the pasture & a very small portion of the Turnips.* That heavy rain from the north east continued into the first week of September and *brought down the rivers very high, as much as I have seen; obliged to swim the Cattle from the Haugh morning to the upper Grounds to pasture all the day.* A *Sowe* of hay was so greatly damaged on the upper part that much of it had to be taken down and spread out on the field to dry.

After the *sowe* had been repaired with the help of all the shearers as well as the other farm workers, the weather improved sufficiently for the harvest to be started on 8th September.

Heavy rain from the north east, described as almost constant on 9th and 10th October, *wett the fields & roads as much as I ever Remember of.* An east wind on 15th

October brought more severe flooding. Frost was mentioned on three nights at the end of October and on the 31st *snow & High Gale continued all night and this day at Intervals till 2 p.m., the snow lying several Inches deep, and on the Hills and rising grounds very Considerably deep. – which must be of Immense dammage to the uncut Corns as well as to those not yet got in. This weather is uncommonly Inconvenient to me who hath so many Houses new & old in no wise prepared for so early winter.*

The snow did not last and in spite of rain of varying intensity recorded on 14 days in November all the grain was safely gathered in by the end of the month.

December 1789 began with mild, dry weather from the south west followed by a characteristic mixture of frost and fresh, with snow on two days.

1790

Frost at the beginning of January 1790 was hard enough to stop ploughing, but fine fresh weather returned from the southward on 6th January and continued for the next three weeks. On 21st January there was *very slight frost in the night, day Extreamly pleasant and even warm weather.* A south-westerly gale with sleet on the low grounds and snow on the hills came at the end of the month, but the winter so far had been open enough to enable the *Labouring* to be *very far advanced.*

The south-westerly weather returned at the beginning of February and lasted the whole month, giving on the 5th *fine dry & spring weather* and on the 24th *fine spring weather from S.W.*

The *remarkable fine & constant good weather* continued into March, but by the middle of the month, as a result of the colder nights and the continuing drought, the *vegetation* was *quite Chequed & Rye grass turning Black.* On 31st March there was *Bleak weather from N.E., and little or no growth in the ground for weeks by past.*

The cold weather continued into April 1790 with some snow and hail from the north east, particularly on 10th April when there was a *Considerable fall . . . which lay till near noon.* A shift of the wind to the south east brought milder air and showers of rain; on 23rd April . . . *vegetation* was *appearing to Recover fast.* For the most part May was warm with an adequate amount of moisture until the 24th when cool, dry weather from the north east emphasised the fact that this had not been a good season for rapid growth.

June began pleasantly enough, but until the 17th rain was recorded on no more than four days and then only in the form of light showers. At last the drought was broken when, on the morning of 18th June and on the following afternoon, heavy rain prevented work out of doors.

The rain which fell on nine days in July was heavy with some thunder on 10th July. Towards the end of the month oats and barley were laid and had to be raised. The weather at the beginning of August was dry enough for the haymaking to be completed with no delay. Thereafter very heavy rain on six days with a *very strong Gale of wind from the S.W.* on 24th August *wet the grounds much and Layed the crop very much.* Northerly winds brought a cold, wet start to the autumn, but on 10th September the cutting of barley was begun. After a week of welcome dry weather, there followed through September and into October changeable weather with cold nights and as many wet as dry days. Stronger winds and heavier rain were experienced towards the middle of October and on the night of 22nd October there *came on very heavy rain from Eastward which Brought down the rivers very high by noon full as high as for a year by past – went over the Dykes into the old water Track amongst the stooks therein.*

Rain continued off and on with the result that on 2nd November *the Rivers* were *full as high as for 18 Months by past & fields very wet.* On 3rd November there was a *Great deal of rain from N.E.* and on 6th November *Early this morning Came on a very Severe Gale of Wind from S.E. attended most of the day with Excessive Heavy rain & stormy – that no person Could work without doors, rivers very high & no Building or windows Could keep out the rain. Miserable weather.* Few entries in the Diary were made on a Sunday, but on 7th November there was the following remark: *last night or this morning the River Bervie full as high as for Eight years by past, but Except some of the Banks Broke a little, no other Dammage done by the Flood – N.Esk very high.* There followed a mild and dry spell, but it was so gloomy that on 15th November James Badenach wrote: *this day frosty but sunshine for first time this 10 days.* The frost continued for several days so that ploughing on most farms was stopped, but at Whiteriggs ploughing on Kavels was continued. There was a mixture of frost and fresh with rain on seven of the remaining fifteen days of the month.

Winter began appropriately on 1st December with *a Considerable fall of Snow & Sleet from S.E. which lay all day,* but it was succeeded on the following day by rain from the north east. The frost was hard between 17th and 20th December when the turnips had to be dug out of the ground for the first time that season and *the River* was *Closed in many places.* Ploughing was stopped again by frost on 28th and 29th December; on the 30th with *fresh weather from Southward,* there were *three ploughs at Haugh on Pease land for Oats & Grass seeds – in general Land would not Labour for Frost & wetness.* Indeed very strong frost on the following day stopped ploughing *though we tryed it.*

1791

With rain and strong winds recorded on ten out of the first eighteen days of January 1791, it was not surprising that on 19th January the river Bervie was *Brought out very high.* There was less precipitation during the remainder of the month, but *thick snow on the Hills & Braes* was observed on the 21st. The strong winds continued until 31st January when there came *Yesterday P.M. and night one of the most severe Gales of wind from WNW for Years by past.*

Cold, stormy weather continued into the first week of February, but by 6th February there was *very pleasant Spring weather . . . from Southward.* For the rest of the month there was a mixture of frost and fresh. In general the mildness of the winter was confirmed by the entry on 25th February; *finished all tillage for Oat seed the earliest I ever had before.* Spring came early enough for the sowing of oats and pease to begin on 6th March. A south-westerly gale on 15th March prevented sowing on that day, but thereafter work went ahead and on 24th March came *The Best Spring day of this season from S.W.*

It was on 4th April 1791 that James Badenach observed at Johnston, where his father lived, *the Eclipse of the Sun very distinctly from 1 P.M. to 3 D° Apparently about*

$\frac{2}{3}^{ds}$, *but little sensible diminution of daylight.* The generally fine weather continued and on 20th April there was *Rhime in the night but day very pleasant dry and warmer to the senses at least than for two summers by past.* The month ended with rather cold, bleak weather from the north east.

The cold north-easterly wind continued into May bringing snow on 6th May followed by warmer weather from the south. On 17th May *Cold Stormy weather from Westward with showers of Hail and rain at Intervals* was succeeded by *a very strong Gale from N.W. Braking down many of the Garden stuffs.* Changeable weather continued for the remainder of the month with white frost on the mornings of 24th, 25th and 26th May.

The first ten days of June were warm and dry, but on Monday, 13th June came the climax to a cold snap at the week-end: *Yesterday and this a very severe & Cruel gale of Northerly wind Blasting every kind of vegetable growth.* This northerly airstream with a *considerable fall of rain* on the 15th was continued on 17th and 18th June *Blasting everything.* The rest of the month was equally unsettled with strong winds, mostly from the south west, and rain in varying quantities recorded on nine of the remaining twelve days.

For the first fortnight of July dry weather from the south west was followed by dry, cold weather from the north west; the cold and drought was said to be *Blasting every kind of vegetation.* Light winds from the south east brought warmer weather, but rain was recorded on only seven days from 1st July until 29th July when *This morning a little before 6 o'Clock Came on a pretty strong Gale & Heavy rain from S.E.*

Warm, dry weather continued throughout the month of August. Rain was mentioned on eleven days, mostly as showers apart from 17th August when there was *in the night a Considerable deal of rain from N.E.* As a result of those favourable conditions the harvest was begun on 29th August, earlier than in any of the subsequent years.

Good progress was made with the harvest during September; rain, mostly in the form of light showers, was reported on only five days.

James Badenach correctly anticipated a change in the weather when he wrote on 10th October: *Yesterday uncommonly pleasant & warm weather but the Barometer for days preceeding showed a Change of weather which accordingly came to pass – last night about 11 Came on rain & stormy weather from North, which Brought down the Rivers much & overflowed a good deal of the low Grounds . . .* The very wet and stormy weather which followed continued for the next fortnight from east and north east, filling the rivers and penetrating the houses. There was some snow on the hills on 24th October.

Fair, pleasant weather at the end of October continued into November, which turned out to be changeable with temperatures near normal for the time of year and precipitation, mainly in the form of rain, falling on twelve days.

On 9th December *Cold stormy weather from North W* was succeeded by a fall of snow and *very hard frost.* Part of the entry on Monday, 12th December read: *Yesterday Excessive Hard frost Thr at 11 A.M. open Air at 21 degrees – this day Came on a little after 10 A.M. a most severe storm of snow & drift with a strong gale of wind from the Southward.* In general it was a cold, dry start to the winter with frost of varying intensity mentioned on 17 days in December and precipitation on seven days, four of snow and three of rain.

1792

In January there was a slight predominance of frost over fresh weather with 13 days of the latter. The only recorded temperature, and probably the lowest of the month, was *22 degrees at 11 A.M.* on 12th January. According to the gardener the frost penetrated to a depth of 15 inches. There were four days of snow and six of rain, but amounts were small until 31st January when there was *a very Considerable deal of rain.*

The changeable weather was continued into February, but this time there was more fresh than frost. The cold spell was principally between the 17th and 27th with two to three inches of snow on 21st February.

Heavy rain at the beginning of March was followed by snow and drift from the east and north east, giving a pattern of stormy weather for the whole month. April was a predominantly dry month with showers reported on nine days; there was a warm spell about 11th April followed on the 13th by *rather Chilly weather & Hail from N.E.*

May began with *Cold stormy weather from N.W. with flirts of snow* and by 12th May there was *strong frost & some snow on the hills & even on the Laighs.* The remainder of the month was less cold, but inclined to be wet and stormy.

There was a continuance of the cold, stormy weather during the first week of June. On 14th June it was *rather frosty in the night* and on the following day a temperature of 50° was recorded, probably at noon when readings were usually made. Towards the end of the month there was some pleasant warm and dry weather from the south west.

The rain in July came on nine days in sufficient quantity to promote growth. Temperatures varied and it was not until the end of the month that there was *very pleasant warm weather from the Southward.*

On 1st August the temperature reached *71° perfect shade.* By 6th August the weather was still described as *very warm from the Southward,* but the temperature had dropped to *62° shade noon.* After climbing again to *68° at noon shade* on 13th August, the weather broke on the 16th with the first rainfall of the month, apart from one small shower on the 10th. Thereafter the weather was changeable with five wet days and variable temperatures.

A wet start to September was followed by a brief dry spell enabling harvest to begin on the 10th. Thereafter white frost was recorded on seven mornings with black and white frost *which hath Destroyed the Potatoes stems* on 17th September. The remainder of the month was so unsettled that *From all the Accounts from the different parts of the Country South & North, the Crop* was *in a very Indifferent Condition, much dammaged by the Constant rains, lodging, frosts, and every other Mischief.* A week's respite from the rain at the beginning of October was followed on the 9th by *considerable rain last night & this day at Intervals, Constant rain from the Eastward* on the 10th, and so on for the next four days. After two days of fair, moderate weather from the south west, strong winds and rain returned from the northward culminating on 20th October with *Heavy rain & high wind from the N.E. which deluged the fields & Brought down the Waters very high.* Heavy rain on seven days accompanied the stormy weather on the remaining eleven days of October.

The whole of the crop, *the last in very Bad order,* was led in on 3rd November. Strong south west winds brought dry weather for the first ten days of the month.

Gales from different directions were reported on five days between 10th and 17th November. Rain was mentioned on 10 days during the month, heavy enough to cause flooding on 14th, 21st and 22nd November.

December brought some appearance of winter with snowfall on twelve days. On 13th December there was a *pretty Constant fall of snow all day through. Evening about 5 Inches depth.* The snow never lay for long and rain fell on nine days, leaving only ten dry days in December. This was yet another month of strong winds and gales mentioned on eight separate occasions.

1793

January 1793 began with the stormy mixture as before; alternating frost and fresh was accompanied as appropriate by snow and rain. The greatest depth of snow was *four or five Inches & rather soft* brought on 5th January by *a furious Gale.* By contrast on 13th January it was *very rainy & stormy weather from N.E. but fresh.* Winds were lighter during the second half of the month, but fresh weather again alternated with frost.There was no precipitation after 11th January with the result that on the 28th, the *Ground* was *dryer than for some time past.*

The dry spell was broken on 1st February. During the month there was precipitation on ten days, four with snow which did not last. The frost that was mentioned on six days was of no great intensity; stormy weather was noted on seven occasions.

March came in like a lion and the weather remained unsettled. Cold, dry weather was succeeded on 11th March by *a Considerable fall of snow & very stormy from the S.E.* More snow followed, but there was then a week of warmer, dry weather until 21st and 22nd March when there were strong winds from the south west and enough rain to flood the rivers and *stop all work without doors.* A north-easterly airstream for the remainder of the month brought an alternation of warm and cold dry weather, except on the 28th when the wind backed and brought *Snow pretty Heavy from S.E. and towards evening pretty deep.* Spring so far had been *an Exceeding Backward Season, and every kind of vegetation uncommonly Backward.*

After four wet days at the beginning of April, the weather remained dry and rather cold; on 14th April there was *very Cold & Bleak dry weather from the Northward* with *very strong frost* in the morning followed by *very Heavy snow & high wind from the N.W. which Covered the Country quite white & lay till about noon.* For the rest of the month cold, dry weather predominated. There was no more snow and rain fell on only three days towards the end of the month.

In May there was little change until the 7th when *very pleasant dry and warm weather from the Southward* gave some hope of an improvement. Less warm but *fine dry weather from the Northward* took over a week later and lasted until 24th May when there was a white frost which *Blasted the Ash leaves.* From the 29th to the end of the month the weather was very stormy and very cold from the northward and on 31st May *a very strong frost Black & White Blasted Potatoes shaws and many of the leaves of the Forrest Trees. Observed vegetation much Checkqued & some Bear yellowed.*

At the beginning of June warmer weather with a moderate amount of rain was succeeded once more by cold, dry weather from the north. On 15th June there was at last a *Good heavy rain about Half an Inch in the Water Gage Garden.*

That was the first mention of a rain gauge, the Pluviometer, which recorded a total for the month of *2 Inches & four Tenths of an Inch* (61.0 mm).

The cold remained and the unseasonable nature of the weather was demonstrated by *a perfect storm from the northward Blacking every vegetable* on 19th June. It was noticed, too, that towards Aberdeen *everything* was *very Blasted & Cattle* were *starving. The Inclemency of weather* was blamed for the fact that *several of our people* were *Confined.* On 22nd June *the sick servants* were *recovering of a kind of Scarlet Fever.* On the 24th there was still *no Growth in Grass & but little Vegetation.* At the end of the month prospects for hay, pasture and corn crops were not good.

After a disheartening start to July with *fair & dry weather but little growth from all ye rain Owing to the Impetous winds,* there was a gradual improvement producing on 10th July *Fine dry weather & Extraordinary warm Thr shade 72° at 11 A.M.* The remainder of the month was variable in temperature with rain on seven days giving a total of only 1 inch (25.4 mm) and making the *Grounds Excessive dry.*

The first ten days of August produced *pleasant summer weather* although the only temperature given was no more than 65°F. During the remainder of the month, between showers and heavy rain amounting to four inches (101.6 mm), haymaking was carried on with fair success. The crop otherwise was observed to be *Extreamly Green & Backward over the whole Country.*

An *Eclipse of the Sun 10 deg* was reported on 5th September. Rain and stormy weather prevented the completion of the building and thatching of the hay *Sowes* until the 12th. Pleasant weather, interrupted by heavy rain on four occasions between 13th and 24th September, was followed by a dry spell culminating with the start of the harvest on 30th September. The rainfall for the month was 2.8 inches (71.1 mm).

After a strong gale and heavy rain from the southward on 1st October there was a long dry spell which lasted until the 22nd when there was *Thick fogg & Small drisling rain.* On three of the remaining days of the month small quantities of rain were reported, but the total for October was no more than a quarter of an inch (6.3 mm). A marked change of temperature was recorded on 31st October when there was *Very Strong Frost, Ice near Half Inch Thick.*

A week of cold weather at the beginning of November was followed by strong wind and rain, often very heavy, on twelve days between 8th and 22nd November. After the rain, warm, damp weather caused heating in the stacks to such an extent that *many in the Country are said to be rotting.* The total rainfall for the month was 4.5 inches (114.3 mm).

A spell of pleasant dry weather at the beginning of December made it possible on the 7th to lead home the *Out Barley* which by that time was *Only fitt for the Horses use.* Within a week there was more heavy rain which on 13th December *brought down the rivers as high as for five years by past.* Unsettled weather for the remainder of December produced a total of 4.25 inches (107.9 mm), but it was not noticeably cold until 31st December when there was *pretty strong Frost & all the Hills & Braes* were *Covered deep with snow.*

1794

After two days of frost at the beginning of January 1794 the weather became fresh from the westward and continued so until the 23rd when there was *very strong Gale & High Wind from Westward* with some snow. By 27th January more snow and strong wind from the north west *filled up the roads & Hollows to many feet and rendered traveling of Carriages and also Horses Impracticable almost.* Early on the morning of 31st January there *Came on a Strong Thaw with Showers of rain from the Southward which Carried most of the Snow Except in Blown Places.* The rainfall in January was computed as *four Tenths of an Inch* (10.2 mm), but that almost certainly did not take account of the snowfall, because James Badenach wrote on 28th February 1795, with reference to the gauge, that *it doth not appear that it can be measured of snow.*

On 1st February the ice on the river broke. Frost was mentioned again on the 15th, but it did not amount to much. A month of unsettled weather produced three inches (76.2 mm) of rain and there was *a Considerable vegetation on the Fruit & other Trees* were *well Budded & some of the Early Flowers Blown.*

March was predominantly dry, calm and not too cold, but it was not until the 18th that sowing was begun. Thereafter good progress was made with few interruptions. The *Quantity of rain* was *1 Inch 3 tenths* (33.0 mm).

In April the winds strengthened from the northward bringing cold, unsettled weather. On the 8th three inches of rain fell in 42 hours. The Haugh was flooded, soil and seed were washed away and part of the area had to be resown. The remainder of the month was dominated by winds from the southward bringing drier and warmer weather. Most of the *four Inches and One Half* (114.3 mm) of rain fell during the first week; after 8th April small amounts were mentioned on only six days.

There was a cold, dry start to the month of May giving on the 12th a *very White frost & Injuring the Grass & tender vegetables.* On 15th May it was warmer from the south, but cold, stormy weather returned five days later, reaching a climax on 28th May when *very heavy rain (1 inch rain)* fell. That accounted for most of the total of $1\frac{1}{2}$ inches (38.1 mm) for the month.

The last three days of May were warmer and on 1st June the beginning of summer was heralded with a recorded temperature of *67° Shade at Noon.* In spite of that inch of rain at the end of May, more was needed. The drought continued and the weather was described as *very warm,* although on 16th June the reading on the *Thermometer in the shade at Noon* was only 59°. On 23rd June *a tolerable shower of rain which went about one Inch only into the Ground* and some *showers with Thunder about noon from S.W.* on the following day relieved the situation temporarily, but gave a total for the month of no more than 0.75 inches (19.1 mm). The warm weather continued and was described on 29th June as *Excessive Hot Weather-Breeze S.E.-Therm^r^ 10 A.M. and at Noon 70 degrees which is as high as I ever Observed it in North Britain.*

On 1st July the temperature had fallen to 64°, but the weather was still described as *very warm.* On 2nd July it was 71° and remained in the 70s for six days, reaching a maximum of 73° on 8th July. At 60° it was cooler on 10th July, but rose again to 66° on the 12th when there *never was more Occasion for RAIN.*

On 16th July *Uncommonly Hot weather (Th^r^ 68°) and drought Scarcely Observed so Severe before* was followed on 19th July by *Uncommonly warm weather from South – Therm^r^ in shade noon 76 degrees – much higher than I ever saw in this Country before.* On the night of 20th July the rain came and on the following day there was thunder to the northward. More rain on 24th July helped to produce a total by the end of the month of 1.60 inches (40.6 mm); the last week was cool and dry.

The weather turned cold at the beginning of August with heavy rain and a strong gale from the north west. However, apart from another spell of wet weather between 23rd and 26th August the month was dry and pleasant, giving a total of *Two Inches & One Half* (63.5 mm) of rain. Thatching the hay stacks was finished on 16th August; it was observed that the potatoes were *Scabbed or Spoilt on the surface owing to the rains.*

Harvest began on 1st September and good progress was made until the 16th when heavy rain and strong winds, with little respite until 27th September, caused damage and difficulty. The month ended with dry weather from the north and frost on 29th September *which totaly destroyed the Potatoe Stems.* The rainfall for the month amounted to 3 inches (76.2 mm).

In spite of some rain during the first week of October, the whole crop was brought in and *Thatching stacks* was *all done but one* on 8th October. The remainder of the month was unsettled with again a total of 3 inches (76.2 mm) of rain. Although frost was mentioned on six mornings, temperatures were such that on 31st October *The Pasture Grass* was *as good here as during the Season.*

Frost at the beginning of November was hard enough to make it *stiff for the ploughs.* There was no more frost until 27th November with the result that *the Open Weather Considerably Saved the Scanty Crop of Fodder.* Rainfall amounted to 2.75 inches (69.8 mm).

It was wet enough at the beginning of December to provoke the remark on the 10th that there was *Fair weather all this day for a wonder.* The remainder of the month was dry with frost mentioned on six days; snow on 25th and 26th December was *about 2 Inches depth average.* The frost on the last two days of December was *keen* and *strong. The Quantity of rain this month is 2 Inches* (50.8 mm).

1795

Very hard frost continued into January 1795 with a respite of fresh weather from 5th to 9th January. Thereafter winter took over in earnest, with more frost and northerly winds bringing showers of hail and snow. On 18th January there was the first considerable fall of snow which lay on level ground to a depth of 5 or 6 inches. By the 20th the snow was *now above 8 Inches average.* On 21st there was *a Considerable additional snow, by the best Medium I could take the Depth is about 10 Inches at a Medium average, although the People call it Knee deep.* The temperature on 22nd January was *lower than I ever Observed it this 20 years. The Thermometer Fahrenheit at Noon Exposed on a North Wall no wind or snow was as low as 13 degrees or 19 degrees below Freezing.*

The cold weather continued until the end of the month. Strong winds caused additional drifting and although there was only one further blizzard, it was hardly surprising that the following comment appeared at the end of the month: *The Country is Covered with a Greater load of Snow than I have Observed for many years by past – but by Blowing Cannot be Ascertained.*

At the beginning of February *4 or 5 Inches additional depth* fell on the 4th and on the following day *I do think SNOW average 18 Inches and deeper than for 20 years by* *past.*On 8th February three days of thaw were followed by more wind and snow. On 13th February, with the thermometer at 28°, the weather was *Clear & fair all day – The Usual Adage I Hope will be falsified – Old Candling.*That day was Candlemas Day "Old Style" instead of 2nd February "New Style". The adage to which reference was made went as follows:

If Candlemas Day be dry and fair,
The half o' winter's tae come an' mair;
If Candlemas Day be wet and foul,
The half o' winter was gane at Youl.

On 14th February *the POST Hath been Stopped from the Southward since Tuesday last* (10th February) *and the Mails Obliged to be Carried by Men.* During the remainder of the month there were one or two days with an appearance of fresh weather and on 24th February the road was opened all the way to Aberdeen, but on the 27th *a very stormy night with a Considerable deal of Snow & Blowing . . . rendered the roads that were Cutt little better than before and almost Impassable.* The final comment for the month read: *The Storm is universally allowed to be the most Severe in the memory of man. No 4 Wheel Carriages Have been able to travel for 5 weeks by past & no Appearance when so.* Precipitation was not measured for the reasons already given.

By the end of the first week of March much of the snow had melted and the ice on the river was breaking up. Frost and snow returned on 12th March and it was not until 21st March that there was *a tolerable thaw of snow* and on the 24th *the Snow mostly gone on the Fields.* On the afternoon of 25th March it was possible to plough *for the first time for above Nine weeks.* The quantity of rain for the month was given as one inch (25.4 mm.).

During the first fortnight of April there was so much rain that James Badenach measured the amount in the gauge on 14th April as $3\frac{1}{2}$ inches (88.9 mm). Some cultivation was possible at intervals during the month, but sowing did not begin until 24th April. Rain occurred on eighteen days during April and the total for the month was 5.5 inches (139.7 mm.) – *very great in seed season.*

There was a promise of warmer and drier weather during the first week of May, but on the 8th cold, stormy weather from the north was followed by snow showers, rain and, on 15th May, *almost Constant Storm of Snow & Wind from the N.E.* Warm, dry weather from the south west followed for a week, but cold, north-easterly winds with gusts of *snow & hail* returned on 26th May. On 31st May it was remarked that *the Grass Pasture & for Hay* was *the most Backwards for years – most of that for Hay eat up for want of pasture, and the Country in great strait of Beastials Meat – The Fruit Trees not fully Blossomed nor Ash or Elm Foliated – as Backward nearly as 1782. Quantity of rain 1 & $\frac{1}{2}$ Inch* (38.1 mm.).

After very heavy rain, *near $\frac{1}{2}$ inch in Gauge,* on 2nd June, there came three days later *remarkable fine growing warm weather from the S by E* with a temperature of 62°F. This improvement did not last and the remainder of the month was cold and rather damp. Temperatures were recorded on three other occasions, the highest being 55° on 15th June. Rainfall amounted to 1.75 inches (44.4 mm.) and at the end of June there was still the suggestion that *Everything* was *as Backward nearly as 1782.*

The eight temperatures that were given in July ranged from 57° to 63°, a measure of the continuing cold of that summer, although 60° on 7th July was described as *very warm* as also was 63° on 26th July.The northerly winds were dry as well as cold and growth was checked. The weather was warmer towards the end of the month, but with a total of only 0.75 of an inch (19 mm.) of rain for the whole month, the *Bear Crop looked upon Generally as thin & Scanty* and *Turnips Scarcely ready yet for Thinning.* The grass appeared to be withering.

There was plenty of rain during the first fortnight of August with temperatures at noon just above 60°; a maximum of 71° recorded on 12th August. Haymaking was carried on with difficulty until the 18th; thereafter warm, dry weather ensured that all the hay was finished on 4th September. The *Quantity of rain in August* was *three Inches* (76.2 mm.) – *which fell Chiefly the first 12 days.*

After a spell of dry, warm weather at the beginning of September, it was possible to begin harvest on the 17th when the recorded temperature was 58°. Cooler weather from the north west was observed on 26th September and the dampness inhibited leading. The comment at the end of the month showed how the crops had been influenced by the adverse weather during the year: *The Oats few of them ripe & very green as yet in many places – The Wheat, Pease & Beans & the Bear will not turn out good either in Quantity or Quality – The Bear not whole meal and in the South it is said to be the same. Quantity of rain this month One Inch* (25.4 mm.) *mostly at the beginning.*

The harvest was continued with difficulty during the unsettled weather at the beginning of October. There came a depressing event on the 10th when *the rain & storm brought down the rivers as high as for some years by past & alongst with it some grain from the upper farms on this river* (Bervie). More rain followed, preventing shearing and deluging the fields. On 15th October James Badenach *Observed the Oats Growing that were Cut about a week ago & others uncut sadly smashed.* On the following day he gave a dismal review of the situation and remarked that *rain in the Water Gage is 5 Inches* (127 mm.) *of rain already.*

On 19th October another unusual effect produced by the inclement weather was observed. The oats that had not been cut were growing and this was *a very rare Occurrence though not near the Ground.* By making the best of the occasional dry spells during the rest of October, the bulk of the crop was led home by the end of a month in spite of rainfall amounting to 7.75 inches (196.8 mm.).

At the beginning of November cold north-easterly winds brought *Strong Frost & Considerable snow night & first half of the day* (3rd November) *which lay on the Braes & Hills all day & sorryful to see it Amongst the Growing Grain.* In the neighbourhood the harvest on many farms was far from finished and many were observed to be *leading off the Hook.*The weather between 5th and 17th November was unsettled, but there were some pleasant spells. Very heavy snowfall returned on 18th November blocking the road to Stonehaven *very Early Indeed, and suppose never known to be so early by any person living.* The snow was mostly gone from the low grounds by 23rd November, but it was still lying on the rising ground towards the north. Two days later more snow was followed by several days of strong frost, but the month ended with rain and fresh weather. *The Quantity of rain a little Uncertain Owing to the Water being Frozen, but Believes it about Two Inches & one Quarter* (69.8 mm.).

Unsettled weather at the beginning of December was followed by very heavy rainfall from 11th to 15th December. This time *the river was Higher than for a Dozen years by past & fields deluged with water* and *the Rain in the Gauge in this month already 4 Inches* (101.6 mm.). *Observe most Farmers at Ley Ploughing, stubble Ground being too wett.* Thereafter the rain was more intermittent, but the remainder of the month was as unsettled as the beginning. *The Quantity of rain this month five Inches* (127 mm.).

The open nature of the winter so far was evidenced by the fact that *the Fields* were *now very Green & tolerably dry but labouring in general never more Backward.* The crops, too, were turning out badly; *the Oats want one Third or more in the Yield, and as much in the Meal.*

1796

The weather during January 1796 was variable, but predominantly mild. On the 17th it was *warm & Milder weather than in June last* and on 20th January *Fine fresh weather from S.W. and free of all Frosts for Many weeks by past and the Fields as Green as in April.* Rain was well distibuted giving a total of 3 inches (76.2 mm.) for the month.

Although the first fortnight of February was cold and wet, by the 16th the *Tillage was well advanced although in wett Condition.* On 24th February it was a *Beautiful day, free of Frost,* but the temperature was only *44° Noon shade.* Snow was mentioned three times, lying on the Hills & Braes on 29th February, but not on the low ground. The rainfall amounted to 3.4 inches (86.4 mm.).

Snow and frost continued into March and by the 7th the ground was too hard to plough. There was less frost on the following day and ploughing was soon resumed. On 16th March the temperature at noon had reached 46°. The sowing of pease for the season on 21st March and of oats two days later was followed by cold weather with *a Considerable deal of snow* which did not, however, interfere with harrowing. Very hard frost on 25th March *Frosted some Potatoes unluckily exposed to it.* The month ended with *fine dry weather from N.E. Quantity of rain this month 1.4 Inches* (35.6 mm.).

The weather during April was much more settled than it had been for a long time. A number of temperature readings was given, the highest 58° on 20th April and the lowest 45° on the 11th. The quantity of rain for the month was *Nine Tenths of an Inch* and by the 27th *the Ground* was *very hard.*

An unpromising start to the month of May was a temperature of no more than 40° on the 6th accompanied on the previous night by *Heavy SNOW which by morning was three or four Inches Depth Over all yet by afternoon was mostly gone.* A strong frost on the morning of 9th May *Blasted Some Early Potatoes.* A spell of cool weather with noon temperatures of only 46° restricted growth until the 18th May when there was dry weather with sunshine. Rain which fell at the end of the month gave a total of 2.2 inches (55.9 mm.) for May.

The cool, dry weather continued into June and the *Strong White Frost* on the morning of 10th June *Blasted the Potatoe stems in many places.* Cold, stormy weather lasted until 25th June when temperatures rose to a maximum eventually of 63° on 30th June, producing *very pleasant & warm weather for the first time this season.* In June there was one inch (25.4 mm.) of rain and *the Hay not Half a Crop and the Oats midling good, but the Bear universally thin . . . the Pasture Excessive Scanty & not half its usual Quantity.*

James Badenach's figures showed just how cool and wet the weather was in July. The average of the five noon temperatures given during the first fortnight was 54°. From the 15th to the 29th seven readings showed an improved figure, but no more than an average of 61° and a maximum of 64°. The total rainfall for July was 4.5 inches (114.3 mm.).

Temperatures for the first fortnight of August were similar to those at the end of July, but *remarkably warm weather* gave a maximum for the month of 69° Noon on 18th August. Temperatures for the remainder of the month ranged between 55° and 62°. Apart from a showery start, when most of the 1 inch (25,4 mm.) of the rain fell, and some light showers towards the end, August was a dry month.

There was little change during the first week of September, but on the 7th a gale interfered with work on thatching one of the hay *sowes.* The temperature on 9th September rose to 69°. The warm, dry weather continued, *stinting* the growth of the turnips, but making it possible to start the harvest on 15th September. The change to cooler and wetter weather came on the 20th from an easterly direction; after that the only noon temperatures given were 52° on the 26th and 51° on 30th September. Most of the two inches (50.8 mm.) of rain fell on 20th, 21st and 24th September.

Strong winds and some heavy rain hampered the harvest at the beginning of October. A noon temperature of 46° on 13th October was followed by reports of hard frosts at night. However, in spite of the unsettled conditions, the last of the crop was led home on 21st October. At the end of the month James Badenach was able to observe that, because *drought Quickly succeeded the rains, it Hath been an uncommon fine Month for the Harvest in General.* The total rainfall was 1.7 inches (43.2 mm.)

A considerable fall of snow on 4th November came to nothing, but the first fortnight was unsettled with alternate fresh and frost. The heaviest rain, which produced a total for November of 4 inches (101.6 mm.), came in the middle of the month, particularly on 16th November. A very strong gale and a *skimering of snow from the N.W.* on 30th November, when the noon temperature was 28°, gave a foretaste of things to come in the first week of December.

Several days of frost and snow at the beginning of December culminated on the 5th *with a FURIOUS Gale of Snow & drift which Stopped all work without doors.* The snow was two feet deep *about the houses* and there was little or no travelling on the roads. Fresh weather followed on 10th December and continued until the 20th when temperatures below 32° at noon were recorded on six consecutive days. From 26th December until the end of the month the weather was cold, wet and stormy, mostly from the south east. The *Ice Shoals* had *mostly vanished and the Frost* was *quite out of the Ground* by 31st December. Rainfall amounted to 2.9 inches (73.7 mm.) in December.

1797

There was a mild start to January 1797, but on the 12th a noon temperature of 31° was recorded, followed by *an uncommon severe frost.* The weather soon turned fresh again and continued variable until the end of the month. Snow on 31st January melted quickly on the low ground, but lay on the hills. *Quantity of rain this month is just Two & one Half Inches* (63.5 mm.).

Strong south-westerly winds at the beginning of February moderated sufficiently on the 4th to give a noon temperature of 45°, followed by several days of what was described as *Spring like weather.* On 14th February *Came on a strong Gale from N.W.* and that night *it Blowed a perfect Hurricane with a Clear Hard atmosphere from N.W.*

which did not abate till about four this (15th February) *morning with so hard a frost as to stop ploughing in many places . . . Several of our Tiled Houses damaged.*

Conditions improved to such an extent that on 27th February pleasant dry weather with warm sunshine evoked the following entry: *Yesterday & this a Continuance of very steady fine weather – notwithstanding the Change of the Moon yesterday evening which often Alters the weather.* That was a popular misconception which has survived to the present day. Rainfall for February amounted to no more than three tenths of an inch (7.6 mm.).

The weather at the beginning of March remained dry with occasional slight frost. Sowing pease began on 11th March, although a neighbour had sown some oats on ley as early as 28th February. The first significant rainfall did not come until 25th March and the total for the month was 1.2 inches (30.5 mm.).

At the beginning of April it was cold and stormy with such heavy rain on the 4th that the *whole day* there was *no work without doors.* Cold unsettled weather continued and by 17th April there was little growth of grass or vegetation of any kind. Temperatures rose slightly towards the end of the month with a noon reading of 49° on 30th April. The increased rainfall, amounting to 3 inches (76.2 mm.), was well distributed through the month.

The cold, stormy weather at the beginning of May was emphasised by the recorded temperature of no more than 43° on 11th May. Cold, dry weather continued and restricted the growth of grass, but warmer weather towards the end of the month promoted some growth. The rainfall in May was 1.9 inches (48.3 mm.).

June was a cool month as was evident from the four readings given: 60° on 7th, 52° on 19th, 57° on 28th, and 52° on 29th June, with no suggestion of anything warmer in between. On 17th June *the Grounds* were *growing very dry, and running wells drying up, the Pasture & Hay very much Checked, as is all kinds of Grain Crop & much of the latest Bear not Breered.* There was *pretty heavy rain at N.E.* on 19th June, but that night and on the following morning there was an *excessive strong White & Black frost – wind at N.E.- the Potatoe Stems quite destroyed.* It was observed that, within three miles of the sea, the potatoes were *not touched.* Towards the end of June there was some rain that *wet the grounds more than hitherto,* but the total for the month was only 1.2 inches (30.5 mm.).

The rain, mainly from the east, came in July and temperatures picked up to a maximum recorded 70° at noon on 14th July, with winds from the south. A further spell of pleasant, warm summer weather lasted from 18th to 27th July when more rain and cool weather came in from the east. At the end of July *the Quantity of rain in the Gage* was *four inches & one Half-same as July 1796* (114.3 mm.).

Very heavy rain at the beginning of August produced a total of 1.6 inches (40.6 mm.) in three days and *Brought down the rivers as large as for some years* on the 8th. Some hay had been cut on 31st July before the deluge and most of it was lost. During a drier spell in the middle of the month haymaking was resumed, but wind and rain returned to create further difficulties and damage to other crops. The total rainfall in August was 5.5 inches (139.7 mm.).

More heavy rain at the beginning of September was followed on the 8th by *a Violent Gale from West by North.* This did a great deal of damage to the hay coles and stooks. The struggle to work in the fields against the strong winds and heavy rain continued throughout September. The situation at the end of the month after five inches (127 mm.) of rain was expressed by James Badenach in the following words:

Whiteriggs - Principal Crops 1790 - 1797
Seedtime and Harvest
With Associated Weather

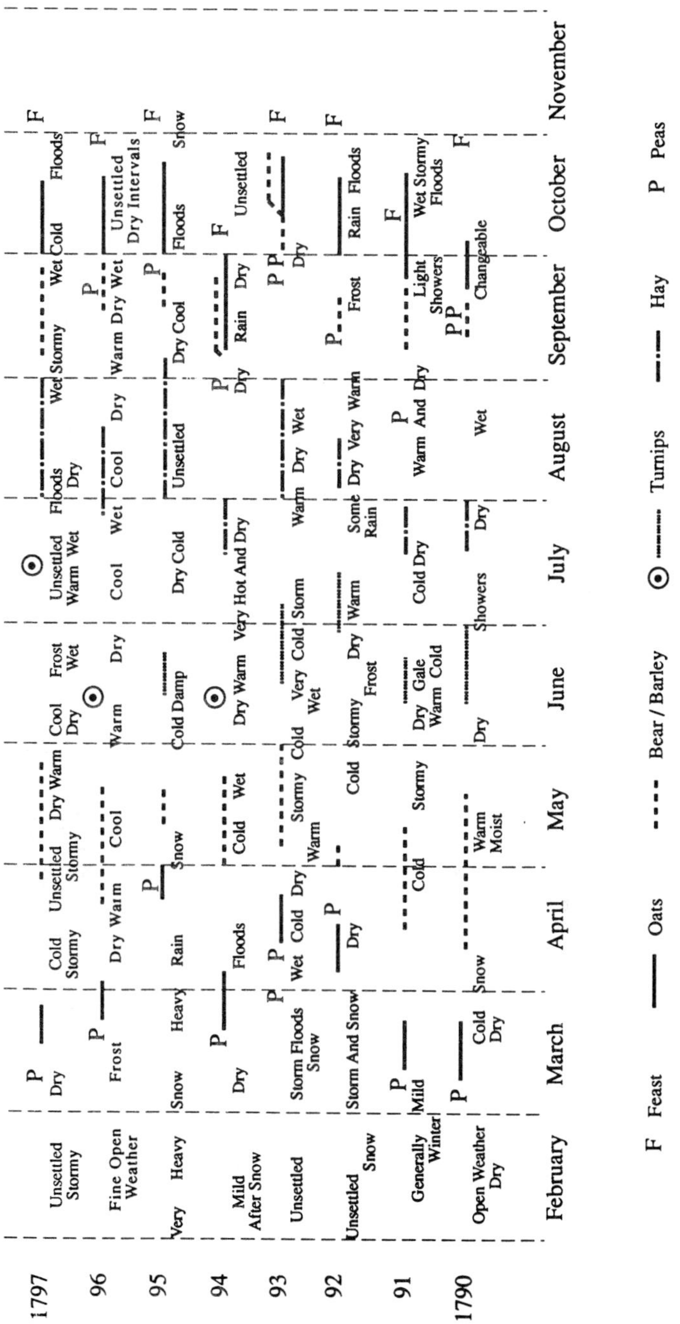

Hear much of the laid Wheat & Oats have little or nothing in them. All the hay is yet in the Fields & much of it in small Coles & useless rotten . . . the Turnips are totally a lost Crop almost every where, 4 or 5 Acres not Equal one Good. Observe some growing in the stooks of Bear &c.- Indeed upon Closs Examination a great deal spoilt.

There was some dry weather at the beginning of October, but strong white frosts did not help and there was little drought. Severe frost on 20th October was followed by a severe gale and a deluge of rain which *brought down the rivers higher than for fifteen years by past-it brought down much Corn, timber Bridges &c.* The month ended with dry, frosty weather, but the rain, mostly after the middle of the month, was 5.25 inches (133.4 mm.). The situation was summarised thus: *I look upon the Crop as much dammaged before shearing, in many places, & Generally as ill got & Bad order as either in 1792 or 1795 years.*

On 1st November the *In Gathering of all the Crop* was finished and it was declared to be *One of the worst Harvests I ever Had & Crop Certainly much Injured beyond Conjecture.* The weather in November was dry with frost at night. At Whiteriggs it was necessary to *turn over One of my Oat stacks into two Small Hollow ones,* but there were stacks in other Yards *smoking like so many KILNS.* On 16th November some were said to be *quite rotten & only fitt for the Dunghill.* A fall of snow on 23rd November gave the country *the appearance of standing winter,* but when the diary ended on 25th November 1797 there was a return to fresh weather.

AGRICULTURAL PRACTICE

Although the rotation of crops on James Badenach's farm at Whiteriggs varied, the most obvious one that could be traced was a sequence of oats, oats, pease or turnips, and, in the fourth year, bear with which grass and clover were sown to provide hay for the following one or two years. The length of time under ley thereafter was not possible to determine, but mention was made of ley coming into cultivation after five, eight and even twenty-five years.

Crops other than those mentioned above included barley, lintseed, and potatoes. However barley did not feature after 1794, lintseed was not grown on a regular basis and potatoes were grown in convenient corners of fields destined for other crops rather than as a rotation crop.

Seeds were almost invariably obtained from the farm; occasionally purchases from neighbours were mentioned. The high cost of red and white clover seed was deplored, but the source of supply was not revealed. In some years small quantities of wheat were grown as winter wheat; beans were mentioned occasionally.

The ground was always carefully prepared for all crops. Lime, obtained by cart from Mathers (see ***TRANSPORT***), was applied to the fields and dung was spread in varying quantities. Turnips were always generously treated and even wheat on Floors in September 1790 was given an allowance of 48 cart-loads per acre.

Sowing began each year if possible at the beginning of March, when yoking took place twice a day. The earliest was 4th March 1790, whereas in 1792 it was not until the first week of April that sowing could commence. Oats were sown as early as possible in the season and if lintseed was being grown it would be next on the list. Bear or barley with grasses and clovers were sown during April, but in 1793 it was well into May before those crops could be sown. June was the month for the sowing of turnips, generally towards the middle of the month, but again in 1792 sowing continued into the month of July.

Cutting the previous year's grass for hay commenced towards the end of July and haymaking continued throughout August. That was a labour-intensive activity involving men and women. The hay was put up into *coles* and *tramp coles. Horse Tramping* was mentioned in August 1791 and 1794 when *sowes* were being built. It was also in August that the lint was harvested and processed. The harvest proper did not usually begin until September, but in 1791 conditions were favourable enough for the cutting of pease to begin on 29th August. The shearing of bear and barley began in September, followed by oats, the harvesting of which could drag on until the beginning of November.

At the conclusion of the harvest the Feast was given; the earliest was on 6th October in 1794 and the latest on 6th November in 1789, when it was 26th November before the fields were finally cleared and the stacks duly thatched. One example of kiln drying was in December 1789 for damaged bear and oats.

An immense amount of time and labour was expended on the turnip crop each year. For the crop of 1790, for example, turnips were grown in two fields, Floors and Rob's Hillock Park. *Ribbing* began in December 1789 for nine working days with two or three ploughs on each day.

In the middle of January 1790 the base of a dunghill at Rob's Hillock Park took two days to prepare and then a month later for five consecutive days in February, two or three carts were engaged in driving dung to the *midding layer.* Thereafter six further cartloads finished the first covering of the dunghill which measured 80 by 6 yards.

For the first fortnight of April dung was driven on to the field at the rate of 24 cart-loads per acre, harrows were *braking in* part of the field and ploughs were in action. All those activities were resumed at the end of April and continued into the month of May.

On 10th May attention was turned to Floors and soon after a roller and a four-horse plough were added to the implements already in use. On the farm there was both a wooden and a stone roller, but it was not always specified which was being used. Work continued during the rest of May and throughout the whole of June, ploughing, harrowing, rolling, gathering and burning all the *quicken* and other weeds.

Sowing began on 11th June and while that was going on, other parts of the fields were still being prepared for seeding. The spreading of dung on the field was the final act before the sowing of the seed which, at Whiteriggs, was done by drilling with a plough. Dung went on at the rate of 28 cart-loads per acre, spread by the women. On 30th June was *finished the whole of the Turnip tillage for season by Compleating the rolling of the last of the Floors & Midding layer Robs Hillock.*

So far 81 days had been spent in preparation of the ground and sowing the crop.

Work on the turnip crop resumed on 19th July when hoeing *(houghing)* and thinning commenced. The *dressing* of the crop was interrupted for a few days of haymaking between 4th and 8th August, but otherwise was continued daily until 31st August, a total of 31 days.

Harvesting the turnips began on 6th October and it was at that time in 1790 that work began on the preparation of the fields for the following year's crop, earlier than in the previous year. From the beginning of October 1790 until 31st March 1791, when the last of the turnips was driven home, carts were collecting turnips on 66 days. That amounted to at least 121 cart-loads. About twenty of those loads were bringing home turnips after carting dung to a neighbouring field.

From 7th December 1789, when preparation of the ground first began, until 31st March 1791, when the last of the crop of 1790 was brought home, work involving the turnips was recorded on 197 days.

Another important crop at Whiteriggs was oats. They either followed a crop of oats *(award)*, or another grain crop, or were sown on land which had been in ley for a number of years. In 1790, for example, oats followed barley on the Bank and on Kavels, while the ley land was on the Haugh and Blacklands. Nineteen days were spent on Bank and Kavels and 32 days on Haugh and Blacklands, with an average of two ploughs a day. Altogether to bring the four fields to the point when seed could be sown, ploughing was continued for 51 days. Cultivation began towards the end of December 1789 and was completed on 23rd February 1790.

Sowing began on 9th March with 8 bolls of oats on the Old Haugh ley and although the area was partly harrowed on the same day, it took nine harrows on each of the following three days to complete the job. The whole of the Bank was then sown with 17 bolls of oats, and again it took nine harrows three days to cover the grain. A roller was also used on that field. Six bolls of oats were sown on the West side of Blacklands ley and harrowed with eight harrows on that and the following day. A four-horse plough and seven harrows were brought into action on Kavels; then it was back to Blacklands with the harrows to complete the sowing of $13\frac{1}{2}$ bolls during the next two days. Thereafter the four-horse plough, the harrows and the roller completed the sowing of 16 bolls 3 firlots on Kavels by 26th March. Twelve horses were working in the field on that day.

During the 16 days between 9th and 26th March 1790 $55\frac{1}{4}$ bolls of oats had been sown, covered by the harrows, and rolled.

Harvesting the oats in 1790 began on 27th September on the Bank. The 16 shearers and 4 bandsters employed that year spent twenty days cutting and binding in the various fields and completed the work on 20th October. Some of the crop was being led while some was still being cut.

The Feast was held on 25th October, but all was not gathered in until 30th October when 27 stacks stood in the stack yard. There were 18 stacks of oats and 7 of barley thatched with straw; the two stacks of pease were thatched with rushes.

From the earliest preparation of the ley on 21st December 1789 to the final thatching of the stacks on 30th October 1790 a total of 87 days had been devoted to some aspect of the crop of oats. That figure compared with 81 days for turnips only to the point where sowing was completed. A further 31 days were required for dressing the turnips; carting them was recorded on no fewer than 66 days. Certainly extra days would have to be added for *dighting* the oats, but overall the turnip crop was the most demanding on the farm.

Summaries of the course of events during the harvests of 1794 and 1795 were made by James Badenach and have been reproduced in Appendix 4 of this volume.

LIVESTOCK

Livestock at Whiteriggs included horses, cattle, sheep and swine with brief references to poultry and pigeons. During the building operations in 1791 a new pigeon house was built at the south end of the new barns. It was an elaborate building with stones and specially designed pigeon-holes brought by cart from Flemington in Angus. On the south east side of the barns a hen-house was built at the same time. Otherwise poultry and pigeons were mentioned only in connection with the provision of dung for the fields.

Swine, too, were seldom mentioned except as providers of dung, although James Badenach must have bred them to the extent that he spoke of cutting *some swine shots* and on 4th December 1792 he *sold and delivered the fatt swine for Aberdeen at £3-12 sh.*

The number of sheep that were kept must have increased during the years of the diary for in June 1791 he *washed and sheared the few sheep.* In May of the following year he bought 6 ewes and lambs and a further 10 ewes in July at Garvock Fair. In general the sheep must have proved rather troublesome, for in May 1793 three sheep which had strayed three months previously were returned to Whiteriggs from Balmakewan some 8 miles away. Payment must have been demanded for their keep as James Badenach lamented that they would have been cheaper maintained at home. In July 1794 he bought a dozen yeld ewes and in the following year 19 wedders and ewes to feed. By 1796 the flock was large enough for the arrangement of the sale of 40 fat sheep to James Christie, Butcher at Arbuthnott. They were to be *lifted in parcels and paid for as lifted.* They were taken, two or three at a time between 16th March and 16th May 1796. (Appendix 5)

Hurdles or frameworks with crossed slats made of fir and sometimes warped with broom were used to provide a temporary pen for the sheep. Those *flakes* were usually set up in the same area as the cattle flakes and moved regularly, often at the same time, but whereas the cattle flakes were taken in during the winter, the sheep remained out of doors in their enclosure. June was the usual month for washing and shearing the sheep, but in 1796 and 1797 they were not shorn until 11th and 10th July respectively. No reason was given, but in both years June was a cold month.

After the erection of new byres in 1790 there was an increase in the number of cattle kept on the farm. At the beginning of November 1790 nine cattle were housed for stall feeding and seven cows in two new byres. A fortnight later ten yearlings were put into the new West byre. In 1791 the cattle were housed as early as 18th October and there were then 19 for stall feeding and 33 other cattle, young and old. That was the only occasion on which total numbers were given, but that was the probable size of the herd from the time of the building of the new byres until the end of the diary.

The milk cows and those newly-calved were put out to grass in the middle of April. It was not until the beginning of June that the cattle flakes were set up on an area of ley that was due to carry a crop in the following year. Within the confines of the flakes there were as many as 24 young cattle. The flakes were moved regularly, the frequency depending on the season and the nature of the ground on which they were set up. In some years the flakes and the young cattle were moved from their original site to another part of the farm to take advantage of some better grazing.

The cattle were put out at the beginning of June and it was usually the beginning of November when the cattle were housed and the flakes returned to the farm for storage or to be used as temporary shelters for the turnips.

There were several forms of sickness that affected the cattle. In 1791 a young stot, six months old, died unexpectedly of the *Quarter Ill*, as it was known locally, or Black-Quarter. This was described by James Badenach as a *pretty general obstruction and extravasation of blood in the different parts of the body*. In other words it was an enzootic apoplectic disease peculiar to cattle and indicated by lameness of the fore-foot and blackness of the flesh. In subsequent years the calves were bled as a precaution.

Fluxes, the nature of which was not disclosed, caused the death of three bull calves in 1795 at a time when there was *in general great deaths amongst them*.

On several occasions cows were described as *hoved*. That condition resulted from grazing animals becoming blown up after eating too much fresh green fodder. Attempts were made to lance or *stab* the swelling, but that was not always successful.

During hot weather, of which there was not a great deal during the period of the diary, the cattle were described as *startling*. In other words they were being so badly stung by flies that they rushed about wildly and had to be brought back to the farmyard.

In March 1790 the cost of good work-horses was £25 sterling. On the farm there were at least twelve such horses, the largest number that was mentioned as being used at any one time. There were never more than four ploughs in action at once, so that it was reasonable to suppose that eight of the horses were suitable for ploughing, while the remainder could be used for riding or driving as well as drawing a harrow, a task that was undertaken by a single horse. A reference to a *highland sholt* which was *excessive ill with a tinge or colick for 14 hours but recovered*, was probably one of the twelve. In February 1793 James Badenach lost the first horse ever in his possession *by a violent colick*. It had been well enough on the previous day during an outing to Brechin, but in spite of receiving *every remedy . . . that I could devise & more than ordinary*, the horse, aged 22 years, died.

There was a smithy attached to the estate and, when necessary, the horses were specially shod for winter work.

There were occasional references to bees and beehives.

SERVANTS

James Badenach himself obviously took a great interest in the running of the farm, but he did employ a grieve who was given a fair measure of responsibility; for example he was mentioned as selling barley in Montrose and buying wedders and ewes elsewhere, as well as being in charge of operations on the farm. In 1797 there was an implied criticism of the grieve when the first *houghing* of the turnips was done in a very slovenly manner, but the blame was placed squarely on his shoulders in November of that year when a stack of oats built on unsound wood had to be taken down and rebuilt. *This Carelessness & Bungling of the Grieve in Managing the*

Corn Yard will Occasion a very Considerable loss of the Oat Crop, being so free in Handling.

It was hard to tell exactly how many regular workers were employed on the farm, but there must have been about twelve. For example on one day in November 1790 there were two carts fetching coals from Stonehaven, two carts driving turnips with two other men filling the carts, 4 men planting trees at Thornylee, and an unspecified number, possibly two, dighting bear. In addition to men and women, children were mentioned in connection with the potato crop. Extra workers were taken on, especially at harvest, when the total work-force reached its peak. On one such occasion there were as many as 26 people working in the fields, but the usual number was nearer 20. Not all the shearing was done by women; *Begun harvest 12 Shearers half men.* The rates of pay for *6 Hired people from Polburn* on 14th September 1789 was *1/ Men & 8ds Wowmen per day & Meat.*

The servants at Whiteriggs were engaged at the Thornton (*Taunton)* Fair and the Garvock Market at the Candlemas and Lammas Terms respectively. There was a smith attached to the farm, but other specialists such as masons, slaters, wrights, a painter and quarriers were engaged and dismissed as required. Wages were not often mentioned, but on 31st May 1790 when *3 Men Begun Quarrying at Dava* one received a shilling and the other two 10 pence per day *with victual.* On 30th January 1793 at *Taunton fair* servants were *still Increasing in their wages at this term about three or four shillings or thereabouts perhaps more, the Shearers Greatly Increased about 5/ each – much against farmers.*

Work varied in quality then as now. The entries on 5th and 6th August 1796 told their own story: *Own People & 4 Hired people at dressing turnips and one Plough preceeding them . . . 10 People as yesterday at the Turnips of which did more than double the Quantity and as well. Unaccountable how they Evade at times & not apparent either at the time.*

Trouble among the servants was recorded once only. It occurred on 24th January 1793 and was described thus: *by a violent Sedition without any reason given 4 Men servants & two Wowmen left their work – but two of the Men Acknowledged their fault & promised to Behave properly were admitted again the rest shall not if Law & Justice is in the land.*

Their health seemed to be fairly good. In 1790 four servants were laid up with measles, In June 1793 some of *our people* were confined as a result of the *inclemency of the weather;* three days later *the sick servants* were *recovering from a kind of scarlet fever.* In January 1795 several of the people were indisposed with colds and it was observed that there was a very great mortality at Montrose and amongst all people in general at that time.

Social life among the servants was seldom mentioned. On Saturday, 13th June 1795, however, allowances were made for a marriage to take place. *Very Early Yoked at the Turnip field, drilling & Dunging so as to Make Half days work from about 9 to 10 A.M. on account of a marriage of one of the Servants thereafter.* Holidays were never mentioned, but there was no work on Sundays and there were several other occasions when work was suspended.

Old Christmas was observed on 5th January each year; that date corresponded to 25th December (Old Style), and was still observed, although it was nearly 40 years since the calendar had been changed. Although it was a day when there was *no work without doors,* there was no mention of a Church service on that day. The people were variously described as *mostly absent, at their several Christmases* and *visiting their friends.*On the second Thursday in July was the Parish Fast before the Sacrament at Fordun. There was no work, *without doors,* apart from any masons or wrights who came from other Parishes. Attendance at Church was mentioned on the Saturday and Monday following, but work on the farm was able to be continued on those days.

New Year's Day was never observed as a holiday, but it was chosen in 1794 as *a Thanksgiving day at Church for the Good Harvest & Collection for ye Soldiers abroad.* The only other occasion on which a Harvest Thanksgiving service was mentioned was on Wednesday, 30th November 1796. *Being a Thanksgiving day for the good Harvest* there was *no work without doors – excepting sending 7 Bolls of Oats to the Mill.*

Appearing in the Diary for the first time in 1794 were *general fasts of the Church of Scotland.* In that year it was on Thursday 27th February and observed to be *well kept everywhere.* In 1795 it was held on Thursday, 26th February, in 1796 on Thursday 10th March, and on Thursday, 9th March in 1797.

MARKETS, TRYSTS AND FAIRS

The buying and selling of livestock, fodder, grain and, regrettably, servants, took place at a variety of centres. They were named variously markets, trysts and fairs, but the activities at each were much the same. Some specialised in a particular commodity and were over in a day, while others lasted two or three days, marketing a wide range of goods and livestock. During the year roups (auctions) usually of either cattle or grain were held on individual farms, a method of selling used on occasions by James Badenach himself. He also sold cattle, grain and hay privately. Each year there were about twelve different markets in the district around Whiteriggs, with March the only month in which there was no market.

For Whiteriggs and the neighbouring farms, the most important markets were the *Taunton* (Thornton) Fair at the end of January each year, corresponding with the Candlemas Term (2nd February) and the Garvock Fair at the end of July, corresponding with the Lammas Term (1st August). In addition to all the buying and selling of produce from the farms, those fairs were attended by farm servants, male and female, who were engaged or re-engaged at agreed wages. On the occasion of Garvock Fair in July 1796 came James Badenach's most outspoken complaint about the system that he hated so much: *Cutting Hay till Breakfast after, all Persons Male & Female at the ACCURSED Markett of Garvock for Hiring Servants which should be abolished.*

The majority of the other markets mentioned in the Diary were also within a five mile radius of Whiteriggs. At Laurencekirk there was a tryst towards the end of April each year and at Auchenblae in 1791. A new cattle market was established at Bervie in May 1792, lapsed in 1793 and was revived in 1794. In May the Rood Market at Montrose was held and cattle markets further afield at the end of May included Glamis and Forfar.

In June cattle were the main interest at Cammock Muir while at Trinity Muir on the following day horses were also on sale. There was a market at Dunsmuir a week later, while at the end of June or, in some years, at the beginning of July, a second market at Cammock Muir was mentioned. Megray Market also took place at that time.

Paldy Fair, lasting for at least three days, was an important occasion in mid-July. It was followed at the end of the month by the Garvock Market described above. In August came the Laurencekirk Fair and the Lammas Muir Market. On only two occasions were markets mentioned during September when the harvest was in full swing. The first was a reference on 12th September 1789 to the fact that cattle were cheaper at Bartle Fair, which was never mentioned again. The second observed simply, on 13th September 1792, that *most of our Hands at Idle Marketing at Bervie.*

The Market or Fair at Drumlithie was a regular feature of the second week of October, with a second market at the end of the month mentioned in 1789 and 1794. A third market at Laurencekirk, was held early in November. There was no fixed date for the last market of the year at Auchenblae. In 1794 that Market, often referred to as the Auchenbale Crofts, was held as early as 7th November, with a second market on 19th December, while in 1789 it ran over into January 1790.

MILLS AND DISTILLERIES

In addition to a malt barn and brewhouse on the estate, occasional reference was made to a distillery, but few details were given. Two other distilleries were mentioned, the first in July 1789 when a total of 72 Bolls and 1 Firlot of Barley was delivered to Mrs Mackenzie at the Woodstock Distillery. On the second occasion 10 Bolls of Bear were sent to the Jacks Bank Distillery.

Oats were sent to the mills for oatmeal during the winter months, with occasional journeys in summer if supplies were running short. The local mills used by James Badenach were the Mill of Garvock, the Mill of Pitrany, and the Mill of Kair. The last-named was used more frequently than either of the other two. It was usual for people from the farm to work at the mill. On 29th November 1790, for example, the grieve himself and two women were at the Mill of Kair.

Andrew Young, the Mill Master of the Mill of Conveth, bought oats, oatmeal, barley, bear and wheat from Whiteriggs on different occasions.

Flax was sent to the mills of Mondynes and Drumtochty; occasionally barley was delivered to Bridge of Leppie for malt for the family.

PRICES

Throughout the diary, usually in the middle and at the end of the month, prices of a range of farm commodities were frequently given, together with some reasons. as the writer saw them, for their rise and fall. Such figures as were available have been reproduced in the following tables (Tables 1 to 8) and graphs (Figures 1 and 2).

In June 1789 the rise in the price of bear was due to its *exportation to France.* In February 1790 there was *a great Stagnation of the Sale of Fatt Cattle and Grain which is very hard on the Farmer owing chiefly to an unjust Embargo on the Exportation of that article.* At the end of March it was reported that prices were unexpectedly low in view of the previous year's poor harvest, deficient by a third and even by a half in some places. Meal from oats was also deficient by as much as a firlot per boll. In spite of that, prices were low and there was no demand for grain or cattle. However, a month later the prices of grain rose and there was a demand for fat and lean cattle when they had nearly all gone. Prices continued to rise until July when the prospects for the forthcoming harvest looked good, but James Badenach had his doubts. Towards the end of the year markets remained dull and he believed that there was *much Scarcity of Cash in the Country* at the time.

At the beginning of 1791 the scarcity of fat cattle led to some demand and grain prices rose. By the month of June there were many cattle available, but not much sale owing to a lack of demand from the South.

At the end of June 1792 the cattle markets were depressed as a result of the lack of grass. Crops on wet land were looking bad and the wheat crop had failed in the south of Scotland where the hay crop was very light. It was little better in the north and the turnips *must be a bad crop.* By September it was clear that the crop had been seriously damaged by rain and storm. The price of meat rose 3^d in 10 days.

Prices were high in 1793 following a bad season and a poor crop. At the end of February the fat cattle were *not Saleable though Beef dear, Owing to the Turnips being Generally done.* It was thought that the French War would depress trade. In December the prices of beef fell because of the shortage of salt to cure it. The price of salt at that time was 32^d per peck.

Whether owing to the War or otherwise there was no demand for fat cattle at the end of February 1794, although by the end of April there was a greater demand, especially for cattle fit for driving. As the year progressed there was an *uncommon demand for Cattle for England & High prices* followed by a great demand for grain. At the end of November the open weather had *Considerably saved the Scanty Crop of Fodder and what* was *of more importance the Great demand for Cattle* had *thinned this Country of them Amazingly.* However, fat cattle were not in demand.

Scarcity in 1795 again led to increased prices of grain. Bear reached 27/ per boll when it fell *owing to the Legislature stopping the Distilleries for a time.* In spite of that, prices in July and November were given as 26/ and 25/ respectively. However, even by the end of October, it was realised that the crop of 1795 was one of the *most destroyed crops.* Matters would get worse because the grain would not keep and the situation would be as bad as it was in 1782 or even worse. In November the distilleries were still stopped and the grain was of poor quality.

In the middle of February 1796 meal was *scarcely to be got* and there was word of *threatened mobbings.* At the end of the month there was *a Great Alarm for a Scarcity in the Ensuing Summer & a County Meeting of Heritors, Farmers & Ministers Consulted on the Business.* In July cattle prices were high owing to the demand from England and on account of the war. There was little or no oatmeal except *of foreign corn.*

By the end of January 1797 the prices of grain had fallen very considerably and there was *a very dull sale of all kinds owing to much Importation from Ireland and other places. Cattle of all kinds above all former prices 10/ per stone.* However in March there was stagnation in the sale of grain, hay and *Fatt Beastials* thought to be due to an *Imaginary deficiency of Money, and of French Inmvasion of this Country.*By the end of March it was the derangement of the London and other banks that stagnated business and generated bankruptcies. The theme continued into April when there was no demand for hay or for any farm produce whatever and prices were falling. That was blamed on *Many Imaginary Conjectures about Political Matters, not properly understood, & deranged Mercantile Speculations of kinds.* Towards the end of the year there was an increased demand for cattle from the south and the scarcity of meal at Montrose was once again leading to an increase in price.

TABLE 1 PRICE OF SHEEP IN SHILLINGS AND PENCE 1794–97

Sheep		
–		
–		
1794	July	3*s*.9*d*. each
1795	July	4*s*.0*d*. "
	Aug	4*s*.6*d*. "
1796	July	7*s*.0*d*. "
	Aug	7*s*.6*d*. "

TABLE 2 PRICE OF BEEF IN PENCE PER LB 1791–1797

Beef		
1791	Feb	3½–4
	Nov	2½–3
1792	June	4–4½
1794	Dec	3½–4
1795	Apr	4½
	Nov	3½–4
	Dec	4
1796	May	6–8
	June	6
	Sept	4½–5
	Dec	5½–6
1797	Jan	5½–6
	July	5–6
	Oct	5–6

TABLE 3 PRICE OF CORN AND FODDER IN SHILLINGS AND PENCE 1790–1797

Corn and Fodder		
–		
1790	Feb	13*s*.6*d*.
1791	Feb	15*s*.2*d*–15*s*.6*d*.
	Sept	20*s*.
	Dec	18*s*.
1792	Jan	21*s*.–22*s*.
1793	Feb	16*s*.8*d*.–21*s*.
–		
–		
–		
1797	Feb	16*s*.–19*s*.

TABLE 4 PRICE OF BARLEY IN SHILLINGS AND PENCE PER BOLL

Barley		
1789	June	14*s*.
	July	15*s*.–16*s*.
	Nov	15*s*.6*d*.
1790	Jan	15*s*.
	Dec	14*s*.
1791	Feb	14*s*.
	Apr	14*s*.6*d*.–15*s*.
	May	15*s*.
	June	15*s*.6*d*.–16*s*.
	Nov	16*s*.6*d*.–17*s*.
	Dec	17*s*.
1792	Dec	18*s*.–19*s*.
1793	Nov	16*s*.6*d*.
1794	June	18*s*.–20*s*.
	Dec	20*s*.3*d*.
1795	Nov	25*s*.
–		
–		

TABLE 5 PRICE OF OATS IN SHILLINGS AND PENCE PER BOLL

		Oats
1789	July	12s.
1790	Aug	15s.6d.
1791	Feb	12s.9d. (Seed 14s.6d.)
1792	Feb	– (Seed 16s.)
	Sept	13s.6d.
1794	Dec	16s.
1795	Jan	15s.6d.
	Mar	16s.–17s.
	July	20s.
1796	Mar	–
	Nov	16s.–17s.

TABLE 6 PRICE OF WHEAT IN SHILLINGS AND PENCE PER BOLL

		Wheat
–		
1790	Jan	*22s.6d.*
	June	*26s.6d.*
1791	Nov	*20s.*
–		
1794	Nov	*21s.–22s.*
	Dec	*23s.–24s.*
1795	Mar	*27s.–30s.*
	July	*40s.*

TABLE 7 PRICE OF HAY IN PENCE PER STONE

		Hay
1789	July	3–3½
1790	Aug	3
1791	Aug	5
	Nov	5½
	Dec	7½
1792	Jan	6
	Feb	6
	Apr	7–8
	Aug	6–1
1793	Feb	8
	May	10
1796	Jan	7
	Mar	7–8
	Apr	8
	July	10–1/
	Sept	7
	Oct	7–8
1797	Feb	10
	Aug	3½–4

TABLE 8 PRICE OF SPINNING IN PENCE PER SPINDLE

		Spinning
–		
1790	Jan	11–12
	Dec	13–14
1791	Mar	13
	Nov	15
1792	Dec	15–16
–		
1796	Dec	16
1797	Jan	16–1/
	July	16–18
	Aug	16–18

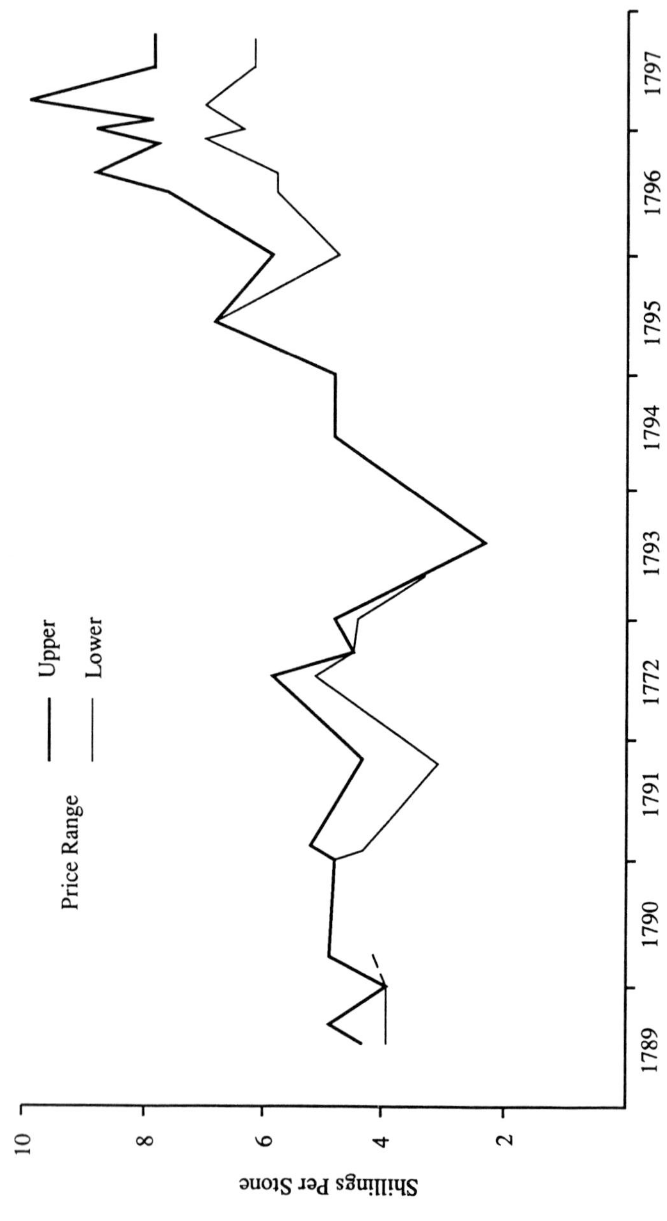

Prices of Fat Cattle

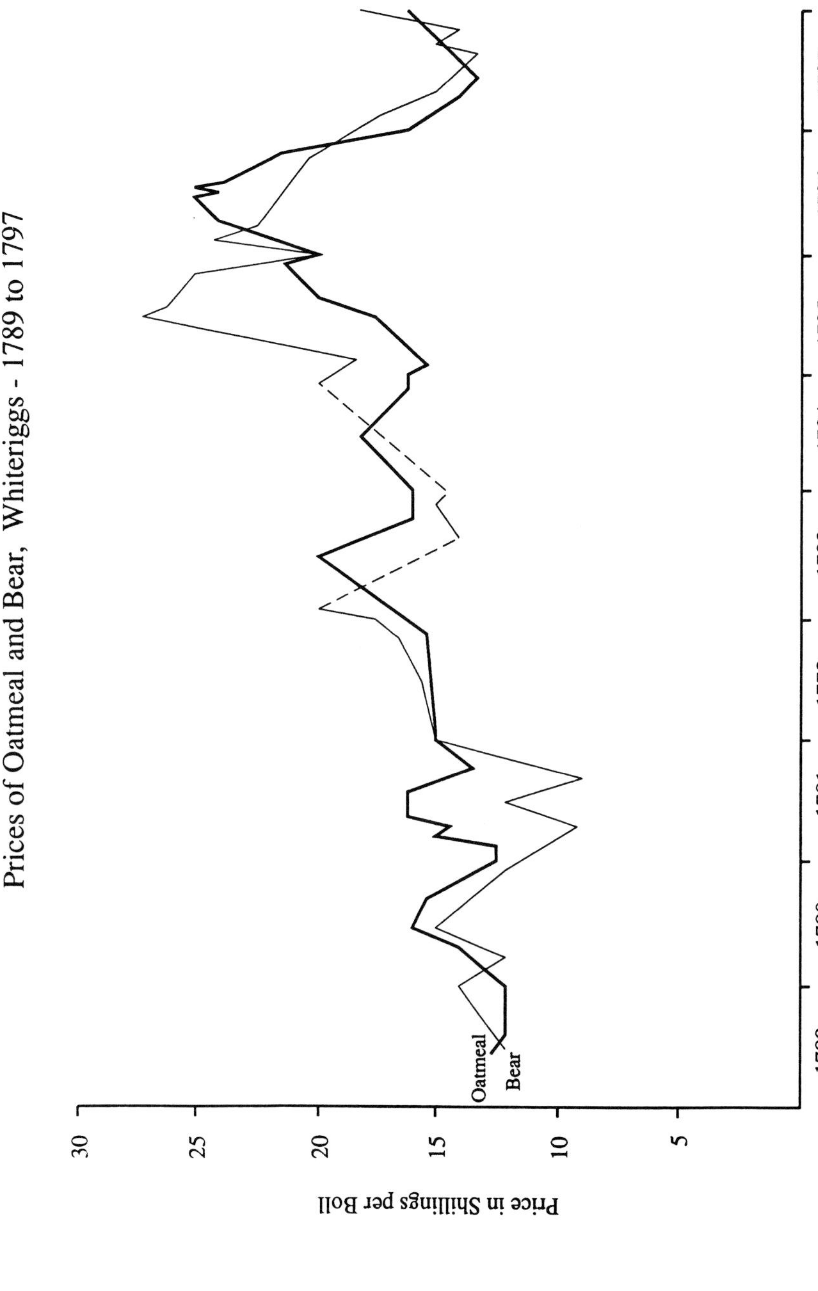
Prices of Oatmeal and Bear, Whiteriggs - 1789 to 1797
Price in Shillings per Boll
30
25
20
15
10
5
1789
1790
1791
1772
1793
1794
1795
1796
1797
Oatmeal
Bear

BUILDING AND RECONSTRUCTION

When the diary began in June 1789 building work was in progress with masons and other workmen already in action. New barns were being built and many journeys to Montrose and Flemington had to be made for wood and slates respectively (See ***TRANSPORT***). By 31st August the masons had nearly finished the cow byres and the barn floors were being laid with clay. In October, as a precaution no doubt, the old West Barn was thatched for a season and the old byres were thatched with rushes, but the slating of the new barns was finished on 3rd November and on the 19th sundry places were searched for *Smithy Culm* to harden the new barn floors which were nearly ready for use. On 18th December the new barns were described as being *Extremely Convenient.*

On 12th December a cartful of *fogg* was collected for *deafening the Noise with plaister lath from the Kitchen below the parlour.*After some more lime had been brought from Mathers for the *plaister kitchen,* the slaters were dismissed for the season on 25th December 1789.

On 27th March 1790 stones were still being brought from the quarry at Dava for the new byres. In April lime and sand were required to point and harl the houses. In addition to those repairs, a cottar house was being built near the Aberdeen road. Throughout June two carts a day, sometimes making three journeys, kept up a constant supply of stones from Dava for the byres and slates were brought from Pitscandly. Five masons and two slaters were engaged on the work. On 19th June the furniture was cleared out of the old byres for the new ones and on 25th June the old byres were taken down. The rubbish was carried off to provide foundations for parts of the new byres that had not yet been started.

By 20th July the roof of the schoolhouse was taken down to make way for a new one of foreign wood and slates. The masons were soon engaged on rebuilding. During August most of the activity that was reported was concerned with the carting of stones for the buildings, but some gravelling of the *High Road through our farm to Aberdeen* was undertaken, as well as gravelling the road to the house. On 30th August the north wall of the old barns was taken down to build part of the new byres and on 7th September the roof couples were fixed on the new byres. Gravelling the roads and slating the byres continued during September.

On 6th October the calves were put into the new byres and on the following day the roof of part of the old barns was removed to complete the new byres. From then to the end of October the masons laid the floors, *causewayed* and slated the byres in which it was then possible to house for stall feeding *9 cattle & 7 cows in two new byres.* On 11th November the masons were dismissed for the season and on the 19th the 10 year old cattle were *put up into the new West Byre.*

In March and April 1791 at least 10 days were spent in driving gravel to make up the approach road to the house from the public road to Aberdeen. Wrights and masons were engaged once more and work on the schoolhouse was resumed. In July work was concentrated on a pigeon-house to be built at the south end of the new barns and later a hen-house at the south east side. Special stones had to be brought from Flemington and Laurieston for the purpose. The pigeon house was finished on 9th August and on 31st August the roofing of all houses was completed, *a welcome business.*

In 1792 masons were engaged as early as the month of March to build a stone dyke on the west side of the garden. A cart which had been sent to Mathers for lime for that job found that no lime shells had been burnt. The journey was not altogether wasted as the cart returned by way of Johnshaven with a load of coal. On the following day a cartload of lime was obtained from the Mill of Garvock.

Towards the end of April work on a new kitchen and a servants' house was commenced. For a second time the journey to Mathers for lime was abortive; on that occasion the cart came back by way of Laurieston with dressed stone for the chimney head. The amount of stone required to build the new kitchen chimney did not please James Badenach when he described it as *a perfect glutton.* The kitchen itself required a great deal of material with rigging stones, slates, lath and lime.

Work went ahead during June on paving the kitchen and fitting up the cellars and milk-house. On 14th June there was *a power of people at the cellars, West Chimney Head taking down &c.* Steps were also made for the new kitchen to the lower stair. On 3rd July all the masons were dismissed except one who was repairing the old kitchen. While they were away, at least two days were spent in driving gravel to the Aberdeen road on the north side of the farm. The return of the masons was not mentioned, but more materials were brought for the kitchen during August and September; at the end of October the new kitchen was being slated.

On 19th March 1793 the building of a cottar house was begun and it was finished on 23rd May, apart from the thatched roof, material for which was collected by *several People at Casting Divots, Holing Broom &c for new Cottar House.*

At the end of May there was a great communal effort in the erection of a bridge over the river Bervie below the farm. The foundations were begun on 27th May and on 29th June the following comment was made by James Badenach: *Men & most others in the Neighbourhood assisting in putting up the Coomb for the new Bridge . . . it appears when sett to be very High & will require much making up at both Ends - will be a Roomy Arch.* The bridge was finished on 5th October. On 8th October a temporary bridge was being built over the Leppie burn.

In July 1793 stones were brought to pave the dung court and gravel was fetched to cover them. In September two slaters were working on the *small Houses little Court.*

Following the building of the new bridge, the first task in January 1794 was the strengthening of a causeway to protect the Haugh bank below the bridge on the east side of the river.

In March more gravel was driven to the area round the house and the well at the house was covered.

In July three carts were engaged in driving gravel to the public road, a statutory requirement. Apparently it was honoured in the breach rather than in the observance for on 31st October James Badenach remarked that the people in the neighbourhood were at the public roads *for the first time for many years.*

In July 1794 more work was carried out on the roads around the *Mansion House.* The cottar houses were thatched in August and the main building activity of the season was the roofing of the long loft. For that purpose *2000 slates at six score to the hundred* were brought from Flemington and timber was collected from Montrose.

In October work was continued on the bank below the bridge and it was not until the 18th November that the slating of the long barn was finished.

During May and June 1795 some general repairs to the houses were undertaken, dykes were mended and work which had started in the previous year to construct *sivers* at Thornylee was continued. In July gravel was again driven to the public roads on four different occasions.

On 10th February 1796 the roof of a cottar house was brought home from Waterlair and it was observed that several thatched houses had fallen. On 14th March two carts were employed all day carrying stones from Thornylee once more to bank the river below the bridge.

On 20th August a great task was undertaken by three men who dug a *canal* 38 feet long, seven feet broad and three feet deep in which to steep the lint. That evening it was filled with water from the Leppie.

There was little activity in 1797. At the beginning of January two cottar houses were thatched and in February work began on the levelling of an old dyke on the west side of Floors to produce *one continued plain of* the *whole Haughs.* From time to time the levelling of other dykes on the Haugh had been mentioned and that was the last. On 22nd May there was road making on the public road to Laurencekirk.

PLANTING

On 1st December 1789 no fewer than 25,000 seedling trees were collected from Stonehaven to transplant into the new nursery which James Badenach had made at Whiteriggs. A week later the nursery ground and a hedge at the north west corner of the garden were planted out. On 14th December after whins had been burnt and cleared on a *sprott brae* towards Kair, the area was planted with *Scotch firs* and *inclosures* were made up. Firs were then planted at Thornylee and in January 1790 broom was cleared for more planting there. On 26th January a patch by the brae at Shepherds Haugh on the east side of the river Bervie was planted and enclosed.

At the beginning of February 1790 work was started on *Inclosing the Cottar Yards with Ditch & Dyke, Hedge row of Thorns behind, and Ash Trees at 4 Ells distance within.* On 18th February 30 *Guin* trees, some of them five feet high, were planted on the west side of the garden within the hedge. On the 19th at the north west corner of the garden, six ash trees, which were destroying the hedge, were dug up and 20 geans were planted *in the new nursery benorth the garden.* It was presumably the same six ashes that were *transplanted for Gates* on the following day.

In March planting of the Cottar Yards was continued with ash and hawthorn; ash trees were added to the plantation on Shepherds Haugh Brae. In May the young trees in the nursery were weeded. Enclosure of the Thornylee planting was begun on 9th and finished on 24th June.

Planting was discontinued until 10th November when four men were engaged at Thornylee on planting *all kinds of Forest Trees in Mixture.* Six men were working there on the following day, but four was the usual number that continued the planting for seven more days. On 30th November James Badenach wrote that *a Good deal of that Plantation much finer ground than either myself or others had any Idea of till Pitting Shewed it.* Between 16th and 21st December the gardener and another man planted larches at Craig Braes *which will render them fine Plantations if they thrive.*

There was bad news from Thornylee on 20th January 1791; hares were destroying the young plantation of trees, particularly the laburnums. In February the gardener prepared another nursery for seedling trees especially for planting the *Bog Atry.* The nursery was planted in March with seedlings and flowering shrubs. In April the area of Shepherds Haugh that had been planted in January 1790 was provided with a paling. In October preparations were made to deal with the area on the north west side of the garden where clearing had taken place in February 1790. In December stones were driven to the west side of the garden where a dyke was to be built in addition to the strip of planting.

In February 1792 the trenching of the ground on the west side of the garden *for planting a belt of a variety of trees as a screen to it* was continued. At the same time ash trees were removed and several were transplanted as gate posts. In March some firs were planted on Shepherds Haugh and a *handsome paling* was put on the house side of the belt of plantings. In April a mason was employed to build the dyke on the west side of the garden, the planting of trees was begun and a sunk fence was incorporated into the scheme. In June the gardener and the women were employed in *hand houghing the strip of planting W. side of garden.*

In June 1792 the not inconsiderable task of draining and planting the area known as the *Bog Atry* was begun. First of all five ditchers were employed in enclosing it. No more was heard until 17th December when *quick* was planted around the *Bog Atry* for hardwood. On three days at the end of the month two men were engaged on ditching to drain the *Bog Atry* for planting.

The work was continued into January 1793, the gardener joining the two hired men already working in the area. From 12th January onwards pitting for planting the *Bog Atry* with oaks was undertaken.

While work on the *Bog Atry* was continued, other activities were mentioned in February including the planting of a hedge of thorns, roses, and sweet briars at Brae head. Larch, horse chestnut and ash were planted along the east dyke of the Bank, the north fence of Back Yards and the Cottar Yards.

By the third week in February the *Bog Atry* was ready for planting with all kinds of forest trees. After five days of work by the gardener and the two men the best part was finished on 9th March. A significant clue as to the possible meaning and former use of the *Bog Atry*before it was drained came in a statement by James Badenach that they had *dunged most of the pitts from it lying thick on the sward.*

Two entries in July spoke of cleaning a hedge on the north side of the house and dressing the belt of planting on the west side of the garden.

In February 1794 the gardener was filling in blanks amongst the forest trees at Waterlair and planting larches at Waterlair Park where they had been neglected or smothered by broom.

At the end of the month the gardener and two men were back again at the *Bog Atry* draining for planting. Willows were planted at the beginning of March after 400 to 500 *setts* had been cut for the purpose from the Red Willow trees at Kair.

Thereafter it was back to Waterlair to clear more whins and broom; a week later the gardener was busy planting larches on the north side of the Corn Yard and Red Willow trees at Thornylee.

Towards the end of March the gardener put *Young Elms 2 Years Transplanted into the Nursery,* followed by 5,000 larches. In July the hedge around the thriving *Bog Atry* plantation was cleaned and weeded, followed in August by the belt of planting on the west side of the garden. In November the nursery hedge was dressed and in December thinnings of fir were brought from Pittengardener for a dead hedge at the north west of the garden.

It was hardly surprising that there was little planting done during 1795 on account of the bad weather. On 4th April alders were planted in the *Bog planting*, but it was not until December that the gardener dressed the hedge around the *Bog planting* and planted ashes at the Loup Yards where half of the oaks had failed two years before. More hardwoods were added to the *Bog planting* at the end of the month.

In January 1796 larches were added to the hard woods already in the *Bog Atry*. That was followed in February by the planting of ash, elm and larch on the east side of the plantation behind the house. The gardener was then given the task of filling up the planting that had been done three years before on the brae at Shepherds Haugh. The only other items mentioned in 1796 were some more trees planted in the *Bog Atry* in March and in October the cleaning of the thorn hedges.

Many larches were planted in January 1797 starting with a row of them at the foot of the *march dyke* on the north side of the river opposite the house. That was above the new bridge and between the river and Shepherds Haugh. Towards the end of the month and at the beginning of February larch trees were added to the seven-year-old firs at Thornylee. More elm and ash trees were added to the *Bog Atry* in February. The only other entry referred to the cleaning of the hedges around the *Bog Atry* on 17th November.

GARDEN

At Whiteriggs in 1789 there was only a fruit and vegetable garden surrounded by hedges, but its position in relation to the house was not mentioned. Work on a flower garden, variously described as *before* and *benorth* the House, was begun on 19th February 1790. Apparently an experienced gardener was appointed in 1791. Before that year reference was made to *the gardners*, but from 24th February 1791 the singular was used more frequently and on 30th March the following entry mentioned the gardener by name: *sowed Onions, Carrots, Beet &c in Broad land, which W.Smith my Gardner Saw Practised last Season about London to be Hand Houghed.*

In 1789 garden peas were in flower on 19th June when the apples and small fruit appeared good. *Collyflower* was planted on 25th June and on 8th July some strawberries planted that year were ripe. Green peas were ready on 21st July, but at the end of the month the gooseberries and raspberries were not *ready for use.* Although it was *esteemed a bad year for apples and other fruits,* gooseberries and *small fruits* were available on 15th August.

In 1790 cabbage was planted on 15th February and shortly afterwards work was begun on the new flower garden in front of the house. By 5th March the gooseberries were foliating. Work on the flower garden continued, while in the vegetable garden apples were grafted and on 15th March *cabbages, coleworts &c* were planted followed by onions, carrots and savoys. On 20th April were sown *several small Garden seeds Lettice, Spinage &c.* By 25th May the cherries, geans and apples were beginning to blossom.

On 15th June there was *a dish artichokes uncommon early.* On 30th June some strawberries were ripe and the roses were in flower. By 6th July it was evident that there would be *no fruit of apples &c* that year. Peas, potatoes and strawberries were ready by 14th July. On 30th July seeds of spring plants for the following year were sown. It was not until 16th August that the gooseberries and raspberries were ready. All the soft fruits and the potatoes had been a good crop, but on 31st August it was again remarked that *there will be little or no apples &c this season.*

The garden hedges were clipped in October and dung was driven into the gardens in December. All the fruit trees were *covered round about the root with dung;* dung hills were made up in the garden preparing for the new *Hot Bed* mentioned in March of the following year.

On 24th February 1791 the gardener transplanted some of the best turnips to the garden for seed. On the following day five large pear trees, about seven feet high, were planted on the north side of the garden *to the West Plott.* On 1st March beans were sown and potatoes planted. A week later *curleys and cabbages* were planted. On 24th March *collyflower & some Salads* were sown in the new *Hot Bed.* On 30th March the *onions, carrots, beet &c* were sown by *my Gardner W.Smith* as quoted above. On 9th April annuals were sown in the flower garden.

By 16th May the garden produce was very backward; *none of the apples, plums, pears, cherries* were *yet fully blossomed.* Even by 31st May the fruit trees were not fully blown. A month later cabbages and all kinds of salad had been ready for some time. Strawberries were ripe on 13th July; it was 18th August before the spring plants were sown. Onions *and other garden stuffs* were *pulled up* on 30th September. In November the garden hedge was clipped. During December the gardener covered the strawberries, artichokes, raspberries and all fruit trees at their roots with *Pease Holme & Dung.* The gooseberries were pruned and dung was barrowed into the garden.

In 1792 work in the garden was first mentioned on 15th March when some fruit trees were planted. On the 16th and 17th garden peas and beans were sown. *Small seeds of kinds* were sown on 31st March, followed specifically by onions and carrots on 9th April. On 9th May the gean, strawberries, some plums and pears were in flower. On 31st May the potatoes and standard apples were in flower as were the peas on 7th June. On 13th June James Badenach *observed the Caterpillar amongst Gooseberries, picked of the damaged leaves & Inside & Cut off all new Shoots.*On 14th and 16th June *collyflowers* and *curleys* were planted; the borders were *weeded and dressed.* Strawberries and green peas were ready and the spring seeds were sown late in July.

In 1793 the early spring flowers were backward. Even by 15th February, there were neither snowdrops nor crocuses. At the beginning of March the gardener and two others were making dunghills in the garden and sowing the first peas. Cabbage was planted on 6th April and on the 10th a number of small seeds were sown; on 17th April onions and carrots were specified. In spite of the women and others picking the caterpillars off the gooseberry bushes on 14th June, the bushes were *quite stripped* by 5th July. However it was observed on 13th July that the caterpillars *had stopped in their ravage.*

The first garden peas were sown as early as 12th February in 1794; on 17th March they were above ground and the gooseberries were foliating. On the same day red cabbage was planted. On 20th March some small seeds were sown, followed by onions, carrots and *principal seeds* on 28th March. By 31st May the gooseberry bushes were being ravaged by caterpillars which were attacking the small fruit. By 30th June there were no more caterpillars on the gooseberries. A few peas and strawberries were ready on 1st July. On 8th July day-labourers helped the gardener to weed and align the rose planting on the west side of the garden. The plants for the following spring were sown on 4th August and *coleworts* towards the end of the month. The garden hedges were clipped in October. On 13th December some gooseberry bushes were transplanted.

At the end of January 1795 dung was brought into the garden. As a result of the appalling weather, it was not surprising that the next entry about the garden was not until 27th March when the strawberries were *dressed* and the *small seeds* were sown. On the following days early cabbages and *curleys* were planted; the ground was prepared for sowing onions and carrots. *Garden pruning* took place on 3rd April. Between 13th and 16th April cabbage was planted and carrots, onions and small seeds were sown. By 25th April the gooseberries were in flower and the strawberries by 31st May.

On 20th June the cold north east wind blasted the leaves of the standard apple trees, pears and plums. The caterpillars were spreading. The main crop of *collyflower* was planted on 25th June. By 30th June the peas were podding. On 13th July it was so dry that the *garden things* had to be watered. *Curleys* were planted on 24th July and the spring seeds were sown. On 31st July the garden peas and strawberries were only just ready on account of the drought, but it was not until 13th August that the gooseberries were ready. As usual the hedges were cut in October and the work was continued into November.

Spring plants were already appearing on 9th January 1796; by 17th January the bushes and flowering shrubs were budded and the artichokes were growing. On 31st January the *Daphne mezereum, Helleborus, Hamamelis* and bulbs had all flowered. During the third week of February the gardener was pruning the fruit trees and *dressing the espalier trees.* The *hot bed* was placed in the garden at the beginning of March. On 9th March the gardener was *laying the coleworts for seed;* two days later the *small seeds,* but not the onions, were sown. Early potatoes and cabbages were planted in the nursery on 18th March. On the 23rd the onion ground was prepared, on the 26th more cabbages were planted and on the 31st the onions *and almost the whole of small seed* were sown.

A month later, on 30th April, the annuals were sown in the flower garden. *Collyflower* was planted on 12th May and on 27th May *caterpillars &c* were attacking the bushes and trees to such effect that by the end of the month *the fruit Trees were mostly Blasted leaves & flowers – and the small fruits Gooseberries Excessive rich but begun to be Injured by Caterpillars.* However the lilac was in flower, the garden peas were *offering to bloom* and the onions and carrots were good.

It was a late and rather miserable season; by 19th June few swallows and no butterflies had been seen, ash trees were not in leaf and on 28th June James Badenach did not observe a good crop of anything but gooseberries. On 30th June the artichokes were the size of goose eggs; the fruit trees were blasted, but *other garden articles* looked good. The roses were *offering to blow,* but there were no peas and no strawberries yet.

Curleys for spring use were planted on 6th July and on 8th July the *spring plants* were sown *rather early, but ground in season.* Garden peas were ready on 14th July. No date was given for the ripening of the strawberries which were not ready on 10th July. The next and final entry for 1796 referred to the gardener dressing the fruit trees on 13th December.

In January 1797 it was seen that several of the fruit trees in the garden had been barked and *much damaged* by hares. Spring flowers appeared at the end of the month. During the first week of February the gardener root-dressed the espalier plum trees; snowdrops were out and the pears and other trees were in bud. The gardener *planted or made up early cabbage, curleys &c and furred them up.* The Daphne flowered on 21st February, about three weeks later than in 1796. Early potatoes were planted on the 25th and three days later some crocuses were in flower.

Onions, carrots &c were sown on 29th March; a month later the hawthorn was *not foliated,* though the *Imperial Crown* had just flowered. On 2nd and 3rd May the gardener *planted about ½ Boll Potatoes by the Dibble.* At the end of May it was reported that the caterpillar was *ravaging the gooseberries much, much earlier than we observed before.*

The *collyflowers* were planted on 5th June; thereafter between 8th and 18th June the caterpillars and other insects were destroying the gooseberries and fruit trees. The gardener did his best to *sweep* them off the gooseberries and although, on 17th June, they were said to be checked, the final verdict was that the fruit trees and the gooseberries were *quite destroyed by Caterpillars and other insects.*

On 27th June there was a hopeful note when the peas were podded, but further gloom followed. On 15th July the gardener was *houghing the potatoes devoured by crows.* On 16th July there were neither peas nor strawberries. Sowing of spring plants took place on 21st July and on the following day potatoes were earthed up. By 25th July peas and strawberries had been enjoyed for several days. Some fruit must have been had from the gooseberry bushes, because on 31st August it was said that the *gooseberries* were *nearly finished.*

On 14th September the onions were taken up and tied to pea sticks in the garden. The storms on 21st September broke the espalier frames. The last entry about the garden on 21st November was a sad one: *the Gardner hath dressed all the Hedges & most of his work is done for the present. He removes &c.* Presumably he left James Badenach's service at that point, but, as the last entry in the diary was on 25th November 1797, it was impossible to tell.

TRANSPORT OF COMMODITIES

There were five principal commodities that had to be fetched by cart to the farm from a distance. They were wood, coal, building stone, slates and lime. Four was the usual number of carts in use at any one time, but if more than five carts were required, others had to be borrowed from neighbours and friends.

Wood

When the diary began on 1st June 1789 work had started on the new barns. Timber for that purpose was obtained from Montrose, a day's journey involving a round trip of about 26 miles. Between 21st July and 3rd September a total of 16 cart-loads was collected on five separate journeys. At the beginning of December 1789, although not strictly at that stage to be categorised as wood, 25,000 seedling trees of various kinds were brought back from Stonehaven as part of a return load.

In June 1790 two carts brought home eight fir trees purchased at a roup at Kair, just over a mile north of Whiteriggs. As building progressed more timber was required from Montrose and three journeys involving eight carts in all were made in August and at the beginning of September.

In 1791 two carts were sent to Montrose on 30th July for wood, but earlier in the month one cart had been sent all the way to Finzean on Deeside for birch wood. It was more than twenty miles by way of Cairn o'Mount, up a very steep track and over the summit at 1488 feet (484 metres).

In July and August of the following year (1792) expeditions to Finzean were more ambitious. Three journeys were made with three flat carts on each occasion to collect birch and sawn fir wood. On the first trip the carts returned shortly after 9 a.m. on the day following their departure. The second journey was made two days later and on that occasion it was after 3 p.m. on the following day before they returned. The third journey was made to collect the remainder of the fir wood and they returned by 8 a.m. on the following day.

On 15th September an unspecified number of carts and people were sent to Pittengardener, a neighbouring farm only about a mile away, to collect thinnings of fir wood *about 18 year planted at 16^{ds} per two Horse Cart load, about 10 feet long & thickness of Man's arm.*

No journeys for wood were mentioned in 1793, but in December 1794 one cart was sent to Pittengardener *for some Thinnings of Firr for a dead Hedge N.W. Corner of Garden.* In August that year three carts were sent to Garrol Wood, about ten miles away, to fetch some fir wood that had been bought there. Later two separate journeys were made to Montrose with two carts on each occasion to collect timber for the Long Barn which was then under construction.

In January 1795, when there were some eight inches of snow on the ground, two carts were sent to Stonehaven for coals and *Oak Garrons* (wooden beams). On 22nd August one cart collected wood for the wright from Inglismaldie Forest about seven miles from Whiteriggs and a fortnight later four carts brought 22 fir trees from Garrol Wood. A return trip was made with two carts to Inglismaldie towards the end of September for birch wood, *moderate priced.*

In 1796 the only expedition for wood was in April when three carts were sent to fetch fir from Garrol Wood for flakes.

One cart was sent three miles or so to Arbuthnott for *2 Ash Trees bought there,* in February 1797. It was for thinnings that two carts went to the neighbouring farm of Pittarrow in April and in July three carts collected fir wood from Garrol Wood for farm use.

Coal

Coal was constantly required and the principal harbours from which it was collected were Johnshaven (6 miles), Stonehaven (10 miles), and Montrose (13 miles). On 28th October 1796 James Badenach recorded that in collecting coals from Johnshaven *it takes 7½ hours Carriage.* Once a cart was sent to Bervie (6 miles) for English coals and on three occasions coals were collected from Gourdon.

Sometimes coals were mentioned as being Scotch or English, but the different varieties were available at any of the places mentioned. The price of English coal was given by the barrel e.g. 1/9 a barrel in 1790; the price of Scotch coal in the same year was given as 8/5 and 8/6 with no indication of an amount, but, as it turned out later, that was the price per boll. *Great* coal and *Smithy* coal were also specified. The number of carts involved was usually mentioned, but the amount of coal was seldom given. When possible coal was brought as a return load after the carts had carried grain or some other product from the farm to the harbour concerned.

In 1789 two journeys were made to Johnshaven with an unspecified number of carts and seven to Stonehaven between July and December with a total of 17 carts.

In 1790 16 separate journeys were made to Stonehaven using a total of 26 carts and only four to Johnshaven with six carts. July and October were the only months when coal was not collected. For the year a total of 32 carts made 20 journeys; a single cart that was used on each occasion covered a distance of about 368 miles.

In 1791, with journeys in every month except September and October, Stonehaven was visited seven times with only one cart each time. Johnshaven was visited 13 times by a total of 28 carts, pease and barley being delivered there on several occasions. English coals were collected on two journeys to Montrose with 6 carts. Journeys for the year amounted to 15 with 34 carts.

In 1792 there were no collections of coal in July, October and December. There were two visits to Montrose with three carts each time, five to Stonehaven with a total of nine carts, five to Johnshaven with nine carts and one to Gourdon with two carts. Total journeys were 13 with 25 carts.

In 1793 February, May and October were the months when coal was not collected. There were four journeys to Johnshaven with a total of 10 carts, three to Stonehaven with four carts, seven to Montrose with 11 carts and when one cart delivered four bolls of wheat to Mill of Conveth, six barrels of English coal were brought back as a return load. Altogether there were 15 journeys during the year using 26 carts.

In 1794 May, October and November were the months without the collection of coal. During the first three months coals were brought back from Montrose where barley was being delivered and there were three other journeys giving a total of 12 for the year with 21 carts. Amounts were mentioned on 6th June when two carts carried 7 bolls of bear to Montrose; two bolls of coal were carried as a return load. On 29th March during one of the six visits to Stonehaven using 12 carts in all, no coals were on sale and James Badenach was obliged to a Mr Low for lending him some. Two journeys were made to Johnshaven with two carts on each trip. Journeys rose to 20 for the year using 37 carts. At the end of the year coals were said to be plentiful but dear at 8/ per boll.

In 1795, the year of the snow, no journeys were made in search of coal during the months of May, June, July, October and December. Eight journeys between February and April were made, often under appalling conditions, primarily to carry oats to Montrose, returning with a load of coals. For example on 28th February two carts were sent to Montrose at 4 a.m. with three horses and a load of 8 bolls of oats to each cart. There were no coals available of any sort. On 3rd March three carts with eight horses were sent with 14 bolls of oats, but they returned with only half a boll of coals. Again on 14th March 14 bolls of oats were delivered, but only six firlots of coal were brought home. Thereafter conditions improved and larger quantities of coal were obtained, 12 bolls on 24th and 12 barrels on 28th March.

Altogether that year 10 journeys were made to Montrose by 29 carts. Stonehaven was visited three times by a total of eight carts. On two of those journeys three carts carried 240 stones of English coals, 80 stones per cart, and on one occasion some displeasure was implied that the carts were *twelve hours absent.* Four visits were made to Johnshaven by 10 carts in September. On 9th February three carts had been sent to Gourdon for two bolls of Scotch coals at 8/3 per boll and on the 23rd one cart was sent to Bervie for 72 stone of English coal. Altogether 19 journeys were made with 51 carts.

In 1796 it was March before coal was required; other months without journeys for coal were June, August and December. Three journeys were made to Montrose with a total of seven carts, four to Stonehaven with 12 carts and eight to Johnshaven, probably with 19 carts, but on one visit the actual number was not specified. It was on that occasion, 12th July, that the carts were sent for coals for the smith and, for the family, English coals at 7/6 per cart. The total number of journeys was 15 with 38 carts.

In 1797, like 1789, an incomplete year in the diary, the only months in which journeys for coal were made were July, August and November. Two were made with four carts to Montrose, four to Johnshaven with 10 carts, one to Stonehaven with four carts and one to Gourdon with four carts. The carts which went to Stonehaven left home at 7 a.m. and returned by 5 p.m.. They were absent on that occasion for fewer than twelve hours. A starting time was not given for the journey to Johnshaven on 15th July, but the carts were back home before 2 p.m. Eight journeys were made altogether by 22 carts. On occasions porter was brought home with the coals and on 5th November 1796 there was brought with all the coals for the season *a fine Fanners from Stonehaven now at 3 10 S^r new & best construction.*

Lime

Rich deposits of lime shells were available on the coast at Mathers about nine miles south east of Whiteriggs by way of a steep track over the Garvock Hills. Milton, at the same location, was also mentioned. In the earlier years of the diary the lime was used mostly in the building works, but later on it was obtained for use on the fields. For a round trip of some 18 miles four carts which were dispatched on 8th July 1789 to collect 16 bolls of shells from the lime quarry at Mathers did well to return by 9 and 10 a.m. Three or four bolls per cart seemed to be the usual load. Altogether that year five journeys with a total of 11 carts were made to Mathers.

In 1790 12 journeys were made to Mathers using a total of 20 carts, all the lime for building. In the following year there were seven visits with 11 carts, again for building.

In 1792 there were several occasions when carts were sent to Mathers, but lime was not available. On 22nd March one cart was turned away because no lime shells had been burnt. On the following day a cartload of soured lime was brought from Mill of Garvock for the garden wall. Carts which failed to collect lime sometimes called at Johnshaven for coals on the way home. On 16th July two carts were left at Mathers overnight and returned with lime on the following day. Of the 22 journeys that were made to Mathers in 1792 all but five of them were for the land in general and the Haugh in particular. As many as 48 carts were involved. When no lime was available at Mathers on 21st May, a cart was sent to Stonehaven, but never again.

In July 1793 no fewer than 10 journeys were made to Mathers with 31 carts. The first visit on 16th July was undertaken by four carts, but one broke down and had to be replaced. Thereafter three carts were sent on each trip. All the lime was for the land, especially Gibson's Park. Two other journeys were made in September and October respectively, once more with three carts on each trip.

In 1794 lime for the builders was collected on three occasions with one cart in May, October and November. Lime for the land was sent for with three carts on four separate occasions in July.

In 1795 only two journeys were made, one on 1st May with two carts and the other on 19th May with one cart. Both were made to fetch lime for the builders.

Two journeys in 1796 were made using two carts in April and three carts in July. Both were for lime for the land.

In 1797 three journeys were made with four carts each time to collect a total of 48 bolls of lime shells.

Slates

Slates were required in large quantities during the building operations between 1789 and 1791 in particular. Those so-called *slates* were sandstone flags which were available in certain districts of the Sidlaw Hills. Those used at Whiteriggs came from the north side of the Hill of Turin in the parish of Aberlemno, Forfarshire. The thinnest of them were from about half an inch to an inch in thickness. They were formed into plates of about 14 or 16 inches broad and 18 or 20 inches long. They were of a greyish-blue colour, but after long exposure they assumed a grey or brownish colour. As they were very heavy, they required a much greater strength of wood to support them than slates and had to be laid in plaster lime or fog *(sphagnum polustre)* which had to be frequently renewed, and after all seldom made a roof that was either wind or water tight (Headrick, 1813).

The first recorded journey was to Flemington with two carts on 8th August 1789, but there must have been an impressive sight a week later when 27 carts, four from Whiteriggs and the remainder borrowed, set out at 9 p.m on the evening of 14th August and returned by 8 p.m. on the following evening with *as near as could be counted 2688 (slates) which is nearly one Hundred per Cart, but as the sizes are extreamly different, the smallest number was often the Heaviest Draught.* The round trip was more than 40 miles.

Another trip with 12 carts was made in September, but in October two carts were sent no further than Laurencekirk to borrow 436 slates; these were returned two years later to Jo Scott Smith on 19th October 1791. The slates from Laurencekirk were probably conventional slates because only two carts were required to carry that number, although three carts were used to return them, with four slates added. One of the reasons for not using conventional slates was to avoid paying duty tax on them (Headrick 1813).

In 1790, mainly in July, 15 journeys, using a total of 52 carts, were made to collect slates. On 3rd July 10 carts, five of them borrowed from neighbours, fetched 1009 slates; one cart was borrowed on 13th July, but otherwise only carts from Whiteriggs were used. Three journeys were required in 1791 with a total of 15 carts, four of which were borrowed; one journey was made with 10 carts, six borrowed, in 1792 and finally in 1794 15 carts, a dozen borrowed, were mustered to collect slates for the Long Loft.

Building Stones

Most of the building stones were obtained locally from a quarry at Dava, a round trip of about three miles. Quarriers were employed by James Badenach, who tried a new quarry at Arthurs, on his own land, in June 1789 and again in June 1790, but found it unworkable. During 1791, when specialised building work was undertaken, stones had to be brought from Flemington, the source of the *slates* and from Laurieston, near Mathers, about 10 miles away. Further journeys were made to Laurieston in June and August 1792, and in July 1794.

PART TWO

Transcription of the Diary of James Badenach Esq of Whiteriggs 1789–1797

March 1790

Wedy 17th — pretty strong frost white till 8 A.M., Nine Harrows at Bank & a Roller. Whole day which by evening Made it in good order — Pm Sowed Six Bolls Oats West side Black lands and dighting Oats for seed in Barns &c. Jobs —

Thursdy 18th — White frost pretty strong till 10 AM — 8 Harrows at Black lands, and a four Horse plough at the Kavels all day — Others at Garden, Barns &c

Friday 19th, Ther at 45° — very thick fogg & Closs till PM: but little or no frost — 8 Harrows and a four Horse plough at Kavels all day — leading some Turnips nearly shot, at Garden — and at sundry Jobs little grass

Saturdy 20th — Foggy & strong draught from SW. afternoon — One plough at Kavels, & all Harrows at Black lands nearly finished with including Old Town park 18 1/2 Bolls Oats — at Barns, Garden & Jobs

Monday 22 good with Horses £25th — yesterday strong draught from SW. night frost with much Hoar till 8 AM. — this day very strong draught from South — 8 Harrows finished the Black Lands Harrowing fine. Sowed three Bolls Oats West side of Kavels — four Horse Plough on Do at Garden for Onions, Carrots & other &c &c.

Tuesdy 23 Onions & Carrots — Hoar frost in the night — day fair very strong draught from the Eastward — One Plough, Six Harrows & a Roller on South side Kavels very Hard — Sowed Early 4 Bolls Oats on Do — Vegetables guide Chequered & Rye grass burning Black — seed

Wedy 24th — Early this morning Came on pretty Closs rain from the Eastward which Continued pretty Constantly the whole day — Sowed 4 Bolls more Oats on Kavels A.M. Twelve Horses and Sowed on same all day — Ploughed

A Diary of Agriculture, the Weather and Rural Affairs at Whiteriggs – June 1789

Mondy June 1st Thr 54° This morning a Great deal of rain which Brought down the rivers much & wet the grounds Immensely – no work without doors till P.M. when ploughs at giving the Second fur for Turnips in the Haugh. Masons, ditchers and other Tradesmen at Building Barns &c. Jobbs –

Tuesdy 2d Fine pleasant weather and drought from S.W. 3 ploughs at Stirring for Turnips, and others at Sundries – Barns dighting Bear. Garden, Quarry, &c.

Wedy 3d Fine pleasant weather & Great Growth – P.M. Sowed all of Thornylee with 11B 1ft Bear and dunging and ploughing of the Cottar land for the same,at Barns &c.

Thursdy 4th a Considerable deal of rain last night from Eastward day more or less such – ploughs at Stirring Turnip ground & finished whole Bear seed Cottar land 3 firlots – pasture & Hay grass looking very well – fields very wett.

Fridy 5th at Second furr Haugh for Turnips and driving stones and other matters for the Building – rain P.M. observed the vegetation more advanced towards Montrose.

Saturdy 6th Bleak cool weather from N.W. with some skiffs of rain, at ploughing. Barns &c. was informed that the Bear Seed was not near done towards Montrose

Mondy 8th Fine dry weather & pleasant from Southward – 3 men Sowed about 15 acres of Thornylee with Grass Seeds. P.M. Harrowed & Water furred the same in part, – Sundries.

Tuesdy 9th a Considerable Shower of rain night & morning, finished Harrowing & Water furring Thornylee after Braking in the ground for Turnips – Sold all my Barley to Mr Mackenzie for 14/ per Boll.

Wedy 10th good dry weather but Grey from S.E. – very Early sent three carts for lime to Mathers and one D° for coals to Johnshaven. P.M. Set up the Flakes on the Sward ley Haugh at Barns & Sundry Jobs.

Thursdy 11th Hazy dry weather & growing from S.E. – ploughs at giving 2^{d} furr for Turnips Haugh & Gibsons park &c. Rolling Thornylee – leading Sand, Water, Stones &c

Fridy 12th Thr 63° very pleasant Summer & Growing weather from S.E. – 2 ploughs at 2^{d} fur for Turnips & one cart leading Sand Water, & Stones for the Masons. Barns & Jobs. the Pasture is now very fine – Hear cattle of all kinds Sold very high yesterday at Trinity Muir.

Saturdy 13th real Summer weather from S.E. and very warm for the first time this

Season – 3 ploughs for 2^{d} furr for Turnips which finished – leading Stones and at Barns & other Jobs.

Mondy 15th Thermometer 64° Thick foggy weather from S.E. in the night time, but day Clear & very fine warm weather-great growth in the ground-Pease beginning to Bloom in Garden-begun cutting Grass in the field for the Horses Suppers, Still give a kind of two feeds of Corn beside fine Pasture.

Tuesdy 16th Thr 64° very warm weather from N.E. all day & Checkquing the grain on light farms & Spotts Considerably – Harrows & ploughs all day preparing for Turnips. Carts driving Stones – at Barns & Sundrys &c.

Wedy 17th Thr 66° very warm weather from N.E. & Crop looking fine all Harrows at preparing for 3^{d} furr for Turnips – driving Stones, and Busy at Building Barns – dighting Oats in the East Barn & Sundry Jobbs.

Thursdy 18th Thr 68° at 1 P.M. Closs foggy weather and very warm 3 ploughs at 3^{d} fur for Turnip in the Haugh – Trying a new Quarry on Arthur's, which does not promise real Success – every part of the Crops of grass & Corn looks well.

Fridy 19th Thr 62° This morning about 5 A.M. Came on a Smart Shower of rain from N.E. which lasted till about 10 A.M. when fair and good deal of drought from S.W. 3 ploughs at the Haugh which nearly finished – at Sundries – Pasture grass very fine & would require a Dozen more Cattle to keep it down – but prices are so high that it is Impossible to Buy to any advantage – prices of Grain of all kinds risen for export to France so that little will soon be left in this Country. pease Garden well bloomed Apples & also small fruit appearing good.

Saturday 20th Some small rain but of no Consequence & dark growing weather from N.E. – all Harrows at Haugh for Turnips – and other people at Sundries. Some people in the Neighbourhood begun to sow Turnips, but very few.

Mondy 22d Yesterday very pleasant warm weather from S.W. this day rainy & cold. Stormy weather from Northward – all Harrows and two ploughs at foot Haugh for forth furr for Turnips – at Barns dighting Barley & lofting Oats. Quarry Stones driving – and at Sundries &c.

Tuesdy 23d last night & this morning a good deal of rain & strong storm of wind from Northward which hath Blasted & Damaged vegetation considerably – no work of Masons without doors – but working at House work – flited the flakes before Dinner & after 2 ploughs at 4th furr for Turnip Haugh & two Carts driving Dung on D^{o} at rate of about 30 cart load per acre – driving Stones to Building Barns.

Wedy 24th Cold Bleak weather from N.E. and drought – the storm hath Blasted the clover a good deal & checqued the growth of the young Trees – the apples sett & small fruit promises well as doth all the Garden Crop – Bleak & Cold.

Thursdy 25th Cold Bleak weather from N.E. – P.M. some showers of rain and milder weather – all Carts and men wewmen at Dunging the Haugh for Turnips not finished. a Cart leading off clodds & other weeds from same. planted Collyflower

Fridy 26th very pleasant and Growing weather from S.E. – with drought- all carts driving Dung Haugh which finished what was Meant for Turnips there – ploughed the Cart Tracts again & carrying off weeds gathered off the Land & Making up an old Water Tract with them, leveled all the walls of the Barns &c Jobs.

Saturday 27th Frosty dews in the night and days warm and Droughty which hath checqued the growth of most vegetables Considerably – Early yoked the plough to Drill the Turnips Haugh and P.M. same. Men, Wewmen &c Employed about them Sowed about $1\frac{1}{2}$ acre covered & Rolled the same – Dismissed the Masons & other workmen for a fortnight.

Sundy 28th Thr 44° Cold stormy weather from N.W. and checking all kind of vegetation – observed the Crop towards Bervie very Bleat and looking very Backward Indeed – Hay Grass & Wheat thereabout very good as also Beans particularly some Drilled ones Belonging to my friend Walter Sime, Esqr, Collector of the Customs in Aberdeen.

Mondy 29th a strong storm of wind & Exceeding Cold from N.W. 2 four horse ploughs at 3^{d} furr Moorpark for Turnips – and others at sowing & Drilling Turnips & spreading Dung in the Haugh and at Sundries.

Tuesdy 30th more moderate weather than for some days past and towards noon came on showers of rain from S.E. – which was pretty smart about 4 P.M. – all Ploughs at the Turnip ground, Haugh & North park – nearly finished the first of about five acres or about one Half of what is to be of them this season. The Grass of all kinds very fine but no possibility of Buying Cattle to eat it being from 4/ to 4/6 per stone – most of the Bear sold for foreign Exportation to France at 12/ per Boll – Meal from 12/ to 13/.

Diary for July 1789

Wedy 1st very agreeable and growing weather all Day – ploughs and Harrows at North park for Turnips & finished all of the Haugh by Rolling – at sundry Jobbs – Beastials sold high at Cammock.

Thursdy 2d agreeable pleasant growing weather from various the whole Harrows all day at North park which finished – lead home some In holding fire – it is said Cattle sold very High at the Markett at Megrie near Stonehaven

Fridy 3d Fine pleasant growing weather – yet the Crop in many places looks but very poorly & foul; most of it here looks well – 4 Carts at Driving dung all day for Turnips N.Park. laid it on about 36 Cart load per acre which is rich dunging – Spread & Drilling all day – at the same – Neighbours far Back also with them. People at roads driving gravel from my Haugh.

Saturdy 4th Thick foggy weather all day & considerable growth – one Cart driving Dung from Cottar Houses & 4 Carts driving D^{o} from the farm yard all day – The Gardiner and Hired other people at the Turnips, Sowing, Covering & Rolling on North park of which did last Half – and at Sundries.

Mondy 6th Yesterday night an Immense quantity of rain from S.E. & stormy weather. This day fine, pleasant season – all people & Hired ones at the Turnips N.Park. Dunging, Drilling &c.

Tuesdy 7th Fine pleasant weather and Drought from S.W. – all people & others at the Turnip N.Park which finished nearly whole of – the rain hath Bettered the Crop with the warmth.

Wedy 8th a good deal of rain in the morning from the N.E. – sent 4 Carts to the Lime Quarry at Mathers for 16 Bolls shells which returned at 9 and 10 A.M. – a Cart driving sand & water all day to sour the same – P.M. Flited the Flakes & Jobbs. Observe the Turnips sown on Saturdy last Breering. observed some ripe strawberries planted this year.

Thursdy 9th Very fine weather – being the Parish fast for the Sacrament at Fordun no work without doors – observe the Crop towards Johnston middling.

Fridy 10th fine pleasant weather from S.E. two ploughs at East End of Haugh for fallow.Very much run with Thistles but not otherwise foul – at driving stones from the Quarry & Jobbs. observed the Crop but middling about Fordun.

Saturdy 11th a good many showers in the night, day some also – Two ploughs at the Haugh Fallow, one Cart driving stones from Dava. people at the Church – repairing & making Entrance off the Aberdeen road at the Cottar Houses &c.

Mondy 13th Some showers of rain from N.E. and variable weather – 2 ploughs at 2d Furr for Fallow at E. end of Haugh overrun with Thistles of about five feet High – 2 Carts driving rigging stones from Dava, and at Sundries – pasture very fine.

Tuesdy 14th Fair weather and Drought all day from S.E. – 2 ploughs all day at Haugh Fallow & two Carts leading stones from the Quarry – Sheep sold high at Paldy fair.

Wedy 15th a Brisk shower from S.W.in the Morning & same about noon – 2 ploughs at Haugh & two carts leading stones D° – P.M. 5 carts with 24 Bolls of Barley to Woodstock Distillery Mrs Mackenzie at 14/ per Boll. at Sundry Jobbs – the Bank Barley shooting & pease field Blooming – weeding Oats.

Thursdy 16th fine warm weather various – 2 carts at driving stones all day – ploughs finished 2d furr of the Haugh fallow – at Paldy fair not many Cattle and those that were good sold very high from 4 to 4/6 per stone when fed – not a great sale; about 6 P.M. by Thunder showers in the Hills the Water of Bervie came down immensely & overflowed much ground and did considerable damage.

Fridy 17th fair weather till past 10 A.M. when came on a Heavy rain with Thunder – at Quarry & driving gravel &c to mend the road to our House – and at Sundries.

Saturdy 18th very pleasant weather and dry – at 4 A.M. sent five carts to Montrose for foreign & sawn wood for the Barns & they only returned at 8 P.M. weeding Corn Haugh – in Oats some shooting. (28 pair couplings)

Mondy 20th Dry weather till 11 A.M. when came on Excessive Heavy rain till near 1 P.M. 9 Harrows at E.End Haugh and stone cart at Quarry – flitted the Cattle flakes – and at Sundries at Barns &c Jobbs. little rain about Bervie

Tuesdy 21st fair and pleasant weather from Eastward – sent four carts to Montrose for wood – one Cart driving Gravel to mend roads &c about the House – weeding Thistles. Green pease.

Wedy 22d Fair and Cold drought from Northward P.M. – 2 ploughs at 3^{d} furr Haugh – one cart at stones all day, one D^{o} driving Gravel to mend roads here.

Thursdy 23d Cold Bleak weather all day from Northward – 2 ploughs at Haugh fallow, and cart at stones driving & P.M. 3 Harrows at Haugh. Sundry other Jobbs.

Fridy 24th Very warm & pleasant weather – 2 carts driving stones all day – ploughs and Harrows at Haugh Fallow – a Cart brought Coals from Stonehaven – & Sundries.

Saturdy 25th frosty Rhime for two nights past day fair & warm – sent 4 Carts to Montrose for wood – 3 Harrows at Haugh fallow – at Sundries – Cattle selling High.

Mondy 27th Yesterday fine weather till 3 P.M.: a considerable shower of rain for two hours. this day fair & warm from S.E.: 5 Hay cutters at Goulds Hillock a fair crop – 2 Carts at Quarry & two ploughs.

Tuesdy 28th very fine dry weather from S.E. – 2 ploughs going all day Haugh fallow & two Carts driving stones from Quarry – at Sundries – Sowed spring seeds in the Garden.

Wedy 29th Strong dew in the night day fair about noon Came on several showers which continued more or less all day – 4 Carts to Montrose for wood – and a Plough at Fallow Haugh.

Thursdy 30th Fine weather – but some slight showers P.M. – Carts &c Ploughs till Brakefast – then all hands at Garvock Markett – few cattle & dull sale owing to no demand from Southward – Engaged the servants about usual rate – observed Cattle of from 10/ value to £7:10 S^{tr} in the Markett. Observed the Crop in Garvock better than could be supposed in such Backward season.

Fridy 31st This day fair & warm weather various – our people and Hired ones to the amount of 12 Gathered and Coled Goulds Hillock Hay of about $3\frac{1}{2}$ acres into 22 hand coles about 40 stones each – in tolerable condition – On the coast of the Mearns not half the Quantity of rain as with us – and there is said to be much more in the North Country than here. prices of grain rising Barley 15 to 16/ Oats about 12/ Oat Meal 12/ – Wheat none but in the South about 32/ per Boll – the Bank Barley filling and pease the same – the Oats not much out of the Hoses. Gooseberries & rasps not yet ready for use. Esteemed a bad year for apples & other fruit – Hay from 3^{ds} to 3^{ds} $\frac{1}{2}$ per stone.

Diary for August 1789

Saturday 1st a small shower of rain Early in the Morning day fair – Stone Carts & Harrows morning at fallow after Brakefast our own people & Hired ones to the amount of 11 gathering & coling of Hay Goulds Hillock which did part of, and stopped by rain at 4 P.M. Sundries.

Mondy 3d Yesterday fair weather till past noon when some showers, but slight –

this day fair weather but little drought till 3 P.M. some Showers from the Eastward which stoped the Hay Coling – morning & P.M. Horse & hand houghing first sown Turnips haugh looking well – 5 men cutting Hay Bridge Shade.

Tuesdy 4th This day fair weather and drought from Eastward – till Brakefast at Turnips Horse & hand Houghing after D^{o} all hands Gathering & tramp coling of Hay Cowfold – Tradesmen, Masons, Wrights &c.

Wedy 5th Heavy dew in the night day fair about 10 A.M. when rain from Eastward – having more of D^{o} stopped a good deal work without doors – sent 10 B^{ls} Barley to Woodstock – turning Hay Bridge Shade.

Thursdy 6th little drought till near noon when good, at Turnips till after Brakefast when at Hay Bridge Shade – in Small Colls – and at Sundry Jobbs –

Fridy 7th very pleasant & fine hay weather from Westward – all people and Hired ones to amount 10 all day at tramp Coling Bridge Shade Hay in good order – others at Sundries – The crop looking well but late.

Saturdy 8th Dew in the night day very pleasant and strong warm Drought from S.W. – 14 people at Hay coling which finished in fine order – 2 Carts to Flemington for Slates 13r^{ds}14 score & returned by 9 P.M.

Mondy 10th Yesterday and this, exceeding warm weather & dry – this day 9 Harrows at Haugh fallow – all day – one cart at the Quarry P.M. – others & hired ones Horse and hand Houghing of Haugh Turnips – Cutting Hay for seed.

Tuesdy 11th Thr 66° Heavy fogg in the night – Day Extremely clear & Hott, Harrows & plough at Haugh Fallow all the day, others & 2 hired people at Turnip dressing, hay &c.

Wedy 12th Weather as yesterday – only P.M. dark and some appearance of rain from E. A.M. people at Turnips, and P.M. at pulling & Watering of Lint from Back Yd.

Thursdy 13th Very pleasant and dry warm weather from the Eastward. Covered the Lint – 9 people Threshing and Coling of Hay for seed Calf ward Haugh – others at Sundries.

Fridy 14th pleasant warm dry weather – people finished threshing of Hay about 6 or 7 Bolls & Coled the same, after at Turnips dressing Northpark – tradesmen as usual kniting the Barn Couples and at Sundries &c. Hear that the price of Cattle fallen in Lammass Muir.

Saturdy 15th Thr 67° very agreeable and pleasant dry weather all day – all spare people at Turnip Horse & hand Houghing Auld North Park – 4 Carts to Turin for slates which set of last night at 9 o'clock-and returned by 8 evening. 27 Carts Brought as near as could be counted 2688 which is nearly one Hundred per Cart, but as the Sizes are extreamly different, the smallest number were often the Heaviest Draught – Sundries. Gooseberries & small fruits ready.

Mondy 17th Thr 66° uncommonly fine weather – very warm. Own people and Hired to number a Dozen at Dressing north park Turnips – two ploughs at taking away first furr from D° – after in P.M. a three year old Quey Clover Heved, stabbed her on nearer side from two to three inches, much Air, and Grass fermenting I took out with my own Hand, but to no Effect as after the Cow fell & was stabbed she recovered a little & swelling subsided – but as she fell in a Ditch on the wrong side, all was to no purpose & we were Obliged to Bleed & dress her –

Tuesdy 18th Thr 67° Uncommonly fine & warm weather – a Dozen of people did out first dressing of Turnips before Dinner – after that Begun leading in the Hay into a sowe 16 feet Breadth by 70 length – Intended with a Considerable Grow – find the Coles by no means Hoary towards the Bottoms –

Wedy 19th a very Closs & Heavy Dew with a portion of Haar – day exceeding warm & Hott – Early yoked three Carts & two Horses to Draw in whole small Coles Entire into a Sow – 17 people in Employd –

Thursdy 20th Thr 64° This day fine dry weather and pretty strong Breeze from the S.W. which Incomoded our Building of the Hay Sowe a good deal. Continued with our own, Hired & neighbours people about 22 for Gather & Build but could not finish y[e] whole.

Fridy 21st fine agreeable weather from S.W. – all people & Hired finished a very large Sowe of Hay in fine order, against Evening – at Sundries – the Crop Beginning to fill – Cattle lowering in price and Grain on the raise.

Saturdy 22d very pleasant dry weather from West – but at night a smart shower of rain for a short time – in many places an Excessive deluge of rain – finished the Sowe of Hay dressing & dighting the seed saved putting her up – Carts driving stones to cow Byre.

Mondy 24th Yesterday fair & pleasant weather from W – this day same, ours & other people Second dressing of Turnips Haugh, and wrights & others putting up the Couples 45 in the Barns – driving stones.

Tuesdy 25th This day fine weather & stormy from Westward – people at 2[d] dressing of Haugh Turnips which look well, others fliting the Flakes, leading stones & other various Operations, too much for a Farmer.

Wedy 26th Fine pleasant and dry weather from Westward – Two Carts to Montrose for wood, and others leading stones from Dava – Houghed out Haugh Turnips and leading in Rushes, Sprots &c for Thatching Hay

Thursdy 27th This day fair & Bleak weather from S.W. – dighting & sent 20 Bolls Barley to Woodstock – at Laurence fair a number of Cattle, but the prices so high that as far as I could observe not much sale.

Fridy 28th Fair weather and strong drought from S.W. – dighted & sent 20 Bolls of Barley to Woodstock Distillery Mrs Mackenzie – leading stones from Quarry at Dava, reding up Barns for claying the Floors.

Saturdy 29th Fair weather and dry till two P.M. came on a very heavy shower of rain which lasted more or less all the afternoon – Brought 4 Carts full of Peats from Moss of Inch Break at one shilling per Cart – after at Smithy, Cutting Horse Suppers – One Cart thrice this day to Dava for stones to Cow Byres

Sundy 30th The rain continuing heavy till midnight yesterday – and Brought down the River pretty high by day light, the grass on the water sides being cut formerly, not much damage hath been done by the flood – it hath however Injured some of the pasture & a very small portion of the Turnips &c.

Mondy 31st This day fair weather till P.M. some rain but after 3 P.M. came on Heavy rain & Course weather from N.E. – Sent 18 Bolls 1 firlot of Barley to Woodstock P.M. – and one Cart driving stones from Quarry – Masons near finished Cow Byres in the Shades, and at Sundries on the Farm. The prices of Cattle very High from four to five shillings per stone either to feed or to Keep – The Turnips pretty good but to late in Sowing here to be what might be Esteemed a first rate Crop. Some little Harvest of Cottar Land, but nothing else yet in this Corner – the Oats in particular though not so here very late and unless very good weather cannot ripen this season.

Diary for September 1789

Tuesdy 1st The Heavy rain from N.E. continued all night & a good part of this day, which Brought down the Rivers very high, as much as I have seen at the highest a little before noon, but subsided much towards evening, Obliged to swim the Cattle from the Haugh morning to the upper Grounds to pasture all the day – little or no work without doors – at Sundries.

Wedy 2d Closs foggy weather and very Bad weather at this Season about 3 P.M. came on very Heavy rain from the N.E. which lasted most of the afternoon –

Thursdy 3d Fair weather this day and but very little drought – sent 2 Carts to Montrose for remainder of wood, and three others carried down 18 Bolls oats to Mill of Garvock to Mill & Brought home stones from Dava – at Sundries.

Fridy 4th Fair & drought from Northward all day – at Sundries. The Hay Sowe by the late Immense rains very much wet & damaged on the upper part & obliged to turn a Considerable part of it down & to carry out & spread same on the field & to turn up more on the rick – our Neighbours around in like condition –

Saturdy 5th Strong Breeze from the Northward all day & drought except some skiffs – turning & working at Hay rick. 4 Carts own, 5 Pittarrow & two Arthurs & 1 Bridge of Leppie for Slates to Pitscandly for Slates – one of ours Broke although a new one – exceeding disagreeable season.

Mondy 7th Yesterday fair weather and some drought from the S.W. this day rain early morning & more or less all day – Making ropes for Hay Sowes, flited the flakes & cutting rushes &c for Hay Thatching & Jobbs.

Tuesdy 8th This day mostly fair but some small showers P.M. after Brakefast begun Harvest by cutting up the Pease full ready on Kavels, rich of grain but not very Fodery – some carts at Johnshaven for coals.

Wedy 9th Strong Gale from S.W. in the night, day fair till past two P.M. several light showers of rain, all Hands shearers Included at turning down carting out & in Hay to Rebuild the Hay Sowe which finished by night in Midling order & Jobbs.

Thursdy 10th Mixed weather & some light skiffs from S.W. at Hay Thatching all Men which nearly Compleated – & Built the remainder into a stack – finished Cutting up Pease & Begun Barley

Fridy 11th White frosty Rhime in the night day fair & for the first time like Harvest weather – all people at Bank Barley rather too ripe but good crop, laying to the fur to the Haugh Turnips also good crop – The Harvest now almost general. Sundries.

Saturday 12th very strong drought from S.W. all day, all our shearers at Bank Barley rather too ripe – finished Horse Houghing Haugh Turnips – Sundries – Hear Cattle were cheaper at Bartle fair.

Mondy 14th Yesterday a Strong Gale from S.W. morning – P.M. some showers of rain from D° – this day very heavy dew – but all our own people & 6 Hired people from Polburn 1/ Men & 8^{ds} Wewmen per day & Meat at the Bank Barley the best part of which finished.

Tuesdy 15th last night & this morning very strong frosty dew, which Blackened the Potato stems & partly of Clover – our people at shearing Barley which finished by about noon- & after at Kavels Barley good crop. Brought home clay from Brownies leys for Barns.

Wedy 16th Strong gale from the North all day – at Shearing the Kavels Barley – turning & leading of Pease in fine order, laying Barns with clay and at a variety of Jobs. The Slater one Man Begun dressing them.

Thursdy 17th all day a strong drought from Northward and cool weather. three Carts led in all the Pease before noon – after partly led in three stacks of Bank Barley moderate sized stacks, in a new Corn Yard N.W. Head of the Bank.

Fridy 18th Fine pleasant dry weather – after Brakefast finished the 3 stacks of Barley – & Thatched all the five Pease ones – wewmen & P.M. others at shearing the Kavels Barley – at Sundries – put down 2 Bee hives rick.

Saturday 19th very Early morning came on a strong Gale of wind & rain from N.E. which lasted for 24 hours and laid & Damaged the crop considerably – no work without doors – at Sundries within.

Sundy 20th Variable weather but little or no drought – Observed the crop to & from Aberdeen very late in many places, & unequally ripened but cutting it up in no condition – the Oat crop will certainly prove but very Indifferent in most farms.

Mondy 21st Fair weather from Northward – people took up the Lint; dressing N.P. Turnips, at Barn floor laying Barn floors with Clay – finished cutting up Kavels Barley. Th^{r} 134 – and at Sundries.

Tuesdy 22d Fair and some drought from S.W. people at the Barns floors, at Turnips dressing too late and at sundry other Jobbs – Some people leading &c.

Wedy 23d Fair & considerable drought from Northward – 3 ploughs laying to a furr to Turnips, morning – after at Cutting rushes, sprots &c for Thatching & P.M. fliting the Flakes Haugh. Barn floors &c.

Thursdy 24th a slight shower morning, at Sundries all day till dinner time when Begun shearing Oats Back Yards full ripe and good crop – Horse Houghing Turnips.

Fridy 25th Fine dry weather from S.W. but dark – 3 Carts leading in Bank and Kavels Barley into 3 stacks which finished by night – 6 wewmen shearing oats Back Yards – Sundries.

Saturday 26th a good deal of rain last night & frost Early A.M. day fair & pleasant weather from Southward – wewmen shearing Oats E. Haugh – Men Thatching Barley stacks all day. Put the Cattle on the fields which were rid – 2 Carts for lime.

Mondy 28th Yesterday about noon Came on a Heavy rain from the Southward – which lasted till 3 P.M. – this day fair but frost – all shearers cutting up Rushes, sprots &c for Thatch and at sundry other Jobbs on the farm.

Thursdy (Tuesdy) 29th Early this morning high wind & much rain from S.W. – day fair & drought, morning cutting up Oats E^{t} Haugh in Gayte, & after Brakefast begun cutting up E^{t} side of Thornylee Bear, a good crop, & remarkable fine head & fine appearance of new grass – Slaters dressing stones &c.

Fridy (Wedy) 30th Frost in the morning day fair and though Coldish very fine harvest weather from S.W. – all shearers at Floors dry till Brakefast when at E^{t} side of first sown Bear Thornylee – Thatched & sorted the large stack of Barley of which was somewhat suspicious, but it is quite well.

Observe most people cutting up Oats & Bear by no Means ripe, owing to Emulation and the season having pretty much advanced – the Pasture is still very good, and Bestials are thriving accordingly – all our Garden crop in general Exceeding rich more than Common. the Price of grain not settled though not risen of late – the Cattle Marketts not fixed either though no sale at the latest ones at Dundee. Have not yet got many Turnips – Manufactures middling.

Diary for October 1789

Thursdy 1st very early this morning came on very Heavy rain and storm of wind from the S.E. which lasted severely till near 9 A.M. when it was scarcely fair through the day – about sundry Jobs, Causewaying a Cow Byre &c. The river was very high at $\frac{1}{2}$ past 11 A.M.

Fridy 2d Strong drought & Rough cold weaather from Westward – all the shearers at Cottar land Bear 15 Th^{r} – after at W side of Robs Hillock park good crop – at Sundries.

Saturdy 3d variable weather-shearing dry, & wet alternately all day on the Old Town Hillock park fine crop Oats-a Cart to Stonehaven for coals & Sundries – several showers of rain & Grounds as wett as any time this season – a growth in Everything.

Mondy 5th Yesterday fair weather though a little shower P.M. – frost in the night pretty strong – all shearers Early shearing Old Park Oats in Gayte, and after 11 A.M. stooking-a little after 4 P.M.Dismissed by heavy rain from S.E.-Observed several leading-set up down shocks.

Tuesdy 6th last night Came on much rain from S.E. which stopped all work of harvest all the day – people at laying Byre floor, and leading rushes &c P.M. flited the Cattle flakes Haugh Grounds Excessive wet &c.

Wedy 7th In the night and almost whole day almost Constant rain from the Eastward, & rivers much out.Scarcely fordable. all people at Jobs about the Town, laying the Byres, riding of Earth &c about new Barns, into the Dung Court – and at Sundries. Very Bad Harvest weather.

Thursdy 8th Variable weather – people at Sundries till Brakefast & after at Auld park Oats in Gayte, at Drumliathy Markett a good many Cattle but being high prices but little sale – Observe little Oats Cut Down towards it.

Fridy 9th last night, this morning, and day almost Constant rain from the N.E. – rivers much out & falling in by turns through the day – sent two Carts for straw to Johnston and at sundries about Byres sorting, causewaying &c. I observe the Oats in stooks & Growing sadly smashed put 4 Cattle for Turnip feeding in New Byres all night.

Saturdy 10th last night, and this day after 2 P.M. constant and Heavy rain, which hath wett the fields & roads as much as I ever Remember of – Shearers Gayting from Brakefast to Dinner time – the corns to cut & those cut in a very sorry condition.

Mondy 12th Yesterday fair weather & drought from S.W. but about 7 P.M. came on Heavy rain & wind – this day fair but no drought. All the people at shearing Oats Robs Hillock Oats in Gayte – at Sundries.

Tuesdy 13th last night & this morning heavy dew, but day fair and some drought from the South-all Hands at Robs Hillock park Oats-Gayte till P.M. when Bound & stooked it – Saw several of our Neighbours corns in very Bad order. Every sheaf almost wet & grown- Set up all ours, and on the whole not half so Damaged as I expected but owing to so much rain there will be very Considerable loss in Shearing after this.

Wedy 14th Fair & rough weather from S.E. – Shearing & Binding till 10 A.M. when came on very Heavy rain which lasted till one P.M. when abated – all the people at sundries. Thatching the West Old Barn for a season. Sent a Cart to Stonehaven for Coals & Sundries Hair for slaking lime – to the new Barns.

Thursdy 15th Early this morning came on Excessive Heavy rain & high wind from the Eastward which Brought down the Rivers much out & as high as for months by past – no work without doors – rained whole day. The fields much poached by the Cattle – at Sundries.

Fridy 16th In the morning closs foggy weather day fair & very warm sunshine – Thatching the old Byres with Rushes till near noon when Begun Shearing out Robs Hillock park and after at Back Yards Bound & stooked – at Sundries.

Saturdy 17th This day fair weather and but little drought from Southward – all shearers at Back Yards Oats. Great loss by growing in shearing & other causes more than the seed and also Damage in the Fodder – there must be but a very Indifferent Crop of Oats & the pease on Coast.

Mondy 19th Yesterday rain till 10 A.M. from Eastward – rest of the day fair & some drought – this morning a white frost but day fair & Drought from South – Early flited the Flakes, after Brakefast at Shearing of Oats Floors &c in stook & in Gayte – Observe a good deal of the Oats Cut up in stooks, those in Gayte & uncut up Greenly Growing.

Tuesdy 20th last night & this morning very Heavy rain from the Eastward – people shearing out Floors Oats & after dinner at Dickies Haugh, but stopped by rain & storm before 4 P.M. – other people at Sundries. Sent a Cart to Mathers for 3 Bolls lime shells.

Wedy 21st last night a strong Gale of wind from Southward and morning skiffs of rain – after Brakefast all shearers at the Back Waters partly Gayte & partly stooked – some of it & the first cutt Oats very much grown – 2 Carts to Stonehaven for Coals.

Thursdy 22d Strong frost White & Black in the night – day fair and pleasant warm weather – Our people at shearing out Back Waters Oats – and after wewmen at Dickies Haugh D° – the Men at setting up all Gayte & opening up the North Head sheaf in gayte – Observe our Neighbours opening the whole stooks into Gayte – & soon leading it certainly in Bad order.

Fridy 23d Closs foggy weather & small rain from the Eastward, all day shearers at shearing out the Oat Crop by Dickies Haugh : Cutting up Sprotts &c for Thatch to Stacks covering – and thatching Byres &c – Jobbs.

Saturdy 24th last night dark weather – day some wind S.W. with slight skiffs of rain at Intervals – Early begun Binding last cutt Oats in Gayte, and leading home same, Continued to Lead in two stacks of same on Timber in very tolerable Condition – Sundries

Mondy 26th fine dry weather from Southward – all Hands Early begun Binding up Gayte and three carts Leading of the same into 2 stacks on Timber in good order.

Tuesdy 27th fine dry weather from Northward – Early begun binding Gayte, Back Waters, Haugh, and in different fields in tolerable good order, got in best part of three stacks on Timber for Security &c.

Wedy 28th Frosty dew from N.E.: day dark weather – Early Carts driving Gayte from Auld park & after at Haughs D° and P.M. begun Braking & wyling of stooks of which in some from three to 5 sheaves to open – sent 2 Carts to Laurencekirk for 436 slates not long Mead

Thursdy 29th Strong Black frost also quite White till 10 A.M. Early yoked 3 Carts to lead in Oats, from Different fields picked by but some of them not very Judiciously done – at Drumleathy Markett 2^{d} time fine Cattle and of these no sale – Observed the Harvest but Backward towards that place.

Fridy 30th Strong & Cold Gale from the Westward all night – Early begun shearing West side of Thornylee Bear into stooks and Men setting out & picking first shorn D° by no Means Damaged, & also Begun leading of what was dry – but about noon Came on severe storm of sleet & after of snow which Covered all the fields white & lasted till 5 P.M. people at Barn threshing, rope Making &c.

Saturday 31st The storm of snow & High Gale Continued all night and this day at Intervals till 2 P.M. the snow lying several Inches deep, and on the Hills and rising grounds very Considerably deep – which must be of Immense dammage to the uncut Corns as well as to those not yet got in. This weather is uncomonly Inconvenient to me who hath so many Houses new & old in no wise prepared for so Early winter – It is generaly suposed that a great part of the Oat Crop is lost. Considerable part of the Pease D^{o} – and prices of grain is rising accordingly – Cattle very much fallen in price – the spinning likewise fallen

Diary for November 1789

Mondy 2d Yesterday fine weather but little drought, this day fair & dry – Early shearers at Thornylee Bear one half of it uncut will be lost by leading & after dinner lead home some gayte Oats in but Midling order, and at sundries – Harvest Backward as here in the Hollow of Angus, rain Evening.

Tuesdy 3d last night came on rain from Eastward, and which Increased all this day so as to prevent all work without doors, except thatching three stacks of Oats, rolling at the Barn floors – obliged to swim the Cattle through the river to the flakes Haugh – at Sundries – finished Slating ye Barns.

Wedy 4th This day fair weather in general though some slight skiffs P.M. people at Clearing out Rubbage about the Barns, and flited the Cattle flakes – the rivers last night very high, and the Grounds as Wett as for ever remembered – the roads very Bad.

Thdy 5th frosty dew in the night – after Brakefast begun & finished cutting up Thornylee Bear in ill order and after finished all shearing by the patch of Oats on the North side of the river & Carted it over to the South side of the Corn Yard in Gayte, at Laurencekirk Markett a good Many Cattle but no sale.

Fridy 6th This morning came on a severe storm of wind and rain from Northward which prevented all work without doors. at Barn floors rolling & chaffing lead home 2 cartful Turnips from Haugh – Brought in the whole Cattle to Shades & Byres – Evening – the river very much Out nearly as much so as this season. Gave our home gathering feast.

Saturday 7th This day various heavy showers & fair weather from the Northward, took in some Barley to Thresh for Malt to selves & threshing at same, dismissed the Shearers, Quarriers & others for this season.

Mondy 9th Yesterday cold weather from the Northward – and P.M. rain from same quarter, this day fair weather from the North, set up all down gayte, and at a variety of Jobbs on the farm – much Harvest work to do yet.

Tuesdy 10th last night & this morning very strong frost – people own Thatched a stack Oats, and after cutting Rushes &c for doing more of them, leading Turnips for the Cattle 8 of them stalled for good, Milch Cows & Calves &c. Observe several farmers ploughing but grounds very wet and in no condition for it – no drought.

Wedy 11th This morning a strong frost & fields white till 11 A.M. – people at

Barns threshing, Corn yards Eddering and at Digging Potatoes field – leading Turnips, two Carts, Housing the Carts, in Shades &c Jobs. rain P.M.

Thursdy 12th last night and this morning very Heavy rain from the S.E. which is very Miserable Indeed – people at Barns, threshing & drawing for Thatch to stacks – Holed up all the Potatoes & Housed them. Sundries.

Fridy 13th last night frost, this day fair till 11 A.M. when came on rain & severe Gale of wind from S.E. till half past 4 P.M. which drove every Creature within doors, and brought the rivers much out at night, at sundries about the farm &c.

Saturdy 14th The storm continued last night till this morning when drought & strong Gale from S.W. – morning set up the unfortunate Gayte of Bear Thornylee and after it yoked 4 Carts single horse, and lead in a stack of West side Bear into a hollow stack fit only for Horse Grain – Observed a considerable part of the crop out towards Drumleathy.

Mondy 16th Yesterday fine dry weather – but from the rains last night everything wetter than for some days preceding – this day fair till 4 P.M. some showers – 2 Carts to Stonehaven for coals, and two D° leading Home Turnips from Haugh – Others at Threshing, drawing straw & thatching of stacks &c Jobbs – am Credibly Informed that the Crop at Morphie, Dun, Kirkside &c is Infinitely in worse order than any here about, Having a great part Oats and that on Deeside, Doors & other places it is in like Condition – Obliged to stab a Cow hoved by Turnips at the point of Death.

Tuesdy 17th This morning a Heavy rain from 6 A.M. to 9 – at sundries about the farm – sent for the Lint 6 St & one half lint weight from Pitnamoon – afternoon again more rain from the N.E. – It is Certainly one of the worst seasons ever remembered by the oldest persons.

Wedy 18th last night an Immensity of rain & this day last part of it little fair weather – the rivers out Considerably – sent 4 Carts to Stonehaven for Coals – other people at sundries about Corn Yard & at a variety of Jobbs – Many houses falling by the rain and for want of being kept dry – at a roup of Pittengardner & Pittarrow farms about three out of 4 Dozen Cattle sold tolerably – little or no ploughing as yet.

Thursdy 19th Showers at Intervals through the whole day from N.E. & very disagreeable weather Indeed – people Housing Coals and D° all the deals sawn out for the Byres next season Sent to sundry places for Smithy Culm to Harden the new Barns floors – Have now got them near ready for use.

Fridy 20th last night and all this day a very severe gale of wind from N.E. attended with very heavy rain which brought down the rivers, and deluged the whole grounds – no work without doors, at a variety of petty Jobbs within Doors – two men Threshing Barley Constantly. Observe the Oat Gayte growing.

Saturdy 21st last night & most of this day Cold Stormy weather from N.E. – and a great deal of rain, thatched out the stacks – finished threshing stack of Barley for Malt to selves &c – begun threshing in the new Barns very damp & in but bad condition.

Mondy 23d Yesterday small showers but on the whole Drought from N.E. – this day mostly fair & Considerable D° – lead home four cartful of Thornylee Bear from Gayte in no bad order into the new Barn & one man threshing it D° – dighting Barley for Malt.

Tuesdy 24th last night and this day very strong gale of wind & drought from Northward, every Body at Binding & leading home the first Cut Bear on Thornylee in tolerable order into a Hollow Stack and turning over the Broad Band in Bad order.

Wedy 25th last night some skiffs of rain & Rough weather with white frost in the morning – day fair, but little drought from N.W. – turned over a Half stack of Oats into a small one. lead Home the Bear Cottar Town & part Broad Band into a small stack &c. Jobs.

Thursdy 26th frost in the night day fine pleasant weather from N.W. Brought home all the Thornylee Bear into new Barns & settled the Crop, people Threshing Out some of the worst of the Oats by Buffing & then giving it to Cattle. Observe most people have got in their grain.

Fridy 27th a remarkable fine night & day of drought & fine weather from N.W. – with slight frost. all people at Barns threshing, Leading home Turnips & other Jobs – towards Stonehaven Observed the Crop mostly inn.

Saturdy 28th very fine dry weather from the Westward – at 4 A.M. sent 3 Carts to Marykirk for Corns & fodder to a Neighbour farmer Removing from that to Gyrets Myre – at Barns and other Jobbs – roads bad –

Mondy 30th Yesterday very strong frost so much as would Have stopped ploughing – this day the frost went early off – and appearance of rain or snow but it Keeped off – all people Chaffing & Dighting of Barley for Stonehaven till 7 P.M. : price about 15/6 Montrose Measure – Believe most of the Crop is now got in the Mearns, though much must be in very bad Condition – prices of Grain not fixed – Bear from 13/6 to 14/6 & Barley about 1/ more per Boll. Cattle both lean & fatt are rather fallen in price but as yet by no Means low – Butcher Meat from 4/ to 4/6 per Stone, the Manufactures rather dull yet Business is still Carrying on.

Diary for December 1789

Tuesdy 1st Sent 4 Carts to Stonehaven with 12 Bolls of Barley – and Brought Home 2 Bolls Scotch coal & 12 Barrell English D° – with Iron – and about 25,000 seedling Trees of the various kinds to Transplant into a new Nursery Made here.

Wedy 2d Strong rough Bleak weather from S.E. – sent one Cart with Barley to Mill Pitranean and one D° leading Turnips – all other people at new Barns dighting of the Damaged Bear & Oats to Kiln dry – three Gardners at Nursery dressing – no ploughing yet.

Thursdy 3d Frost last night, but moderate day fine weather from S.W. yoked two ploughs to put up E^{t} Haugh fallow into rib – One Cart carried 12 sacks Bear wet to Mill of Garvock to dry so as to preserve for Meal or Horses – 3 Gardiners planting in nursery forest trees of kinds. took in large stack Barley.

Fridy 4th fair weather till about 8 A.M. when a considerable shower & high wind from S.W. – some more at the Intervals – three ploughs at the Fallow E[t] End of the Haugh in very fine order – took in part Barley S Bn.

Saturdy 5th Dark Blowing weather from S.W. – 3 ploughs finished putting in rib the fallow – The Horses get Birled Bear and Damaged Oats P.M. – Observed most of the Oat crop Hutted & Stacked in the fields towards Montrose, Morphie, Balmaledy, Pitcarvie &c & pease in many places Drumlangan &c lying Out.

Mondy 7th Yesterday Blowing dry weather from S.W. – this day the same – three ploughs at the Haugh Floors ribbing for Turnip – P.M. Employed at sundries about the farm – roads in many places Bad Still planting in the nursery 2 Men &c Jobs.

Tuesdy 8th very strong drought from S.W. – three ploughs at the Haugh Floors ribbing – P.M. at Sundries about Barns and other Jobbs – Planted out the Nursery ground and a Hedge N.W. Corner of the Garden.

Wedy 9th Remarkable fine weather from S.W. – 2 ploughs finished Floors and after at West side of Robs Hillock Park ribbing for Turnips – One Cart Brought about 12 Bolls of dryed Bear from the Mill of Garvock for Horses – P.M. at Sundries.

Thursdy 10th Pleasant weather from S.W. but Hazy & soft – 3 ploughs at Robs Hillock ribbing for Turnips – P.M. at Sundries about Barns Lofts &c Jobbs.

Fridy 11th Fine dry & soft winter weather from S.W. – 3 ploughs at Robs Hillock Park as before – P.M. at sundries about Barns & other Jobs – Considerable vegetation as yet – The Christmas rose flowered &c.

Saturdy 12th Strong frost & Hoar also till near noon, pleasant 2 ploughs at Robs Hillock park – One Cart leading Turnips from Haugh and a cartful of Fogg Common Confervei Linnaei – for deafning the Noise with plaister Lath from the Kitchen below the parlour – dighting old Oats for Miln.

Mondy 14th Yesterday fine dry weather from the S.W. – this day the same with a very slight frost in the night – 2 ploughs ribbing at Robs Hillock park – one Cart carrying old Oats crop 1787 to Garvock Mill & leading home Turnips afterwards, Burning & Clearing of whins &c on a sprott Brae towards Kair, planted same with Scotch firs, & Making up Inclosures on the same. Sundries.

Tuesdy 15th This morning about three came on closs rain & stormy weather from Eastward & Northward which continued all the day – Two Carts to Mill of Garvock with old Oats for Meal for family – & one Cart to Johnston with & for sundries – people Employed all day dighting Oats for the Mill – No work in the Neighbourhood without doors –

Wedy 16th last night & this morning a very severe gale of wind from N.W. which damaged the Houses both Tiled & otherwise, and also other Things – 2 ploughs at Robs Hillock ribbing – 2 Carts leading at Turnips Haugh – at Mill 32 B.

Thursdy 17th last night & this morning very heavy rain from the Northward which Brought down the river considerably & wett the lands very much, little work without doors, at Houses repairing roofs – a Divering 4 year old stott died – at Sundries.

Fridy 18th Strong gale from S.W. – three ploughs at finishing ribbing Robs Hillock park, and after at the Back Yards ribbing for Barley – Chaffed out a stack Barley South Barn – find the new Barns Extremely convenient – at Sundries.

Saturdy 19th last night & this morning strong gale from the Westward, & pretty keen frost, 2 ploughs at the Back Yards ribbing – & one Cart to Mathers for 3 Bolls Lime Shells for Plaister Kitchen &c – others leading Turnips – Planting Firs Thornylee.

Mondy 21st Yesterday a little snow & frost – this day fair weather and rather fresh from S.W. – ploughs did out all the ribbing for the season, and begun furing or water furring the Bank for Oats – and at Sundries about the Town – planting Firs.

Tuesdy 22d Fine fresh weather and strong drought from S.W. which hath dryed the grounds much – ploughs at furing the Bank & Kavels full length for Oats – lead Home 2 Cartful of Turnips & Jobbs.

Wedy 23d a slight frost in the night, day fair till about 11 A.M. – rain from Southward, which continued the whole day after – 3 ploughs seed fur at Bank and Kavels – people at the Mill of Garvock – Slaters at repairing of Tiled lofts & at Sundries.

Thursdy 24th last night & this morning a very strong gale of wind from West – one plough Bank – 2 Carts Brought Home about 32 Bolls Meal of old Oats & Girnelled same – one Cart leading Turnips Haugh. P.M. took in part Barley stack – repairing lofts & Jobbs.

Fridy 25th a strong Gale from N.W. & very stormy weather with some showers of sleet – 3 ploughs at seed furr on the Bank & Kavels – Sacking oats for the Mill old – riding up lofts for Barley – Slaters at Sundries – dismiss Slaters.

Saturdy 26th Frost in the night, and betwixt 8 & 9 A.M. came on a pretty heavy fall of snow which lasted till about 11 A.M. about $2\frac{1}{2}$ inches deep – 3 Carts driving Dung to Haugh fallow – others at Sundries – sent 10 Bolls old Oats to the Mill of Garvock.

Mondy 28th yesterday morning early a strong Thaw & fresh wind from S.W. which carried off all the Snow – and this day a very strong drought from S.W. – Two ploughs at Bank & Kavels – 2 Carts at Turnips – P.M. taking in the other part of Barley stack.

Tuesdy 29th This morning early came on Heavy rain & wind from S.E. which lasted mostly till evening – no work without doors – people at Barns & lofting Barley into long loft & at Sundries &c.

Wedy 30th This morning came on Heavy rain & very high wind from Southward – the rain moderated but the wind continued all the day – 2 ploughs begun ploughing at the flaked ley Dickies Haugh ley (25 years old) and two Horse plough at Black lands ley-& Jobbs (11 to 12 acres ley)

Thursdy 31st Fair weather all this day, but dark & Gloomy – 2 ploughs at Haugh Ley, and one ribbing the Haugh lately Turnips in good order – at variety of other matters – Planting Firrs &c.

Price of Grain by no means settled Bear about 13/6 – Oat meal $8^{ds}\frac{1}{2}$ to 9^{ds} per peck – Fatt Cattle about 5/ per stone – Beef & Mutton about 4/ per stone – The

Ribbing in the Country pretty well advanced, but little seed furr done as yet the fields being very wett – The roads at least by ones very Bad & Deep – all Sorts of Manufacturers Complain of a very Indifferent Markett & profits.

Diary for the Month of January 1790

Fridy 1st a Considerable Black frost all day, three ploughs at Haugh Ley, one of them ribbing late Turnip Ground. in ploughing a Hirst of Gravel never ploughed before Broke a plough Beam & some Swingle Trees – at the Crofts of Auchinblae Bought a Milch Cow – Cheap.

Saturdy 2d Hard frost to stop ploughing – 3 Carts driving dung to E^{t} Haugh – others at Barns & Sundry Jobs – about Noon came on very Heavy rain from S.E. all day -

Mondy 4th Yesterday fine pleasant winter weather – this day same – 2 ploughs at E^{t} Haugh & two Carts leading out Turnips from Haugh. Brought Home 10 Bolls $\frac{1}{2}$ peck Meal from 10 Bolls Oats – at Barns & other Jobbs.

Tuesdy 5th Frost in the night, day very pleasant & agreeable weather from S.W. – being old Christmass, no work without Doors all the servants being mostly absent.

Wedy 6th Fresh weather and Drought from SS.E. – 3 ploughs at E^{t} Haugh Ley very stiff work – others at Barns Chaffing Barley – Our ground is now pretty dry ribbing the Turnip ground – Grass & vegetation yet not checqued.

Thursdy 7th Fine fresh weather from Southward and drought – 2 ploughs at Haugh Ley, most of it very rich and lead Home Turnips in return – two Horse plough at Black Lands ley in good rich order – others at the Barns Chaffing Barley – Gardners Sundries.

Fridy 8th very fine winter weather and drought from Southward – 2 ploughs at Haugh ley & one at Black Lands D^{o}. Wewmen and others Dighting Barley & Sundries. Jobs. There is now a Considerable Quantity of Redland.

Saturdy 9th Wind S & weather as yesterday – 2 ploughs at Haugh Ley and Brought Home Turnips – One Cart to Stonehaven for Coals – @ 8/5 – people at Barns &c Jobbs – the fields are now much drier – Grass Green and most trees and shrubs Budded – &c

Mondy 11th Yesterday fine weather & drought from S.W. – this day the same till 5 P.M. came on rain from S.E. 3 ploughs at Ley all day – dighting & lofting Barley and at sundries – the Embargo being taken off the Exportation of Victual it will now rise.

Tuesdy 12th The rain mostly ceased by 9 P.M., but Blowed a strong Gale after Midnight from the S.W. & drought-3 ploughs- Have Broke several Plows & Graith in the Ley – at lofting Barley, &c Jobs.

Wedy 13th very fine pleasant and droughty weather from S.W. – 2 ploughs going – one Cart Leading Broom from Thornylee, and one Cart leading last of the Turnips from the Haugh – P.M. at South new Barn Chaffing Oats. Gardners & others at Sundries.

Thursdy 14th This day very pleasant dry weather & light wind from S.W. – 2 ploughs at Haugh Ley – and Ribbing after Turnips D^{o} – one Cart leading Turnips from Gibsons Park and another leading Broom from Thornylee to clear ground for Planting it – at Barns & Sundries.

Fridy 15th pretty strong frost night and day, 2 ploughs at the Haugh Ley, and one plough at ribbing out the late turnip ground Haugh for Barley – Making a Sheep Cott at Goulds Hillock of Cattle Flakes warped with Broom – Sundries.

Saturdy 16th frost in the night & day but not so Hard as to stop ploughing of Haugh ley two ploughs One Cart leading Turnips from N. park and Others at Braking down an Old Dyke for a Midding Layer for Robs Hillock park Intended for Turnips – No price yet fixed on for Grain Common Bear from 12/6 to 14/ Oat Meal from 12/ to 13/4 – Fatt Cattle not worth above 4/6 per stone – Spinning 11^{ds} to 12^{ds} per spindle – all kind of vegetation yet continues, most Trees & Shrubs being yet in the Bud – No Haddocks or few yet on our Sea Coast. Hare Hunting saw about 24 of them.

Mondy 18th yesterday Hard frost which would have stopped Ploughing – this day fresh weather from S.E. Carts driving dung to East Haugh Fallow. Making up a Dunghill for Garden, and at a variety of other Jobbs – levelling down the East dyke of Kavel's park for a Dunghill Robs Hillock park Intended for Turnips

Tuesdy 19th Fresh weather from Eastward & fair – 2 ploughs Haugh ley – one at stuble ground Old Town park – others at Barns dighting Oats South Barn & other Jobbs.

Wedy 20th very agreeable winter weather from the Eastward – 2 ploughs at Haugh Ley & Old Town Hillock park – Others and all Hands Dighting Oats for the Mill – Sundries.

Thursdy 21st very slight frost in the night, day Extreamly pleasant and even warm weather – 2 ploughs at Haugh ley – and the two Horse plough at Old Town planted Clump. Levelling down Braes & ploughing the last Old Water Tract, the dressing of these & Hirsts are very troublesome.

Fridy 22d Slight frost & very pleasant weather – 2 ploughs nearly finished all the old Ley Haugh – including the latest Old Water tract, very stiff work & levelling the Braes on each side of it with spade & plough – 2 Horse plough as yesterday – P.M. all Hands at Barns.

Saturdy 23d Uncommonly fine weather – 1 plough at Haugh ley – and one at North Clump park – 2 Carts to Stonehaven for Scots Coals dighting & sacking Oats for the Mill. They are however not near so good as could be wished.

Mondy 25th Yesterday slight frost & fine weather – this day frost pretty Hard, but not to prevent Ploughing – accordingly 2 four Horse ploughs – and One Cart driving Turnips – at Barns dighting and sacking Oats for Mill &c Jobs.

Tuesdy 26th Frost last night, and day fine winter weather rather fresh – 3 ploughs at Oat seed furr – others at sundry planting patch by Brae Sheperds Haugh East side of the river & Inclosing it – at Barns & other Jobbs -

Wedy 27th last night a Considerable deal of rain from the Eastward, day fair – all people at the Markett Taunton fair – Servants wages rather high & Scarcely to be got at present. Engaged however most of ours – no kind of Markett for grain.

Thursdy 28th last night strong Gale from S.W. and fair weather till about 1 P.M. came on sleet & rain from S.W. veering round to N.E. – which wetted the grounds very much & Brought down the rivers Considerably. One plough at Black lands Ley – sent 22 Bolls of Oats to Mill of Kair – others leading Broom from Thornylee to rid for planting – at Barns, forming Dunghill layers &c

Fridy 29th a fierce gale & very stormy weather, with sleet here but snow on the higher grounds – all day same kind of weather from North – 3 ploughs at Black lands ley till 1 P.M. when loosed them – others at Barns, chaffing Oats – and at sundry Jobs – preparing some a firlot of wheat by stale Mortar & Quick lime. Jobs.

Saturdy 30th Cold Bleak dry weather from N.W., but fresh – two ploughs at Black Lands ley – and two Horse Harrowing a firlot of Wheat sown in the new Broke up ley Haugh at rate of 3 firlots per acre & Harrowed the same Compleatly – P.M. lead home two Cartful of Turnips from North park – Barns & Jobbs.

Sundy 31st Strong windy weather, saddning the fields Considerably. The Labouring though as far back at Martinmass as in any season preceeding is now very far advanced owing to fine open Winter. at present no fixed prices for Grain – Bear about 13/ Barley 15/ Wheat 22/6^{ds} per Boll, and fatt Cattle & Swine very low priced & very little demand – there will be little or nothing Made by them this season – Manufactures of all kinds very dull in this Country – Meal 12/ – no farms sold – Snowdrop flowred – Most trees in Bud

Diary for the Month of February 1790

Mondy 1st Strong drought from S.W. and an appearance of Spring the Larck singing two ploughs at Black lands ley – One Cart to Mill of Kair with Oats – & leading Broom from Thornylee – at Barns dighting Oats for family Meal.

Tuesdy 2d very pleasant fair & mild weather – 2 ploughs – 2 Carts driving Home Turnips, and one Cart driving Earth from an Old Dyke to fill up an old Water Hole in the Haugh Ley – others dighting Oats & Sundries -

Wedy 3d Exceeding pleasant weather and some growth in the ground, three ploughs at Stuble ground & Ley and at Barns and other Jobs. two days ago begun Inclosing the Cottar Yards, with Ditch & Dyke Hedge row of Thorns behind and Ash Trees at 4 Ells distance within – roads Bad.

Thursdy 4th very pleasant spring Weather, 3 ploughs all day, others at Barns, and a variety of Jobbs vegetation evidently on the Increase &c – Took in first of our Pease West Old Barn -

Fridy 5th fine dry & spring weather – Two ploughs at the Bank & the ley

Blacklands near finished, One Cart leading Turnips & one leading Broom & cut Willow Trees Dickies Haugh – another Cart filling up an Old Water Pitt with Earth E[t] Haugh – & Jobbs

Saturday 6th Remarkable fine weather and many fields quite seed Coloured – 2 ploughs seed furr & two Carts to Johnshaven for Smiddy Coals to the Smith – lead home some Turnips P.M. – at Barns & Sundry Jobbs.

Mondy 8th Yesterday very pleasant weather and Ground seed Coloured – this day same till about 1 P.M. Came on rain which Continued Mildly through the Evening – people at Ploughs & two Carts leading Turnips-Gardner digging for Potatoes Garden & transplanting Turnips D° for seed.

Tuesdy 9th The rain Continued all night & till 8 A.M. when fair till after dinner rain slight – 3 ploughs at stuble ground – others at taking in Pease P.M. – People at Mill of Kair – 22 Bolls Oats gave 18 Bolls fine Meal or better 13 pecks – at Sundries -

Wedy 10th Very pleasant warm weather S.W. – 2 ploughs at Bank. One Cart driving Oats to the Mill & Bringing back in return Meal – at Barns & Sundries -

Thursdy 11th a white frost in the night which continued till near noon – 2 ploughs at Bank & one Cart driving Oats to the Mill – at Barns & Sundries.

Fridy 12th Blowing weather and drought from S.W. – all day 2 Ploughs at Bank & Kavels, and two Carts leading Home some Turnips, and after to the Mill with Oats for Meal – at Barns dighting & sacking Oats, and Chaffing Bear, South Barn, at Sundries

Saturday 13th very strong gale at times from the Westward with much drought & drying the Ground greatly – 3 ploughs at finishing the Tillage of Bank – at Barns dighting Bear and at Sundries on the farm – a few drops of rain only – at the Garden &c. (Dr W died suddenly)

Sundy 14th This day fair weather in general but some skiffs of rain from S.W. Vegetation increasing but very slowly – Observe the Tussilago in flower – the fields quite dry & partly seed Coloured.

Mondy 15th Strong drought from S.W. and the fields & roads very dry at this season – 2 ploughs at Bank & Kavels – One Cart to Stonehaven for Coals – and one leading Turnips. at Barns & other Jobs – at Garden planting Cabbage &c

Tuesdy 16th last night several showers of rain & stormy from the S.E. which Brought the river a little down, this day fair and strong drought from S.W. – 3 ploughs at seed furr, at Barns dighting Bear, P.M. at Sundries. Prices of Grain not fixed or scarcely saleable & fat Cattle the same, very unpromising to Farmers.

Wedy 17th Slight frost in the night, and day fair & drought from S.W. One plough at Kavels and three Carts driving dung for Haugh Fallow – Sundries.

Thursdy 18th very Pleasant weather and much drought from S.W. three Carts driving for Turnips Intended on Robs Hillock Park, on an Old Dyke leveled East side of Kavels – levelling down an Old Dyke also on the Floors for the Midding layer for Intended Turnips on it – Planted in Garden West side within the Hedge 30 Guin

Trees some of them five feet high – at Barns & Sundries. Crocus & Tussilago flowring.

Fridy 19th Strong Breese from S.W. and fair weather – 3 Carts at driving Dung – and one Plough at Kavels – Digged up 6 Ashes on N.W. corner of Garden destroying the Hedge, and other under Trees & Shrubs – Planted 20 Guin Trees in the new Nursery benorth Garden – Making flower garden.

Saturdy 20th This morning came on slight rain at 8 A.M. and continued till noon, when faired – 3 Carts at Dung driving to Robs Hillock – One plough at Kavels – at Sundries – Vegetation farther advanced about Stonehaven than here &c transplanted six Ashes for Gates.

Mondy 22d Yesterday and this uncommon fine dry & pleasant weather from S.W. and ground seed coloured – 2 Carts to Stonehaven for Coals – 2 Carts driving Dung and one Plough at Black lands Ley – Braking down an Old Dyke Floors for a Dunghill layer. Making out the Flower Garden before the House.

Tuesdy 23d White frost in the night day very pleasant weather with drought from the S.W. – 3 Carts driving Dung & Turnips-and one Plough at Kavels P.M. at finishing last Dunghill layer Floors. Observe some of the Laburnums leaved &c. Crocus flowred & Dandelion. Lapwings.

Wedy 24th Fine spring weather from S.W. – 3 Carts driving dung to Robs Hillock park – One Plough at Dickies Haugh for Pease – P.M. at Barns dighting Thornylee Bear to loft – at Garden and Sundries.

Thursdy 25th very Brisk Gale from S.W. and strong drought – 3 ploughs at Haugh for Pease, at the Mill of Kair with above 36 Bolls Oats for family use – P.M. lofting part Thornylee Bear Old Distillery.

Fridy 26th Strong Gale from S.W. and some light skiffs of rain – 2 ploughs Haugh for Pease and two Carts driving Meal from the Mill of Kair 31B 3p. Some under 14 pecks per Boll. lead Home 2 Cartful of Turnips – at Sundries.

Saturdy 27th very strong Gale from the Westward and drought – 3 ploughs at Haugh for Pease and at Sundries, Barns, Garden, and at this time the grounds dryer than for fifteen months by past – Several people sowing Pease & Oats this week.

Sundy 28th More Moderate weather various – Some drissling rain Sleetish in the after part of the day – am Informed the Grass as well advanced in this Neighbourhood as about Dundee, Arbroath & other Coast Towns. There is a great Stagnation of the Sale of Fatt Cattle and Grain which is very hard on the Farmer owing chiefly to an unjust Embargo on the Exportation of that article – Hear of some Farms Comiston &c being sold for 12/ Bear & Meal – Manufactories of all kinds much over done and dull. Corns & fodder worth about 13/6 per Boll – Spring early as 1779.

Diary for the Month of March 1790

Mondy 1st Closs foggy weather & warm – 3 Ploughs at the Haugh for Pease and

P.M. at Barns and at Sundries Jobs about Pease dighting &c. Observe several Neighbours yoking twice a day.

Tuesdy 2d mild warm & foggy weather as yesterday. Ploughs at the Haugh for Pease; at Leading some Turnips nearly finished, at Barns P.M. dighting and sacking Pease for seed – and at Sundries.

Wedy 3d White frost in the night, day very pleasant and even warm weather – 3 ploughs at Haugh for Pease all day – Yoked twice for the first time Middle of day at Sundries – Grass rising.

Thursdy 4th Very Pleasant & more appearance of Summer weather than any part of last season. One plough at Back Waters all day – Yoked early sowing Pease on Haugh from 13 to 14 pecks per acre & partly Harrowed the same with 8 Harrows whole day – at Barns dighting Pease, taking in Hay and Sundries Garden – at Flower Garden.

Fridy 5th Thick foggy weather in the night, day very pleasant & warm weather from Southward – 9 Harrows A.M. at the Pease Harrowing – & 4 only P.M. – with two ploughs at the Back Waters nearly finished for Pease – at Barns dighting Bear & taking in seed Oats for Barn. Gooseberries foliating.

Saturdy 6th remarkable fine & constant good weather – five Harrows at Back Waters sown with $3\frac{1}{2}$ Bolls Pease & partly Harrowed the same – two Carts to Stonehaven for English Coal at 1/6 per Barrell. Several people Have sown Pease & Oats this week – Grass a Bitt -

Mondy 8th Yesterday very strong drought from S.W. – this last night a very Heavy fogg & white frosty dew – day very cold & stormy from S.W. – 9 H finished the Harrowing the Pease & others Braking the Old Ley Haugh for Oats – a Cart driving on Earth on a Hirst Haugh from a neighbouring Dyke & at Barns dighting & sacking of seed Oats. Sundries. An Ox Hoved with Broken Bear.

Tuesdy 9th very strong drought & stormy from S.W. some drops of rain P.M. Early sowed five Bolls of Oats on Old Haugh ley – P.M. three More & partly Harrowed the same – at Barns dighting sent up the Cattle Flakes from Haugh to the Over Burns. Planting the Cottar Yards with Ash & Quicks. Grafted some Apples.

Wedy 10th Fine dry weather but variable wind shifting – the appearance of rain or snow but kept off – 9 Harrows at Haugh ley all day – One Cart driving Earth all day to stannery parts Haugh ley – and one Cart driving Clay & Gravel for Flower Garden walks benorth the House. at Barns & Sundries.

Thursdy 11th Strong drought & very Cold from various Quarters – 9 Harrows at Haugh ley nearly finished in good order – two Carts at driving Earth to Gravelly stanners & gravel for the Flower Garden walks. Sent for five Bolls seed oats from Polborn.

Fridy 12th Strong winds & drought from S.W. – till P.M. some skiffs & moderate – 9 Harrows all day finished the whole Haugh ley a Most arduous Business at least triple ordinary work – 2 Horse ploughs at Blacklands ley – at Barns – Flower Garden.

Saturdy 13th Some light showers of rain in the night, day fair & drought – sowed all Bank & partly Harrowed same with 17 Bolls Oats five of them being from Jas Croll Polborn – at Barns.Planting Ashes Shepherd's Haugh Brae.Garden &c

Mondy 15th Yesterday very stormy weather from S.W. – this day moderate & very pleasant – 9 Harrows at the Bank all day – at Barns & Sundry Jobs, planting Cabbages Coleworts &c small Matters.

Tuesday 16th Strong frost white & Ice on Dubs till noon – day pleasant – the frosts at night & drought in the day have stopped vegetation – 9 Harrows & a Roller on the Bank all day very Cloddy – a sudden drought after a wet ploughing hath left the fields very stiff & ill to Harrow – Hear the Viscount Arbuthnott hath sold his farms Bear & Meal about 12/ per Boll – fatt Cattle about 5/ dull – spinning 11^{ds} per spindle – Grass rising – Garden pease well up, Gooseberries foliated.

Wedy 17th pretty strong frost white till 8 A.M. – Nine Harrows at Bank & a Roller whole day which by evening Made it in good order – P.M. sowed six Bolls Oats West side Black lands and dighting Oats for Seed in Barns &c Jobs -

Thursdy 18th White frost pretty strong till 10 A.M. – 8 Harrows at Black lands, and a four Horse plough at the Kavels all day – Others at Garden & Barns &c

Fridy 19th TH ab^{t} 45° very thick fogg & Closs weather till P.M. but little or no frost – 7 Harrows and a four Horse plough at Kavels all day – leading some Turnips nearly shot, at Garden and at sundry Jobs – little growth.

Saturdy 20th Foggy & strong drought from S.W. afternoon – one plough at Kavels, and all Harrows at Black lands nearly finished with Including Old Town park 13$\frac{1}{2}$ Bolls Oats – at Barns, Garden &c – Jobbs.

Mondy 22d good work Horses at £25 str Yesterday strong drought from S.W. night frost with much Hoar till 8 A.M. – this day very strong drought from South – 8 Harrows finished the Black Lands Harrowing fine – Sowed three Bolls Oats west side of Kavels – four Horse Plough on D^{o} – at Garden for Onions, Carrots & other &c &c

Tuesdy 23d Hoar frost in the night – day fair & very strong drought from the Eastward – one Plough, six Harrows & a Roller on South side Kavels very Hard – Sowed Early 4 Bolls Oats on D^{o} – Vegetation quite Checqued – & Rye grass turning Black – Sowed Onions & Carrots.

Wedy 24th Early this morning Came on pretty Closs rain from the Eastward which Continued pretty Constantly the whole day – Sowed 4 Bolls more Oats on Kavels A.M. Twelve Horses Employed on Same all day – Ploughed up an Old Earth Dyke & West Avenue in above field.

Thursdy 25th This day fair & very pleasant warm weather variable – finished all tillage for Oats, Sowed 2 pecks Lintseed very Old Haugh ley – at the Garden 2 Men – the rain hath Met the Old Sap -

Fridy 26th White frost in the night & pretty hard morning – Sowed out whole of Oat seed being 16B:3f on Kavels and nearly finished Harrowing it – two horse Ploughs at steering some parts of last years Fallow – in Garden at a variety of Sundries – Growth again.

Saturdy 27th Fine pleasant dry weather variable Harrows finished the whole of Oat seed in fine Order, & Begun Braking in the fields for Barley – leading stones for new Byres & at steering Gibsons Park for Bear a clean furr – at Garden planting Savoys – Took in a stack Barley for Seed.

Mondy 29th Yesterday fine dry weather but no growth – this day same. Harrows Braking in Back yards – late Haugh Turnip land for Barley – and leading Stones from Dava &c for Byres – Steering Gibsons Park for Bear & Jobs.

Tuesdy 30th Cold Bleak weather from N.E. all day but no frost – 2 Carts driving Stones from Dava all day for new Byres – one Plough steering after Turnips as yesterday – & six Harrows nearly finished Braking in all the land Intended for Barley – and at Sundries – at the Garden & levelling old Dykes useless before.

Wedy 31st Therm^r ab^t 46° Bleak weather from the N.E., and little or no growth in the ground for weeks by past – 6 Harrows finished Braking all the lands Intended for Barley A.M. P.M. 3 Carts driving out Dung for Turnips on Robs Hillock Park, one plough at Gibsons Park for Bear as before, Brought Home two Cartful being the Quarried stone from Dava, at Barns taking in Hay & Sundries – The late Crop is likely to have cut very ill in most places one third deficient, in others near one half, in Bolls, beside the great deficiency in Mealing Oats, near one firlot per Boll, yet surprising prices are low & no demand for either Corn or Cattle of any kind, which is unaccountable, the Oat seed not finished in this Corner as yet, Grass scarcely a Bite, Scarcely any field flower Blossom.

Diary for April 1790

Thursdy 1st Th^r 50° Cold Bleak dry weather from the Eastward and little or no vegetation – 3 Carts driving Dung as the preceding day – finished first covering Dunghill 80 yards by 6 D° – at Barns & other Jobs &c.

Fridy 2d Weather as yesterday – Carts driving dung for Turnips to Robs Hillock park; Settled some Marches betwixt me & Alexander Smart of Waterlair, tennant at Polbourn, regarding commonty of the Brae of Waterlair up to the Wards &c Jobbs.

Saturdy 3d Cold Bleak weather from S.E. and no Growth in the ground. 3 Carts driving dung all day to Robs Hillock and Plough finished steering fur for Gibsons Park – 3 Harrows P.M. Braking in E^t side Robs Hillock Park very foul for Turnips – taking in Barley, Hay &c.

Mondy 5th Yesterday strong drought but Milder, this morning a Keen frost both Black & White – 3 Carts driving of Dung to Robs Hillock Dunghill all day. P.M. begun the seed fur Barley Haugh after Turnip but too dry – at Garden & at Barns & Sundries.

Tuesdy 6th last night and this day Cold Bleak weather and P.M. some showers of Hail from N.E. – 2 ploughs at Second furr E^t side of Robs Hillock very foul & a Plough at seed fur for Barley after Turnips – & at Barns dighting seed Barley – at Sundries.

Wedy 7th last night several showers of Hail which lay on the Hills & Braes all day

– 2 ploughs at the steering for Turnips, and one plough at seed fur for Barley, at Barns dighting seed Barley sold. no kind of growth in the ground – fields Bare.

Thursdy 8th Cold Bleak weather from N.E. – Ploughs at steering of E[t] side of Robs Hillock park and seed fur for Barley Haugh – & P.M. two Harrows Braking in before sowing being very Cloddy owing to the drought – &c

Fridy 9th Cold & stormy weather with flirts of snow & hail at Intervals whole day – lying on the Hights – One plough at steering for Turnips – & two at the seed fur Haugh – Braking down a dyke Back yards for Potatoes – all vegetation totaly stopped.

Saturdy 10th last night & this morning a Considerable fall of snow from the N.E. which lay till near noon – Carts driving Dung A.M. P.M. begun sowing of Barley – three Bolls after Turnips Haugh in fine order – 6 Harrows & two ploughs at same work taking down an old Dyke for Potatoes &c Jobs.

Mondy 12th yesterday morning frost so strong as to freeze water in the House Portico – & some snow which vanished A.M. This day strong frost morning day very cold from N.E. 2 ploughs at late Turnip ground finished – & 6 Harrows at same P.M. sowed further 14 pecks or 3B 14p on said ground – two Carts driving dung all day at rate of 24 Cart load per acre.

Tuesdy 13th Sharp frost white & hard – day very Cold & dry from the N.E. – 4 Carts driving out Dung as the day before & three Harrows the Ground for Turnips Robs Hillock – at Barns & Sundries. Vegetation quite at a stand & pasture failing.

Wedy 14th Hard frost in the night with heavy showers of Hail & Snow at Intervals through the day – 4 Carts at driving Dung to Dickies Haugh all day – 3 Harrows at Braking as before – at Barns dighting Barley for seed, Houghing at Nursery ground, and at sundry Jobs – Gave the Tennants at Arthurs & Sutty Wells each as much Rye grass quite full of Plantain as Serve an acre each on Condition of leaving no ground unsown w[i] produce of same or such after the ordinary Croping & this Gratis &c. also spreading dung Haugh P.M.

Thursdy 15th a Considerable fall of snow from N.E. and very cold stormy weather – 4 Carts driving dung all the day finished out all the Dunghills on Haugh – & 3 Harrows at Robs Hillock for Turnips – begun Threshing Oats N. Barn for Cattle Fother &c Jobs.

Fridy 16th Strong frost & stormy weather with Hail at Intervals all day – Carts nearly finished dunging Fallow & 3 ploughs at seed fur on D° P.M. – at Barn & at Sundries – no sale for either Corn or Cattle -

Saturdy 17th Frost both Black & White and day with some showers of Hail & sleet. 2 ploughs at Barley seed fur A.M. & P.M. sowed almost 2 Bolls Barley West side Fallow and most Harrowed D° – 2 Carts to Johnshaven for Coals and to Milton for Lime shells to point the House – returnd yoke of Lime.

Mondy 19th yesterday strong drought but Milder – this day fair weather and the wind shifted to the S.E. – 3 ploughs all day at seed fur fallow – Others spreading dung & Barns dighting seed Barley sold & for own use – at Garden & Sundries. Sold two of my stall fed Cattle under 5/ per stone.

Tuesdy 20th White frost in the night, day fair & Moderately mild Ploughs & Harrows all day at Barley seed – sowed A.M. 2B 3p and partly Harrowed same – at Barns & sowed several small Garden seeds, Lettice, Spinage &c Viogla Matha flowring.

Wedy 21st Cold Bleak weather from the Southward all day & about 3 P.M. Came on light rain from S.E. which Continued – 2 ploughs at E[t] Haugh for Barley and Carts & Harrows at D° – sowed two Bolls Barley P.M. – at Barns planting Potatoes &c.

Thursdy 22d some showers of rain from S.E. with growing weather finished sowing out Dickies Haugh with Barley and very nearly finished the Compleat Harrowing of it with 6B 3f and $\frac{1}{2}$ peck P.M.: 3 Carts dunging at Back Yards -

Fridy 23d Fine showers of rain A.M. and vegetation appearing to Recover fast – Carts driving dung from Dung Court and Harrows at Barley & Ground for Turnips, ploughing the Patch north side river Shepherds haugh for Barley

Saturdy 24th a Considerable shower in the morning early, day fair & drought from the S.E. – at 5 P.M. some rain & stormy from N.E. – three carts driving dung all day from Dung Court – two Horse Ploughed out & Harrowed all strip N.side Water $6\frac{1}{2}$ pecks Barley – after seed furr south side of Back Yards – spreading dung on D°.

Mondy 26th Yesterday morning a Considerable fall of rain & very Cold stormy weather from N.E. – this better but Cold – Carts dunging, Ploughing & Harrowing all day at Back Yards – and Robs Hillock Park – Ground rather Claggy -

Tuesdy 27th wind shifted in to the Westward & more Mild & dry – Carts &c as before – Dunged out the Back Yards & partly Sowed & Harrowed the same with Barley – Observed the Grass much better about Aberdeen & Stonehaven than in the Hollow of the Mearns, but Trees & Shrubs not more so – prices of Victual rising.

Wedy 28th Slight frost night, morning Mild, day stormy & Bleak from S.W. – 3 ploughs all day at Back Yards and Sowed part of D° – planting Potatoes &c Jobs -

Thursdy 29th Slight frost morning, day fair & some drought from S.W. 2 ploughs at Back Yards A.M. and Five Harrows at D° – P.M. 3 ploughs finished the tillage for Barley – at the Tryst at Laurencekirk a good Many Cattle of but Indifferent sorts severals sold pretty high – heavy shower P.M.

Fridy 30th fine dry weather from various points – 2 ploughs at Robs Hillock for Turnip, and Harrows finish the Back Yards sown with 5B & 2 pecks Barley ready for Grass seeds – leading sand for lime to point & Harle the Houses, and at Planting of Potatoes on a Ditch South side Back Yards – and at various Jobbs – Prices of Grain rising Chester Bear & Oat Meal about 13/4 – Fatt Cattle in much demand now they are nearly all gone, lean Cattle also – the Pasture is as Backward as I ever saw it in any former Season & vegetation of all kinds very tardy & Backward : fodder going fast.

Diary for May 1790

Saturdy 1st Fine growing weather and drought from S.W. – 2 ploughs at Robs Hillock park for Turnip Sowed out Back Yards with 56 lb Eng: of red Clover, and $3\frac{1}{2}$

Bolls of Rye grass & Plantain – Harrowed & partly Rolled the same, planting Potatoes – (Sent for a Stone Ornamental)

Mondy 3d Yesterday morning & first half of the day an heavy fall of rain from the N.E. which Brought down the river very High & overflowed several patches Haugh – water at hight by $\frac{1}{2}$ past 3 P.M. – this day fair – 3 ploughs at steering Robs Hillock – P.M. 6 Harrows Braking in the Floors for Turnips – Planting Potatoes.

Tuesdy 4th variable weather with some slight skiffs of rain – finished Braking in the Floors for Turnips A.M. 3 ploughs at Robs Hillock park – leading stones from Land for Building, planting Potatoes &c. Harling & Teething the Distillery Houses &c.

Wedy 5th Stormy dry weather from S.W. – 3 ploughs all day at steering Robs Hillock park, and P.M. a Roller finished rolling the Back Yards sown with Grass seeds. at Repairing Houses & begun rebuilding a Cottar House near the Aberdeen road &c.

Thursdy 6th very Cold Bleak weather and a Considerable fall of rain this morning from the Northward – 3 ploughs all day at Robs Hillock park – Masons & wrights at various Jobs – grass yet but very Indifferent.

Fridy 7th Dry weather from the N.E. and some growth in the ground notwithstanding – 3 ploughs at steering for Turnips all day – at such Jobs as the day before. Rolled some Cloddy parts of Bank Oats &c Jobs

Saturdy 8th fine weather and warmer & More growth than hitherto vegetation accordingly sensibly Increased – 2 ploughs at steering for Turnips – and two Carts one for Coals to Johnshaven and another for Lime shells to Mathers – at Sundries as before.

Mondy 10th Yesterday fine dry & warm weather from the Northward – this Cold Bleak D^{o} & some skiffs of rain from the N.E. 3 ploughs all day at steering for Turnips at Robs Hillock & the Floors, at Sundrys as formerly.

Tuesdy 11th very pleasant weather and strong drought from N.E. – ploughs and Harrows at the Grounds for Turnips – and P.M. Broke in Gibsons park for Chester Bear, the Grass & all other articles vegetating more Briskly.

Wedy 12th Fine dry weather and ploughs & Harrows at Sundries Building at a Cottar House, and at setting up the stone stathels in the Cornyard, and at a Roup of the late Dr Walkers affairs things sold high.

Thursdy 13th Remarkable fine dry weather & warm from the Eastward – 2 ploughs at second furr Floors for Turnips very stiff and Carts driving on Rubbish Dung on Gibsons Park for Chester, at planting Potatoes & sundry Jobs -

Fridy 14th Fine pleasant dry weather and uncommonly good for Barley seed season – finished 2^{d} furr of the Floors for Turnips – and Dunging – & P.M. seed furrow Gibsons Park for Chester Bear – finished also planting Potatoes about 9 pecks total the Grass rising fast, prices of grain rising -

Saturdy 15th Fine growing weaher various Chiefly from S.E. and some skiffs of rain P.M.- all ploughs and some Carts driving dung all day to Gibsons Park – P.M.

sowed nearly two Bolls Chester Bear from Rev Mr Spark on South side of said Park – and partly Harrowed the same – at Sundries.

Mondy 17th Yesterday fine dry weather & much sunshine – this morning a Heavy fall of rain from the Northward – with more or less through the day. Ploughs & Harrows at finishing the Gibsons Park being whole of Barley & Bear seed – The Grass now pretty good. Observed the Bear seed not near done about Arbuthnot, Allardice & the Coast side prices of Grain rising farther Every place There will not be much fodder left -

Tuesdy 18th last night a severe gale of wind & Colder from the Northward – this day more Moderate – Harrows & drag D° at the Floors for Turnips very Lumpy – Sowing the grass seeds on Gibsons park Evening and at Barns and sundry other Jobs &c.

Wedy 19th Moderate weather from S.W. – Sowed out Gibsons Park with 46 lb red Clover & 3 Bolls rich Rye Grass full of Plantain – Harrowed & Water furred same all Horses in Harrows, Drag D° & roller, but Coming on rain about $\frac{1}{2}$ past 10 A.M. softned it much & Made it reduce much Easier -

Thursdy 20th This day Blowing rough weather from S.W. and all Harrows, roller &c A.M. at Floors – P.M. 2 four Horse Ploughs giving 3^d furr to it – and Rolling Gibsons Park after Grass seeds sown.

Fridy 21st fine pleasant weather but windy & some skiffs of rain from Westward – 2 ploughs at 3^d furr on the Floors – Harrowing & Rolling same – finished Rolling out Gibsons Park – Yoked at 5 A.M. & put the Horses on Grass two which it Tinged much –

Saturdy 22d Weather much as yesterday but warmer from S.W. – 2 Carts to Mathers for Lime shells, and one to Johnshaven for Scots Coals at 8/6 – one plough & some Harrows at Floors – at Sundry Jobbs – Building a Cottar House.

Mondy 24th Yesterday dry Cool weather from N.E. – day same – 2 ploughs at the Floors for Turnips, and three Harrows at D° – taking in Hay for Horses and at Sundries, weeding the young Trees Nursery by the Gardiner & others.

Tuesdy 25th This day Cold Bleak weather from N.E. and no Growth – One Plough, 6 Harrows & Roller all day – finished 3^d furr Floors – at Nursery &c – Hear lean Cattle are selling very dear as also sheep – Cherries, Guins & Apples standars only beginning to Blossom – & everything as backward as last year.

Wedy 26th remarkable strong drought Continues from Northward much sunshine but little growth – 3 ploughs at the Floors 4th furr now fine – Others at sundries – Hear many fields destroyed by worms Chiefly Ley Broke up – a worm of about 1 Inch – thickness Goosequill

Thursdy 27th Strong drought from N.E. and sun shine Harrows at Robs Hillock & three Carts to Mill of Halkerton flitting a Cow Herd to this place, and at sundries – Jobs –

Fridy 28th This last night slight frost and day strong drought from the N.E. – 8 Harrows all day at Robs Hillock park very dry – and two Carts A.M for a Cottar

Mans Plenishing from Allardyce – P.M. all the Harrows & Drag D° on a bene field – Barns &c

Saturdy 29th Drought not quite so strong as for some preceding days but Every species of vegetable in the greatest want of Rain – 3 Carts driving Dung from Town to the Floors all day – and six Harrows at the Robs Hillock Park Cross Harrrowing it – Middle of day setting up the Cattle Flakes Over Burns – Barns &c. Pease Garden Blooming & Cherries, Apples &c.

Mondy 31st Yesterday Closs foggy weather and appearance of rain from the Eastward but it keeped off. this day same – It Hath however Mended the Growth of both Corn & Grass Considerably – 3 Carts driving dung all Day to the Floors – and three Harrows & a Roller all day at the Haugh Barley after Turnips too late as it hath Happened being no rain lately & after at Robs Hillock Cloddy – Prices of Grain on the rise, Meal Oat scarce about 11^{ds} per peck and Bear Chester for Aberdeen 15/- prices of lean Cattle are about 4/ per stone not quite so high as last Season – 3 Men Begun Quarrying at Dava this day one 1/ & two 10^{ds} per day with victual

Diary for June 1790

Tuesdy 1st Early this morning a Brisk shower of rain from S.W. – and showers very refreshing at Intervals through the day – 3 ploughs at the Floors all day – at Barns dighting Oats. Garden &c

Wedy 2d Fine growing weather & drought from the S.E. – all Force Ploughs, Harrows & Roller at Robs Hillock park reducing it into fine Mould, & Gathering & Burning all Quicken & other weeds on the ground – & Sundries.

Thursdy 3d Fine pleasant dry weather from Southward – 2 ploughs, 6 Harrows & Roller Robs Hillock park – & P.M. – 3 Carts drove out all dung on farm to the Floors – at a Roup of wood Kair Bought 16 for 18/ Fir -

Fridy 4th last night & this day very strong drought from N.W. – 3 two Horse Ploughs at E^{t} side Robs Hillock park and 5 Harrows at D° & Men Gathering & Burning an Immense quantity of Quicken, finished turning the Dunghill on Floors being all for Turnips & Sundries.

Saturdy 5th Stormy weather and very strong drought from N.W. 3 ploughs at Robs Hillock park west side & five Harrows Gathering a vast quantity of Quicken on the East side of same, at Barns dighting Oats &c.

Mondy 7th Dry weather till Evening when some showers of rain from S.W. – this day same with some skiffs of D° – 4 ploughs at west side Robs Hillock park & Harrows at or following D° – Barns &c Jobs

Tuesdy 8th Early morning some light showers of rain, after fair till 3 P.M. Came on heavy rain which stopped all work without Doors – Harrowing & ploughing at Robs Hillock – 2 Carts Brought 8 Fir Trees Bought at Kair – at Sundries.

Wedy 9th This morning fair weather from S.W. – the rain deep into the ground – 2 ploughs at Robs Hillock park streighting for placing the Dung – 2 Carts for wood

Kair & P.M. yoked four Carts to Driving out Dung to West side Turnip field 28 Cart load per acre – Yesterday Evening begun Inclosing Thornylee planting – at Barns &c Jobbs -

Thursdy 10th Strong drought from various points – some showers before dinner time – 4 Carts at driving out Dung for Turnips all day & all spare people spreaading at same 28 Cartload per acre – was Informed that Trinity Muir Markett was very dull and owing to Prices being excessive high little sale.

Fridy 11th Coldish & Stormy weather from N.W. with a shower P.M. Early Carts driving dung & wewmen spreading same – after Brakefast Begun Drilling on West side of Robs Hillock & P.M. Sowed & Covered about near two Acres for Turnips – P.M. Two Carts to Dava for Stones & at Sundries -

Saturday 12th Cold Stormy weather from N.W. – and drought Checking the vegetation much – a Plough Drilling Turnips all day from 4 to 5 Acres finished by Rolling & in very fine order – One Cart to Stonehaven for Coals & Iron & two D^{o} Driving Stones from Dava three Draught each.

Mondy 14th Very strong drought from N.W. and rather warm weather – 4 Carts driving Dung all day to Robs Hillock one Cart driving stones from Dava all day, after Brakefast flitted the Flakes of Cattle & Sheep. the vegetation much Checqued – at Sundries -

Tuesdy 15th Strong drought Continues from S.W. – 4 Carts driving dung all day to Robs Hillock & One Cart fliting James Law Smith to Hercules Haugh – and one draught of stones from Dava. Wewmen spreading Dung all day – The pasture in General & Hay very poor – a dish Artichokes uncommon Early -

Wedy 16th very strong drought from Eastward & very warm weather – One plough drilling Turnips whole Day – and others at Spreading of Dung and Covering the seed, & P.M. Rolled the whole Sown this day –

Observe the Crop very scanty & thin in general betwixt this & Snaton near St Cyrus – & pasture nearly finished – prices of Grain & Meal rising and growing scarce & it is thought will soon be very high & not to be had – Oat Meal 1/ per peck – driving stones for Byres two Carts from Dava & Sundries – Inoculated two of my Boys at Snaton 10 Miles distance from a Person in them.

Thursdy 17th last night about 11 P.M. a slight shower of rain but of no duration – this Cold Bleak weather but dry – A.M. Drilling Turnips – Wewmen Spreading dung – and P.M. 4 Carts driving dung – tried the Quarry at Arthurs but found it not Workeable – Sundries -

Fridy 18th last night and this day till near noon, almost Constant rain & stormy from Eastward, but not Cold which prevented our Tames yoking in the morning – at Barns dighting & lofting Oats & P.M. Drilling & Sowing Turnips, Spreading dung & two Carts at Dava. first Cut Grass.

Saturday 19th much rain Early this morning & through most of the day, after Brakefast yoked a drill plough for Turnips and three Carts driving dung for two hours, but rain coming on & severe storm of wind from Southward obliged to stop at noon – Heavy rain & Stormy – lofting Oats 42 B^{ls} – Clearing out old furniture of old

Byres for new ones & P.M. sent two Carts to Dava Quarry for Rigging stones &c Jobs.

Mondy 21st Yesterday very warm weather & strong winds S.Wy. This day same & high winds from S.W., drying the grounds too fast, and Braking down trees, shrubs, and other vegetables – 3 Carts driving dung to Robs Hillock all day & four Harrows at the Floors yet pretty rough – spreading dung.

Tuesdy 22d Growing weather from S.E. till 11 A.M. some Thunder & Heavy Showers of rain which cleared up soon & P.M. strong drought from S.W. – one plough at Drilling Turnips whole day – 2 Carts at Quarry thrice and three Harrows at Floors – spreading dung & Covering the Turnips & at Sundries.

Wedy 23d Frosty in the night, day very strong drought from S.W. – 3 Carts driving dung all day from Sundry Places to Robs Hillock not Entirely finished – and two Carts leading stones from Dava three draught a day. Observe our neighbours only begining to sow Turnips –

Thursdy 24th Droughty weather but some showers at Intervals of no duration in the day – 3 Carts driving dung Morning to finish Robs Hillock park – 2 Carts at Quarry, and one plough Drilling. Sowing & Rolling whole day, nearly whole field from 13 to 14 Acres finished in the very best order of any I ever Possessed – finished Inclosing the Thornylee Planting.

Fridy 25th Th^{r} 66° White frost in the night day fair till about 11 A.M. Came on Heavy Thunder showers which lasted at short Intervals till past 6 P.M. & wet grounds much – 2 Carts driving stones from Dava & three D^{o} driving Dung P.M.on Floors field for Turnips – taking down the Old Byres & Carrying off Rubbish to Fund new ones.

Saturdy 26th Early morning some Heavy showers of rain from North and P.M. several showers lighter from same quarter – 3 Carts driving dung on Floors all day, and Carts driving stones from Dava for the new Byres . Wewmen spreading D^{o} – and Men Midle day Carting Found for the Byres &c Jobs.

Mondy 28th Yesterday fair weather from N.W., this day Came on very heavy showers of Hail & rain about half past 11 A.M. more or less so at Intervals till P.M. Drilling the Floors for Turnips, others spreading dung, & sowing what was Drilled Cut up the Turnips for seed in Garden fine Crops & sent three Carts to Pitscandly for slates, flited ye flakes.

Tuesdy 29th Fine dry weather from Southward – the Carts returned past 10 o'clock at night with the slates owing to delays – at spreading dung & drilling & sowing of Turnips Floors all day, the Mould fine & soft, but not neat – at Sundries –

Wedy 30th fine growing weather from the Southward – finished the whole of the Turnip tillage for Season by the Compleating the rolling of the last of the Floors & Midding layer Robs Hillock. Broad Cart Carting out the Fund for the new Byres; turned all the Barley & Oats in the lofts and at Sundries – prices of all sorts of grain high – Oat Meal 1/ per peck Chester Bear 15/ to 16/ per Boll, Barley proportional Wheat 26/6 per D^{o} at Montrose – Cattle Marketts very dull owing to the sellers holding them so high – Hay Crop but very scanty and late; Pasture in General tolerably good; the Barley in the shoot Blade, and Pease near Blooming; the Oats likewise looking well in Many places. Manufactures in general dull, & not as Advantageous as sometime ago – some strawberries ripe – roses Blowing

Diary for July 1790

Thursdy 1st Frosty rhime most night – after 10 A.M. Came on Heavy rain which lasted till 2 P.M. – Carts driving stones from Dava, one to Kesthock for Slates left there last season, riding out found for Byres – Observed the Pease haugh Blooming – & Sundries –

Fridy 2d fine growing weather variable – 3 Carts driving Gravel to the road Back of the Home steading – and others leading water & sundries for five Masons who founded & built part of said work this day – at Sundries – Crop Mending

Saturdy 3d Fine growing and a Considerable quantity of rain through the day – 5 Carts to Pitscandly for slates own and five others tennants &c – Masons & others at Byres Building – Brought home 1009 slates –

Mondy 5th Yesterday fine weather & warm – this day strong drought Northerly – Horses Idle preparatory to sending for slates to Angus – riding up Rubbish from Barns & Byres new ones & five Masons at same Building – observed the Crop about Bervie Earlier & better than here.

Tuesdy 6th Cold Bleak weather from the Northward – 4 Carts for Slates and one Cart driving Gravel for roads about the House, and different tradesmen about sundries – There will be no fruit of aples &c this year here.

Wedy 7th Cold weather from North West till 4 P.M. Came round to Southward & Milder – One Cart to Mathers for Lime shells, and the rest Idle. Men at Carrying out straw from Barns & Built it on stathels and took in last stack of Oats N.Barn, Sundries.

Thursdy 8th Fine summer weather – wind S.W. Fast day at Fordun, the Masons however as usual – In the Evening sent 3 Carts for slates – Grass growing Scanty.

Fridy 9th last night and this day till towards noon a great deal of rain from S.W. – P.M. some strong drought & rough weather from the N.W. – 5 Carts for slates – and others at sundries Barns & variety of sundry Jobs –

Saturdy 10th last night and a considerable part of day heavy showers of rain, and some Thunder – being a Church day, people there – at Sundries – Masons & Slaters all day – Observed some of the rankest Barley & Oats swayed.

Mondy 12th Yesterday a great deal of rain – and this day several showers rendered everything excessively wet the Masons cannot work for it – people at Church – afternoon wind veered to the N.W. – Cold Bleak weather- Sundrys –

Tuesdy 13th at 2 A.M. Came on rain which lasted till past 7 – five Carts last night for slates One from Stanacky returned by 3 P.M., the four others at – Observed some of the Rankest Barley fallow Haugh offering to Lodge – Crop appearing fine –

Wedy 14th Cold Bleak weather from N.W.- but with drought – people at redding out New found North side of the Building Byres, and dighting & sacking Oats for Mill of Kair – delivered 5B 1f pease to the Millar at Garvock – Cutting Grass Auld park same day, Erected Baking Oven in the Distillery – and five Masons, and two Slaters daily Employed – Pease, Potatoes & Straw Berries ready.

Thursdy 15th Fair weather Held from Northward till about 1 P.M. Came on rain at Paldy fair for two or three Hours – Sent Oats to Mill of Kair to Mill & 2 Carts to Quarry for stones – at Markett high prices for Cattle & quick sale more so than last season's – Crop looking very well about Fordun, Drumtochkty & other places – prices of Grain and Linnen Cloth rather in the fall, but there is yet demand for both &c.

Fridy 16th In the night a very strong Gale for a few Hours day fair & Moderate – lead Home 4 Cartful of stones from Dava Quarry – & Evening sent 3 Carts to Pitscandly for slates – the crop not so good about Drumlithie & Glenbervie as Fordun.

Saturdy 17th Very pleasant weather all this morning, day fair and some very light skiffs P.M. from S.W. last night at 10 P.M. sent 3 Carts for slates, this at 3 A.M 2 Carts to Mathers for 8 Bolls Lime Shells for Building the Byres, at Houghing Potatoes in the field which finished – Cutting Grass for Horses Suppers & variety of Sundries – many people have begun Cutting Hay a very light Crop & very bad weather hitherto for winning it – a Cart Broke down with slates at Hair Cairns

Mondy 19th Yesterday fine dry weather from S.W. – this day very good summer weather from S.E. – One Cart for the Slates & others driving stones from Dava, Horse Houghing the first sown Turnips very good – took off roof of the Intended School House West side of Closs to give it a new one of Foreign wood & Slates – & at Sundries.

Tuesdy 20th Strong Gale from Southward in the day & heavy shower of rain at 5 A.M. skiffs & rough weather all day – Carts leading stones from Dava, & taking away first fur from the Turnips – taking down roof of School House & Sundries. a Sow Died.

Wedy 21st Very strong Gale from the Westward all day, which must Damage the grain & hurt vegetation Considerably – Carts driving stones from Dava 6 Cartload, a Plough all day taking away fur from Turnips & six people thinning Turnips P.M. – Masons & serv^{ts} slapping doors & for Chimney West side of the Court for School House, and Sundries –

Thursdy 22d more Moderate weather from Northward – Evening wind veered to the S.E. – Carts driving stones from Dava. Horse Houghing Turnips & Men & Wewmen at Thinning of D^{o} which offer extreamly well. Most people Cutting & Coling of Hay.

Fridy 23d Fine pleasant weather from Eastward – at 1 P.M. sent 4 Carts to Pitscandly for Slates – one leading stones & Rubbish from New School House & Masons at Building D^{o} – Men & Wewmen dressing of Turnips Robs Hillock Park very fine – 3 Men begun Cutting Hay Old Town Park as good as our neighbours,–

Saturdy 24th fine pleasant & warm weather – people dressing Turnips all day – others turned over Hay Cut yesterday P.M. – 4 Carts own & two David Barclays N.Water Bridge returned with Slates by 3 P.M. – Hay Cutters begun Cutting West side of Thornylee Hay very moderate Crop – Sundries.

Mondy 26th Yesterday fine drought from S.W. – this day same but rather Stormy – 4 Men Cutting Hay Thornylee, and ten people mostly after Brakefast Gathered & Coled about 5 Acres Auld Town park into 6 pretty large Tramp Ricks – before Brakefast Flited the Cattle Flakes Over Burns.

Tuesdy 27th Hoar frost in the night day fair till 11 A.M. when Came on rain at first light but Increased to be very heavy by 1 P.M. Stopped all work without doors after dinner at Barns dighting Oats for Meal & at dressing of Turnips very fine, at Mill of Kair &c (30 Bolls).

Wedy 28th Early morning Hoar frost and day drought from the Westward – which raised some of the lying Barley & Oats – all people at Dressing of the Turnips, & others Dighting Oats N.Barn – the Hay Cutters finished Hay Business; Brought 29B 1fir Meal from Mill of Kair this day -

Thursdy 29th Hoar frost in the morning, day several showers of rain laying the Rank Corns – people amongst the Turnips till Brakefast, afterwards all people at Garvock Market, little sale of Cattle, but the Servants rather higher – Sale of victual very dull.

Fridy 30th most of this Day alternate rain & fair weather and a greater vegetation than during the Season but not very faverable for the Crop – as much of the best is lodging – people at Horse & Hand Houghing of Turnips, sent 4 Carts for Slates at Noon – Masons & Wrights Busy – Sowed seeds of Spring plants.

Saturdy 31st Frosty weather in the night as Common, day quite stormy & very Cold from Northward like winter – own Carts & two of my Fathers from Johnston with Slates arrived at 4 P.M. – some rain – 6 people thinning Turnips Robs Hillock best part done, one Plough at D° – For some weeks by past a Considerable stagnation in the sale of Grain & rather falling prices owing at least to the appearance of that on the ground, but for my own opinion I judge the Season promiseth very uncertain, Constant rain, frost & high winds with a very late Crop at present forebodes no good prospect at this day of high return -

Diary for August 1790

Mondy 2d Yesterday showers of rain from diff^t^ Quarters, this day a Considerable deal of same to wet the Grounds very much, at ploughing fallow for Wheat Floors, and others at Turnips dressing – Two Carts to Mathers for Lime – flited the Flakes.

Tuesdy 3d Frost last night day fair till past 5 P.M Came on a very heavy shower till past 6 – finished first Horse and Hand houghing of Robs Hillock Turnips and afterwards at those of the Floors P.M. Turned first Cut Hay Thornylee P.M. leading sand & Giving a fur to fallow for Wheat.

Wedy 4th Cold Bleak weather from N.W. but dry – people at Barns dighting &

sacking Oats for the Mill – 3 Carts to Montrose for Wood to West wing of the House, and P.M. all hands at Turning of Hay Thornylee, and at Sundries.

Thursdy 5th Fine dry weather from various Chiefly S.E. – 15 people all day at Gathering & tramp coling of Hay Thornylee, in good order, of which did about 7 Acres – Sundries.

Fridy 6th Dry weather but Bleak from the Eastward – 15 people all day Hay Gathering & finished coleing of Thornylee Crop of same into fifteen Tramp Coles, Supose about 120 stones strong Hay per Acre, sent one Cart for Lime shells for the Slaters & Sundries.

Saturdy 7th White frost in the night to Damage some flowers – and day fair weather till past 5 P.M. – appearance of rain & dark weather – 2 Carts to Montrose for wood and one Cart to Stonehaven for Coals – other Carts driving stones from Dava, flited the Flakes.

Mondy 9th Yesterday very stormy weather from S.W. & Showers of rain at Intervals – this day rough from N.W.but milder, 2 Carts to Mathers for Lime for the Byres Building and two Carts driving stones – Horse & Hand Houghing the Floors Turnips all day – and at a variety of Jobs. -

Tuesdy 10th Fine dry weather from Southward. Carts driving stones from the Quarry and others at Horse & Hand houghing for the first dressing all the Crop of them – & P.M. begun giving 2^{d} dressing to the first sown of D^{o}. Observe the Crop whitening in some places & in others as here offering to Fill – at Sundries.

Wedy 11th a little frost night day moderate weather & some showers of rain P.M. Carts driving stones – and others at Driving Gravel to the High road through our farm to Aberdeen – others at Dressing of the Turnips promising very fine – at Sundry Jobs -

Thursdy 12th Fine pleasant weather but little drought from S.E.- people at Turnips, and Carts driving stones from Dava, Graveling the road to House and at Smithy and other Jobs – Gooseberries near ripe, the Barley & pease filling & Earliest Oats.

Fridy 13th Closs foggy weather and some rain morning and through the day – flited the Cattle Flakes morning – at noon sent three Carts to Pitscandly for Slates, and other Carts driving Dung from Swine Cotts & Hen House to Floors for Wheat, others Drawing Thatch for the Stacks – 4 Servants laid up with the Measles –

Saturdy 14th last night & this morning Considerable rains – day fair till past 5 P.M. some light skiffs of rain from the S.E. Carts returned at past 4 P.M. with Slates & other people drawing straw for Thatch morning and after pulling & watering the lint a tolerable good Crop – at Sundries – the pasture very Scanty.

Mondy 16th Yesterday afternoon a very Considerable deal of rain & also in the night – Early this morning white frost – day fair & drought from N.E. – at Barns before Brakefast – after pulled out all the Lint & put it into the water before Dinner – Others leading of stones from Dava, and P.M. a plough laying to a fur to Earliest Turnips & others Houghing at same a very fine Crop: Observe the Barley much Hurt by the Contrary winds & much Dirt of kinds rising amongst all the Crop which must much Injure the Crop, Pasture very scarce – Barley filling fast. Goose Berries, rasps &c small fruits a good Crop – as also the Potatoes.

Tuesdy 17th Fair & settled weather from various points – all people at the Dressing of Robs Hillock Turnips, and in the Middle of the day gave another or 3^{d} furr to the old road & Dyke Floors for Wheat – Masons at N Byres –

Wedy 18th about 7 A.M. some showers of rain – day fair till about noon when Came on Excessive heavy showers at Intervals – & loosed all the people from the Turnips – after at the Barns dighting of Oats, and at Rye Grass Seed – 2 Carts driving Dung and Rubbish from the new Slated Houses & Sundries.

Thursdy 19th This day pleasant weather and fair whole day-uncommon, a plough at Laying to a fur for the Turnips, and all other people at dressing of D^{o} – very ill to do owing to late rains — notwithstanding lateness of Crop no demand here at present for either Barley or Oat Meal.

Fridy 20th In the morning a Heavy rain which wet the fields & roads much, all people at dressing of Turnips, and sent two Carts to Pitscandly for Slates, and others at Sundries – Thick & foggy –

Saturdy 21st Strong drought from various – Slate Carts returned by 5 P.M. : one plough laying to a fur to Turnips, and P.M. three Harrows at Floors Intended for Wheat, all other Hands Houghing at Turnips and only two of them dighting of last years Rye Grass seed –

Mondy 23d Yesterday pleasant weather till towards Evening when Came on some showers of rain from Westward and Early this morning a very severe fall of rain & high Boisterous wind from Westward which snodded the Barley a good deal, all people Hand & Horse Houghing of Robs Hillock park Turnips &c Jobs.

Tuesdy 24th last night & this morning a very strong Gale of wind from the S.W. and fair weather – all people at Turnips dressing by Plough & Hand houghs – others Quarrying and 2 Carts lead home some stones from Dava. Deranged the Lint of those who had it spread & Injured ye Corns.

Wedy 25th very strong drought all day from Westward – Observe the Crop well advanced little to the farmers advantage at least a fortnight by the High wind, the Barley whitening & the other Crops Loosed at the root – finished 2^{d} dressing of Robs Hillock park Turnips a very tedious Business & mostly Horse Houghed lead six Cart Load of stones mostly for Houghing from Dava.

Thursdy 26th Fair weather from Westward till towards 3 P.M. Came on Heavy rain for remainder of the day – took out & spread the Lint from 10 to 12 nights in water after at Turnips – at Luerencekirk fair sold two Cattle Beasts at Good prices – also all my last Crop of Barley at 15/6 per Boll – all remaining Oats & a portion of Oat Meal at 15/6 – & at Business.

Fridy 27th last night & this morning a deluge of rain which wet the grounds much and Layed the Crop very much – this day fair, Coldish & raw – Sent two Carts for remainder of slates – others Leading stones, and at hand Houghing of the Floors Turnips & Jobs.

Saturday 28th This day dry weather and rather Bleak from the North[d] – a Cart with 10 Bolls of Meal for Mr Shand Doors at 15/ – & Brought Back Coals – driving stones from the Quarry 3 Cart load – others at Turnips &c Jobs.

Mondy 30th Yesterday raw wett weather – this morning about near 8 a very heavy shower of rain from the S.W. – all people at Turnips Floors, and at taking down the North wall of Old Barns to Build new Byres, and at Sundries -

Tuesdy 31st Yesterday & last night frosty – this day fair and Moderately mild from S.W. – finished 2[d] Houghing of all Turnips – and after Brakefast begun & lead into a Small Sow 9 Ells by 3 – Golds Hillock all the Hay of Auld Town park about 800 stones in fine order – The pasture very Scarce yet the Beastials very well thriven – prices of Grain very dull and little sale – Hay very Cheap at 3[ds] per stone very low – Hay foggage rising well – There will be little or no Apples &c this season.

Diary for the Month of September 1790

Wedy 1st Fair weather and drought but with some showers of rain of no Avail – very Early yoked 4 Carts single Horse to Lead Home Thornylee Hay into a Sowe 13 Ells by 4 D[o]- and very nearly finished it in fine order – at ½ past sent 12 Bolls Oats to Laurencekirk for Mr Jo May Kinaber &c Jobs -

Thursdy 2d fair weather in general but some showers – sent three Carts very Early to Kinaber Mills with 2 Carts of my Fathers with 28½ Bolls of Oats Sold to Jo May and to Bring out remainder of wood for the new Buildings – a Cart for Lime shells to Mathers & finished leading & sorting all Hay from Thornylee.

Fridy 3d In the morning Came on an excessive hard Gale of wind from the Northward with much rain which stopped work without doors till noon, people dighting & sacking Barley Sold 40 Bolls – others at preparing Thatch Rushes & ropes for the Hay – ploughing Floors -

Saturdy 4th Cold Bleak & Stormy weather from the Northward and some rain but after 5 P.M. very stormy & heavy sleetish rain – 4 Carts own & 4 Borrowed to Stonehaven with Barley 40 Bolls sold to Mr Shand Doors at 15/6 per Montrose Boll and Brought on return four Boll Scotch Coal – Thatched the small sowe of Hay and did best part of the Large one &c. Jobs. put Cows on foggage.

Mondy 6th Yesterday flying showers sometimes smart – this day fair & some drought from the Northw[d] – people at Thatching of the Hay Sow nearly finished – 2 Carts driving Dung to the West side of Floors for Wheat – laying to a fur to Floors Turnips promising well – Sundries.

Tuesdy 7th moderate drought from N.W.and pleasanter weather – two Carts for Lime shells to Mathers and others at Driving Dung as yesterday about 48 Carts load per acre – put up the Roof Couples on new Byres & Sundries.

Wedy 8th fine pleasant dry weather from S.W. – 3 Carts to the Mathers for Lime for Slaters, and one plough at ploughing down Dung for Wheat Floors, and at Sundries on the Farm – Severals begun Harvest.

Thursdy 9th Early this morning a smart rain from Southw^{d} which Continued at Intervals till 10 A.M., then fair but little drought – a 4 Horse plough at ploughing down Dung for wheat; 2 Carts twice at Dava for stones, and a Cart driving Gravel for roads by ye House.

Fridy 10th Slight frost in the last night day fair & rough weather from the S.W. leading stones from Dava & ploughed down all Dung for Wheat & at 10 A.M. begun HARVEST on Back Yards Barley 16 Shearers & 4 Bandsters – the whole field not ripe – Observe most of our neighbours also begun.

Saturdy 11th Strong drought from Westw^{d} & P.M. from N.W. – people Cut up Back Water Pease, and after at the summer fallow Barley Dickies Haugh very fine Crop – P.M. Harrows at Floors for Wheat – Sundries.

Mondy 13th Yesterday strong Drought from Westward, this day same – all Shearers at Haugh Barley all day, and in sundry places – there will not be Constant shearing as the Oats are Backward – The Cows & Calves in good pasture Thornylee.

Tuesdy 14th Dark weather but good for the Labouring at Harvest – Cut up all the Pease Haugh & part Barley D^{o} – Observe every person Shearing & many Leading Home part of their Bear, In the general most people seem to be in too great Hurry if the Season do not Turn Out like that of last Season, which Can hardly be Supposed – at Sundries

Wedy 15th Fine pleasant weather from the S.W. finished Back Yards Barley before noon after at Haugh Dickies nearly finished – and Dismissed Shearers for Want of ripe victual till beginning of next week – Barley Crop not Heavy -

Thursdy 16th a slight shower of rain last night, this day very Strong Gale from N.W. which drove down the stooks, and damaged the growing Corns a little – driving stones from Dava 4 Cartful – Harrows & roller all day at the Floors for Wheat very Hard, and at variety of other Matters on the farm – no sale of victual &c –

Fridy 17th Slight rain from S.W. morning day fair and Good deal of Drought – 2 ploughs finished the seed fur for Wheat Floors and Carts driving all the stones Quarried Home, & P.M. Gravelling the road above the House; setting up Couples in Byres &c. Observe the Crop towards Stonehaven as here.

Saturdy 18th Dark Cloudy weather but fine – after Brakefast Begun leading Home a stack of Barley in good order from Back Yards on the Ground, & One from E^{t} Haugh on Pillars but rain Coming on was Obliged to stop after being about two thirds up – & Covered them with Thatch – Sundries.

Sundy 19th last night Heavy rain which Brought down the river very Considerably by noon – Again in the Evening which Increased it more before Midnight -

Mondy 20th This morning fair & drought from Southward – after Brakefast all Shearers begun Cutting up the Bear on Gibsons Park a Middling Crop – but Obliged to stop by 4 P.M. with very Heavy rain & storm from the S.E. – Some of the Back Water Pease overflowed yesternight & Men turning of them – at Barns & Sundries.

Tuesdy 21st rainy weather till 7 A.M. when fine – Carts driving Gravel to Roads, at 10 A.M. Shearers begun & cut up all Gibsons Park Bear Threaves 85 – after at Haugh N. side river opposite the House Thr^{s} 15 – after at Dickies Haugh Barley – and at Sundries -

Wedy 22d This morning a strong white frost, day fair weather & drought all Shearers till Dinner at Haugh Barley – and after 4 Carts leading Home Barley from Dickies Haugh and the Back Yards – finished two stacks & founded 2 others of D^{o}.

Thursdy 23d a small shower of rain but very slight from N.E. – day fair but little drought from Northward – morning Thatched two stack Barley – after Brakefast flited the Flakes to the Nether Burns – after Dinner yoked 4 Carts and lead Home wheat-finished the 3 & 4 stack of Barley.

Fridy 24th a strong white frost morning day fair & drought from the Westward – Cutting Rushes for Thatching the Pease, and 4 Carts Lead all Backwaters Pease into two large stacks tramped 11 Cartful each after Brakefast.

Saturdy 25th Early begun leading Home Haugh Pease in very fine order & rich, but only got two Cartful when stopped by rain which Continued all day – but Moderate. Thatched the two Stacks of Pease with rushes, and the Barley Stack with straw; P.M. Graveling the road to the House -

Mondy 27th Yesterday Closs small rain & very growing weather. This day fair & drought after Brakefast – begun Shearing Bank Oats – & part Men turning the Haugh Pease in fine Condition – Graveling the road &c.

Tuesdy 28th Begun Early this morning to Lead Pease, but stopped by rain by 9 A.M. – after at Sundry Jobs – at 2 P.M. the wind shifted suddenly to the Northward & faired – 3 Carts driving Gravel to the road to the House, and the rivers came out Considerably & fields very wett &c.

Wedy 29th a strong white frost last night & morning – day fair and Moderate drought from S.W. – People after Brakefast shearing at Bank, Old Town Hillock Park & Kavels Oats all which are fine grain – Slating Byres.

Thursdy 30th Moderate weather and fair with slight drought from the Southward – Early shearers at Bank Oats in different places, but by no means full ripe – little before noon yoked four Carts to Lead Home the Remainder of Pease Haugh, which got done by about 8 o'Clock Evening in tolerable order, and a very rich Crop to all appearance turn out as they may – Observed all our Neighbours leading Bear but in what Condition I Cannot Judge, as not six sheaves of twelve of mine are in any Manner fit for it – of late Cattle Markets are dull & prices fallen both here and to the Southward – prices of Grain is also unfixed the new Crop does not seem to have Lowered the price either in South or North Britain as yet – Bear Meal 7^{ds} per peck. In my opinion from the Immense rains for near two months past, I do not Imagine this Crop will reduce the former value of Grain.

Diary for the Month of October 1790

Fridy 1st Moderate drought all day – Thatching at Pease and Making ropes, at 11

A.M. begun leading the Gibsons Park Bear into a Stack, and wyling the E^{t}Haugh Barley into a Stack. Wewmen at sundries, shearing rushes and after at Oats Bank P.M.

Saturdy 2d Mild weather and some drought from S.E. – Shearing Oats Bank, and about noon Begun leading Home the whole of the Barley Crop into 8 stacks, it hath Cost much Labour, and but just in Midling order, but the weather and season have been very Backward for Barley. The Neighbourhood very far Back in the general -

Mondy 4th Yesterday Closs foggy, but as this Broke out sun shine before noon when warm; – this morning begun Cutting up Oats Bank in Gayte, and when dry bound. P.M. at Ley Oats Haugh, and at sorting stacks &c.

Tuesdy 5th Heavy Dew in the morning – day fair & some appearance of rain – however kept fair before Brakefast driving Gravel to Approach to the House and after all shearers at Oats Haugh, and after at North End of Kavel park D° – Sundries. Cutting grass done.

Wedy 6th Strong drought from S.W. all day – Early all Shearers at upper side of Kavels Oats ripe, and after Dinner at part of Bank D° – Beginning of the week Begun Suppering the Cows with Turnips and put the Calves into the New Byres – Suppered D°

Thursdy 7th In the morning some light rain, through the day a good deal more Showers, Early driving Gravel to the Approach road to House, at Sundries, dismissed all shearers to Monday first, and at Drumleathy Markett but a few Cattle and less sale &c -

Fridy 8th pretty strong drought from N.W. – people flitted the Flakes morning, after levelling down a dyke at Goulds Hillock for a Dunghill to Kavels – P.M. drove some dung from Straw Yard to it – at the Horse Markett Drumlithy Sold two Old Horses at good prices, they have wrought to me annually at 12/6 per H.

Saturdy 9th pretty strong frost in the morning, day strong drought and fine winning weather from the N.W. – many people leading, but here shearing – Carts driving Dung & Rubbish for next year Turnip field Kavels to an Old leveled Dyke Goulds Hillock – took down the roof part Old Barns to Compleat the New Byres & Sundries.

Mondy 11th Yesterday in the morning a very strong frost, day very pleasant weather – this morning some showers of rain, day fair but no drought & Consequently no shearing – driving of dung twice a day &c Jobs – Observed the Crop towards and about Bervie latter than with us, at Sundry Jobs.

Tuesdy 12th last night and this morning some rain and high wind from Southward, begun Shearing at 11 but stopped by a storm of wind & Excessive heavy rain till 3 P.M. when Cleared up – people driving dung before Brakefast, at Barns dighting Barley for Mill to selves, and Masons laying Floors of Byres – the River Came out much towards Evening &c.

Wedy 13th Early this morning Begun shearing Bank Oats – and finished them Evening full ripe & some Out, Threaves

Set up all down stooks, and a very strong drought from Westward – Observe several People Leading Oats, and Others Bear, Sundries.

Thursdy 14th Early this morning Came on heavy rain & stormy weather which lasted till 11 A.M. when Cleared up & after Blew a strong gale from the Westward-which shook some of the Oats-Cut up part of the Ripest of the Smithy Park Oats, a very rich Crop – morning at Barns threshing Oats – 4 Cartful of rank ley Haugh Oats for Thatch to stacks –

Fridy 15th Very Early begun Leading Oats from Old Town Planted Hillock Park, and the Bank & N. end of Kavels Oats of which got in by a little after 7 P.M. five good stacks in as fine order as any person Could desire – Wewmen Cutting up Oats Smithy park the most of it the weightiest Crop I ever saw – Sundries.

Saturdy 16th Black frost morning, day very fine dry weather from the N.W. – Early 4 Carts leading in Oats, from the Banks Haugh, and Black Lands new Cutt, all in fine order into 4 Stacks this day by 7 P.M. – the wewmen shearing all day at Smithy park Oats – and a Considerable loss by shake & shearing. Cattle pretty well off in Meat by the Fields being reade &c.

Mondy 18th Yesterday early Came on a good deal of rain from the Southward & stormy weather which lasted & towards late pretty Heavy – this day fair & drought from the Westward – all shearers finished Black Lands & most of the remainder of Kavels fine Oats, flited Flakes.

Tuesdy 19th Slight frost in the night, day fair & pleasant dry weather from Westward. Early begun Cutting up the remainder of Kavels Oats, and after Brakefast at Haugh Oats dead ripe, and parcel of Wheat 5 stooks – Some very few done shearing this day -

Wedy 20th Early Begun shearing the remaining part of the Crop uncut Haugh Oats dry – finished whole in the Evening – P.M. yoked 4 Carts and lead home all the Black Lands Oats, and after further part of Kavels Oats – wrought till between 7 & 8 P.M.

Thursdy 21st This morning about 7 Came on rain from S.E. more or less with thick foggy weather – all people at Barns threshing, and drawing straw – Thatched two stacks of Oats – Making ropes and at Sundry Jobs -

Fridy 22d Morning Came on rain from S.E. pretty heavy and Indeed so last night; all people at Barns, threshing, drawing straw & thatching stacks, also threshing Barley for Malt, & thatch – P.M. flited the Flakes – thatched some more of the Oat stacks & Sundries –

Saturdy 23d last night Came on very Heavy rain from Eastward which Brought down the rivers very high by noon full as high as for a year by past – went over the Dykes into the old water Track amongst the stooks therein, Evening took out same & Gayted them on the Higher grounds – people at Threshing Barley & Oats for Thatch to Stacks; Causewaying Byres &c Jobs – took in Hay – flited sheep flakes

Mondy 25th The feast Yesterday fair Mild weather – this day same & very warm but no drought, Early all people Men thatching the whole of the Oat Stacks not done before viz 6 of them Others at Clearing out Byres, & Causewaying at them – Observed severals shearing Oats & Bear Many people are not yet near Half Harvest, Misfortune.

Tuesdy 26th Heavy frosty dew and everything very wett – all people at flooring Byres, and assisting Masons P.M. yoked 2 four Horse Ploughs on East side of Bank, stiff & rich ribbing for Pease – lead in some Turnips, Hay for Horses &c – Weather warm.

Wedy 27th Moderate & Mild weather but no drought – 2 ploughs at the Bank ribbing morning – after flited the Flakes – Nether Burns – Digging Potatoes, at Byres, and put Horses on new Grass fine Back Yards, also at Graveling and variety of Jobs – Clipping Garden Hedges Batts appeared.

Thursdy 28th Dark Grey weather from Eastward & not much drought One plough Ribbing Bank, one at Back Waters ribbing for Barley after Pease, one Cart to Mathers for Lime Shells, others at Byres, digging Potatoes &c – at 2[d] fair Drumlithie Cattle are no Markett –

Fridy 29th fair weather but little drought from Northward – Early begun Wyling the Oats, Haugh & Kavels and 4 Carts to lead Home the same which did into 3 stacks, One of them Hollow, by evening & Barned some Cartfuls being whole of the Crop for Season – finished with a HEAVY SHOWER as vexatious as in many other parts of the Harvest.

Saturdy 30th White frost in the morning, and day very little drought from the Northward – Thatched the stacks lead yesterday being whole H 27 – and P.M. Two four Horse ploughs Ribbing Bank for Pease & Turnips – Slating Byres, took in Hay for Horses, and Turnips for Cows, Calves &c Observe Many people leading Oats in no great Condition undoubtedly – Sundries –

Sundy 31st This day Small rain & dirty Bleak weather from the Southward – many people had lead home a good deal last night after Dark – There is a great deal of the Crop yet out in the Parish of Marykirk and Indeed in Most farms.

Diary for the Month of November 1790

Mondy 1st Came on rain about 9 A.M. which Increased with stormy weather from S.E. – 2 ploughs at the Bank Ribbing – after flited the Flakes – dighting Oats – Evening heavy rain from East.

Tuesdy 2d Fair weather, warm & Bad for the Corn – the Rivers full as high as for 18 Months by past & fields very wett – 2 ploughs Ribbing the Bank – Housed for stall feeding 9 Cattle & 7 Cows in two new Byres – Observed Mr Barclay of Urie's people shearing Oats at Dava quite green & not a sheaf yet in.

Wedy 3d a Great deal of rain from N.E. – 2 ploughs at Ribbing Bank & Backwaters – leading Turnips from Floors, at Byres & Barns – Cut a Bull & some Swine Shots – rain through the day -

Thursdy 4th Small rain & no drought from any Quarter – 2 ploughs at Ribbing, and one Cart to Stonehaven for Coals & Sundries – at Laurencekirk Tryst – not many Cattle & no sale – Sold a stott, no demand for Grain & Observed great part of the Oat Crop Out betwixt Polborn & about that place, at Digging up Potatoes, Causewaing Byres & Sundries –

Fridy 5th little frost morning – day fair & some drought towards afternoon – 2 ploughs Ribbing at Bank and Back Waters, one Cart Leading Turnips, at Byres flooring, Dug up all Potatoes a fine Crop, but wet – finished Slating for a Season – Yesterday put in all Cattle.

Saturdy 6th Early this morning Came on a very Severe Gale of Wind from S.E. attended most of the day with Excessive Heavy rain & stormy – that no person Could work without doors, rivers very high & no Building or windows Could keep out the rain. Miserable weather.

Sunday 7th last night or this morning the River Bervie full as high as for Eight years by past, but Except some of the Banks Broke a little no other Dammage done by the Flood – N.Esk very high.

Mondy 8th Fair weather from S.E. – but no drought – 2 ploughs Ribbing at Kavels & finished Do Backwaters – One Cart leading Turnips from Floors, at Byres and at variety of Jobs – Hear of much loss on the Coast of Shipping by the Saturdays Gale &c –

Tuesdy 9th Moderate fair weather & Calm – 2 ploughs at Ribbing of the Kavels for Turnips fine season, leading of Turnips, and one Cart Clearing up Rubbish & dung from before the new Byres, and at Sundries &c

Wedy 10th Dark Gloomy weather & very little drought from So – One Plough at Kavels, One Cart leading Turnips Floors, and one Cart driving off Rede Dung from before the New Byres to Golds Hillock Dunghill 4 Men Planting at Thornylee, & Sundries –

Thursdy 11th Droughty weather though no sun for a week by past one plough at Kavels, One Cart for Coals to Stonehaven, One D^{o} driving dung to clear before new Byres, Dispersed Masons for Season – 6 people planting Thornylee, observed most people leading Bear & Oats in very Indifferent order –

Fridy 12th Thick and Dark weather from S.E. Two ploughs at Kavels, and One Cart leading of Dung & Rubbish from about the Houses &c leading some Turnips – at Planting & Pitting for Hard wood Thornylee Park -

Saturdy 13th Dark Gloomy weather but Mild, 2 ploughs at the Ribbing for Turnips Kavels, Lead some turnips and a Cart Clearing out the Cart Shades & Other parts about the Houses, and at Sundries –

Mondy 15th Yesterday mild dark weather – this day frosty but Sun Shine for first time this 10 days – 2 four Horse ploughs at Kavels, lead Home some Turnips – at Barns dighting Oats – & Sundries.

Tuesdy 16th Strong frost in the night day fine & pleasant weather 2 ploughs at Ribbing for Turnips Kavels – at Barns dighting Oats for the Mill, at Planting Hard wood on Thornylee, and at Sundries – the Crop is not yet settled in the Neighbourhood no prices yet fixed for Grain in our Marketts,

Wedy 17th Strong frost all day – One plough at Kavels nearly done, 2 Carts Carried 25 Bolls of Oats to Mill of Kair for Meal to family – and lead Home some Turnips – flited the sheep Flakes & put up Hay Hacke in D^{o} – Sundries &c

Thursdy 18th very strong frost so as to stop Ploughing on most farms – yet continued it here on Kavels – a Cart to Stonehaven for Coals, and 2 Carts driving of Dung to Goulds Hillock dyke Dunghill and at variety of Jobs – Planting Thornylee 4 Men with all kinds of Forest Trees in Mixture

Fridy 19th Fair weather till past noon when Came on a Gale from the Eastward with Heavy rain, a plough at West side of Bank ribbing – and a Cart with Oats to the Mill of Kair twice, also one Cart lead Home 4 Cartful of Turnips from Floors, put up 10 year olds into the new West Byre

Saturdy 20th The rain Continued last night till past midnight and Brought down the rivers very high, day fair & Moderate, Sent 4 Carts to Stonehaven for Coals and at Sundries about the farm – Frost at night.

Mondy 22d Yesterday several showers of Hail & rain – this day strong frost & Cold from N.W.-2 Carts to Stonehaven for Coals to the Smith and 2 driving Turnips two other Men filling at Barns dighting Bear & 4 Men Planting.

Tuesdy 23d very strong frost so as to Stop Ploughing, day some showers of rain from S.W. – 3 Carts driving Dung for Kavels & one Cart at Carting stones to a part of the Floors field like to Brake by the late Floods – &c

Wedy 24th Fresh weather and fair till a little after Midday when Came on rain pretty smart from S.W. – 2 ploughs at Black Lands ribbing across – one Cart at driving stones to Embank the Water at Coley Pool and others at Planting Thornylee – dighting Oats.

Thursdy 25th Early this morning Came on rain from Northward which was very Closs & Heavy most of the day & night, one Plough at Blacklands, and a Cart leading Turnips, one at Stonehaven for Coals and another Leading stones for Banking the Rivers – other people at Planting, and Dighting Oats N. Barn –

Fridy 26th Rough stormy weather & Gale from N.W. – one plough at Black lands – one driving stones to fend the river Head of Bleaching green – lead some Turnips – One Man Grieve & two wewmen at Mill of Kair with 37 Bolls of Oats – Meal 31B 3 pecks –

Saturdy 27th very strong drought from N.W. & no frost – Sent two Carts to the Mill for the Meal, and two D° driving ten Cart load of Turnips fine season – others dighting Oats N. Barn, Girnelling Meal &c Planting

Mondy 29th Yesterday Cold Bleak weather and Hard frost, so as to render Ploughing not good work – accordingly one Cart carried 15 Bolls Oats to Mill of Kair and three Carts driving Dung out & leading of the remainder of Floors Turnips – at Barns dighting and two Men Planting in Thornylee, Sundries

Tuesdy 30th This morning some showers of Hail and fresh weather from the Eastward – 2 ploughs at the Blacklands ribbing for Barley – and One Cart Carried at three draught 17B 3f Oats for Meal – P.M. at Barns and Sundry Jobs. 2 Men planting at Thornylee a good deal of that Plantation much finer ground than either myself or others had any Idea of till Pitting shewed it – Prices of Corn very uncertain as yet Bear & Meal about 12/6 – and good Cattle saleable, but poor ones very low. Believe there is much Scarcity of Cash in the Country at present, and Manufactures but Midling – much of the late Crop spoilt by rain & scarcely yet settled in general.

Diary for the Month of December 1790

Wedy 1st a Considerable fall of Snow & Sleet from S.E. which lay all day – 3 ploughs at Black lands ribbing – lead some turnips – at Barns dighting Bear for the Horses, and at Sundry Jobbs –

Thursdy 2d last night & morning a very heavy fall of rain from N.E. – day mostly fair & Cold – 3 Carts driving dung all day – they & one More lead Home 5 Cartful of Turnips from Robs Hillock Park just begun – at Barns dighting Bear for the Horses &c Jobbs –

Fridy 3d last night strong frost – this day Came on rain, about 11 o'Clock very heavy & Stopped the Carts driving Dung & two people Planting about 1 P.M.: after at Dighting & lofting Bear for the Horses &c. Jobs –

Saturdy 4th This day a severe Gale of wind from N.W. which Tirred West gavel of old Barn and hurt some of the Tiled Houses but little dammage – 3 ploughs finished Ribbing of Black lands & at variety Planting &c. Jobs on the Farm.

Mondy 6th Yesterday very pleasant Mild weather & Calm – this day same but frosty in the night – two Ploughs finished Ribbing west side Bank in good order, which finishes whole Ribbing for the season for Turnips, Pease & Barley 46 Acres. P.M. took in the 4th stack of Oats S.Barn &c.

Tuesdy 7th Soft & Mild weather from the Westward – 2 ploughs begun the seed fur on Dickies Haugh for Oats Having ribbed 46 Acres for Turnips, Barley and Pease, at Barns, & Sundries about Planting.

Wedy 8th Scotch Mist & Slawky weather – one plough at Dickies Haugh, and two Horse plough begun giving a neat Clean fur across to Floors for Barley – ribbing not answering with so strong ground in Droughty Spring, Two Carts leading Turnips from Robs Hillock &c.

Thursdy 9th Fine pleasant weather from Westward – 2 ploughs at Floors, and four Horse D° Dickies Haugh, people at the Mill with 32B 3 pecks Oats, Meal 30B 2 pecks – at dighting Oats, Midding Layers &c Jobs –

Fridy 10th Moderate weather from Westward ploughs as before – two Carts Brought home Meal from the Mill and at Barns & Sundries – planting &c –

Saturdy 11th Frost and some skiffs of snow from Westward – Ploughs going as usual and at Sundries

Sundy 12th pretty strong frost & some little snow but did not Lay here, though Considerably so to the Westward of Brechine – Fatt Cattle more valuable there than in the Mearns –

Mondy 13th Fresh weather & Moderate from Northward, a two Horse Plough Floors, & one plough at the Dickies Haugh for Oats, 2 Carts lead some Turnips, and after Brought Home & Housed all the Cattle Flakes – Planting Larch's in the Brae behind the House, taking in Barley.

Tuesdy 14th pretty Hard frost night & day, wind South W – 2 ploughs at Dickies Haugh, and one Cart at Leading of Turnips – at Barns dighting – two Men Planting Larchs in the Upper Clumps pretty large ones pitted – took in Hay and at a variety of Jobs &c –

Wedy 15th a severe Gale of wind & rain from Southward which lasted till Morning, when Came on another Gale from NNW very furious – two Ploughs at Dickies Haugh, and a Cart driving 16 Bolls Oats to the Mill of Kair.

Thursdy 16th last night strong frost – two Ploughs at the Dickies Haugh seed fur, in good order, though I observed none of our Neighbours Could work owing to Frost but the Haugh was warm, dry & Covered with a fine Stubble, driving home Turnips, & at Barns Dighting Oats for the Mill – Gardner & another Man Planting Larchs at Craig Braes which will render them fine Plantations if they thrive – at Sundries; prices of Grain not yet determined –

Fridy 17th very strong frost & wind N.W. – 3 Carts driving dung all day and one Cart driving Oats to Mill of Kair for family use, at Barns, Planting Larchs Craig Brae, Barns &c – at Auchenblae Markett not above 100 small Cattle or Cows, Bought a ferry one.

Saturdy 18th very strong frost from N.W. and Moderate weather – 3 Carts went to Inchbreak for Peats and returned by 5 P.M. One Cart leading Turnips and Obliged to Dig them up for first time this season – at Barns dighting Oats and at Sundries –

Mondy 20th Yesterday very hard frost, and this day same the River Closed in many places – 2 Carts to the Moss for Peats, and one leading Home Turnips, and sent two Cartful of Oats to the Mill of Kair – and at Sundry Jobs &c –

Tuesdy 21st last night a strong Gale from the Southward with Heavy rain which this day was fresh from S.W. 3 Carts driving Dung & leading of Turnips – at Barns & Sundry Jobs &c –

Wedy 22d very strong wind from S.W. and some rain – 3 Carts driving dung all day – and at Smithy Shoeing Horses and others driving Home Turnips – at Barns dighting up Oats & Sundries –

Thursdy 23d last night a Heavy fall of rain from the Southward & river out carried off all the Ice this quite stormy from N.W. & frost – 2 Carts driving Dung from Cottar Town to Old Town Park for Barley – two Carts lead Home some Turnips After dinner – at Barns dighting Barley and Gardiners driving in Dung to Garden.

Fridy 24th moderate weather and fresh towards afternoon, two Carts to Stonehaven for Coals to Ourselves & Cottars – One Cart driving dung from Cottar Houses to Old Town Clump planted Park – Have got there about 20 Cartload Excellent Manure – and Cut a water tract for Conveying the Dung Court Urine &c through Corn Yard to y[e] Bank.

Saturdy 25th last night strong fresh from S.W. and this day moderate weather from N.W. – 2 ploughs at Haugh for Oats begun Pease land – Streighted from River to North side Haugh, one Cart leading Turnips all day – flited sheep Flakes, Barns & Sundry Jobs.

Mondy 27th Yesterday fresh weather from Westward – this day the same all Ploughs at Haugh seed fur Grass seeds & Oats – lead Home some Turnips – at Mill of Kair with 44 Bolls Oats – Meal 40B:3 pecks.

Tuesdy 28th very hard frost to stop Ploughing – 2 Carts lead Home from Kair 40B 3p Meal, and one Cart leading Turnips & others pulling D° – at Barns dighting Barley – Gardner Carrying in Dung Garden.

Wedy 29th very strong frost and Bitter Cold weather from South – 3 Carts driving dung to Nether Burns ley above 30 Cart load per Acre – at Barns dighting Barley and lead Home some Turnips P.M. Sundries -

Thursdy 30th Fresh weather from Southward, three ploughs at Haugh on Pease land for Oats & Grass seeds – in general Land would not Labour for Frost & wetness – Spreading Dung on Ley – & Sundries.

Fridy 31st very strong frost to stop Ploughing though we tryed it – after three Carts driving Home Turnips all day in fine Condition for it –

The Prices for Grain about 12/ for Chester & 14/ for Barley – Oat Meal from 12 to 12/6 per Boll – Fat Cattle if good about 5/ per stone, and Lean ones about 3/6 D°

The Manufactures pretty Brisk, price of spinning about 13 or 14^{ds} per spindle –

still Employed Planting two Men, and Covered round all the Fruit Trees about the root with Dung – Making up Dung hills in Garden.

Diary for Month of January 1791

Saturday 1st Cold Disagreeable weather from S.W. and appearance of rain – sent 3 Carts to Johnshaven with Meal and Brought out 2 Bolls Coals in return – at Sundry Jobs on the farm, and at Glenbervie

Mondy 3d Yesterday very Cold weather from Southw^{d} – about 9 night wind veered round to the N.W. with a gale, some snow & Hard frost, this day frost – 3 Carts driving dung to the ley Nether Burns, and at Barns dighting of Barley, and Garden, &c Jobs –

Tuesdy 4th This morning about 7 Came on very severe storm of wind & rain from S.E. which Increased towards Evening into a very severe Gale that no Houses Could Keep the rain out, also very dark – lead Home 3 Cartful of Turnips before Dinner – at Grass seeds & Barley dighting.

Wedy 5th about 8 last night the Storm abated, this day frost and Moderate weather from S.W. – being Christmas Day no work with the farm – Could have Ploughed &c.

Thursdy 6th Frost in the night – Morning strong winds & rain from S.W. – 2 Carts driving dung to Nether Burns ley, and P.M. two Carts lead Home Turnips, at Barns dighting & lofting Barley & Bear – Sundries –

Fridy 7th about 7 A.M. Came on a Severe Storm of wind & rain from the East which Continued till noon when Moderate weather – P.M. 3 Carts driving dung and lead Home some Turnips, at Barns &c Jobs –

Saturdy 8th a Good deal of rain last night, and most part of this day – 3 ploughs at Haugh seed furr viz the Old Ley Award – & Evening lead Home 3 Cart load of Turnips – at Barns dighting &c.

Mondy 10th Yesterday moderate weather from the N.W. – and this morning Came on rain & very stormy weather from S.E. with Heavy rain from 1 to 2 P.M. – 2 ploughs at Haugh Award – leading of Turnips – at Barns &c Jobs –

Tuesdy 11th Strong winds Westerly & fresh – 2 ploughs at the Haugh Award fine season, & two Carts leading Turnips from East Side Robs Hillock Park – Sundries.

Wedy 12th Strong wind from S.W. all last night, after 10 A.M. Came on some rain which Increased with High wind to be very severe from S.E. – three ploughs at Haugh, other Men spreading dung long Burns and at Barns, Smithy & other Jobs – P.M.

Thursdy 13th Strong winds in the night and all this day, three Ploughs at Haugh, but obliged to leave it and take to the Bridge shade ley – rather young but the Clover having over Croped itself in the Hay Cropt Obliged to Plough it up – at Barns &c Jobs.

Fridy 14th strong drought from the Westward & rather better weather than for weeks by past – 3 ploughs at Bridge Shade ley in good order – spreading dung and pulling down Dike for Dunghill Layer – Sundries Barns Chaffing & Dighting Oats N. Barn &c

Saturdy 15th variable weather frost & fresh, High winds & rain – One Plough at Bridge Shade ley, One at Floors and two Carts Leading Turnips – at N. Barn Dighting Oats, and at Sundries, taking in Hay – Garden &c.

Mondy 17th Yesterday fair & frosty till P.M. – high winds & a good deal of rain from S.W. – this day high winds but fair from Westward – 2 ploughs at Bridge Shade ley – and one Cart leading home Turnips from Robs Hillock Park – Sold two of my fatt Cows for Montrose for about 5/ per stone – there is a Considerable demand for fatt Cattle but prices not high – few or no farms sold –

Tuesdy 18th slight frost in the night day fair till about 11 A.M. Came on a most severe Gale of wind from the S.E. – attended with a deluge of rain full as severe as on the 4th Inst – Stopped the Ploughs for a little after Miday – after at Barns & other In Jobs – Hear of Oats for seed & Bear for D° sold at 14/ payable at Whitsunday – by Smart Polborn Oat Meal 12/ to 13/ –

Wedy 19th The rain & wind Continued last night till very late & Brought the River out very high – this day stormy & Cold but dry till 3 P.M. when Came on Considerable rain from S.E. which lasted for several Hours, people at Ploughs Bridge shade and leading Home Turnips – at Barns & Sundries.

Thursdy 20th slight frost morning – day fair till about noon Came on rain pretty Heavy about Dinner time – 3 Ploughs at Bridge shade ley and P.M. at Barns dighting & sacking Oats for the Mill – The Hares are destroying the young Plantation of Trees particularly the Laburnums &c Thornylee –

Fridy 21st Cold Stormy weather from N.W. and thick snow on the Hills & Braes – Two Ploughs at the Bridge shade ley – & one Cart leading Turnips – at Barns sacking Oats for the Mill & other Jobs –

Saturdy 22d Frost in the night, day fair till past 2 P.M. Came on rain & stormy weather from S.E. – one Cart to Stonehaven for English Coals, one D° carrying Oats to the Mill of Kair with Oats, one D° leading Turnips and a four Horse plough nearly finished all the Bridge shade ley – taking in Hay – Barns &c.

Mondy 24th Yesterday moderate weather from Southward – this day a strong Gale from S.W.but fair – One Plough finished the Bridge shade ley – & some of Kavels. One Cart Carrying Oats to Mill of Kair, and two leading Home Turnips – at Barns & turning Dunghill.

Tuesdy 25th very strong Gale from S.W. & fresh weather – 2 ploughs at Upper Burns Ley, & one at steering the Floors after Turnip good season – at Sundries –

Wedy 26th This day fair weather & drought from S.W. but Cold – at Taunton fair a vast Concourse of People, but little Trade, the servants either for House work or on the farm very high, and for shearing the same, Engaged the greatest part of mine at somewhat more than last year, though others were advancing them very much – this Business of Increased servt^{s} wages must turn out Serious if not better Prices for farm Produce.

Thursdy 27th Finer weather from S.W. than for Months by past – 2 Ploughs at the Burns Ley, and one Cart Leading Turnips -and which are shooting fast – the Grounds Considerably dried – at Sundries.

Fridy 28th Strong frost in the night & skimmering of snow which did not Prevent two Ploughs at Over Burns Ley, and one at Haugh Award – the Ploughing at Polburn &c Stopped by frost. Turning Dunghill Golds Hillock P.M. &c Jobs.

Saturdy 29th Strong drought from S.W. & P.M. from N.W. – 2 Ploughs at Ley and Haugh Award – and two Carts Leading Turnips – at Sundries – filling up an Old Water Track Over Burns with spades &c

Mondy 31st Yesterday P.M. and night one of the most severe Gales of wind from W.N.W. for Years by past – Damaged very little the Houses &c – This day moderate but much Drought from the same Quarter – 2 Ploughs at Haugh for Oats, and one two Horse D° steering South end of Floors for Barley fine season – feeding Cattle and Indeed all the Cattle on the Farm are for some days getting Plenty of Turnips, as they are offering to shoot & flower, and Have as yet nearly one Half or above 8 Acres remaining- Prices of Grain very dull & Precarious – but of Fatt Cattle there is great demand being scarce & no Fish on the Coast – the Fields now seed Coloured –

Diary for Month of Feby 1791

Tuesdy 1st Moderate weather till about 11 A.M. when the wind shifted from S.W. to the N.W. in a very strong Gale & Storm – damaging roofs of Houses, Corn Stacks, Trees &c Evening rain – 2 Ploughs at Floors, and Haugh, and two Carts Leading

Turnips – People at Mill Kair – much about whole Meal which is by far the best there of season.

Wedy 2d In the night Came on severe Squalls of Snow which Lay all day with Frost – sent to Kair 3 Carts for the Meal, and after at South Barn Dighting of Barley all People, Observed some Ploughs, not withstanding was very severe & stormy – Sundries.

Thursdy 3d Stormy weather and a Small fall of Snow on the Ground – pretty Hard frost – wind N.W. – 3 Carts driving Dung to Nether Burns – & P.M. leading some Turnips and Hay – at Barns dighting of Barley all day- Observed no Ploughing

Fridy 4th pretty strong frost last night – day mild & agreeable, and the more so Contrasted by ye late weather – 3 Carts driving dung to N. Burns ley – and two Carts P.M. leading Home Turnips – digging them – at S. Barn dighting Barley all day &c Jobs –

Saturdy 5th Pretty strong frost in the morning – 3 Carts at driving Dung to Goulds Hillock Dunghill for Kavels Park for Turnips – and others at Turning D^{o} – leading Turnips, lofting Barley –

Mondy 7th Yesterday very pleasant Spring weather – this day the same from Southward – 2 ploughs at the Haugh, and one Cart Leading Turnips & Carrying 4 Bolls Barley to Bridge Leppie for Malt to Family – dighting & lofting Barley – the Gardiner Digging the Nursery for a new sort of seedling Trees for Planting **Bog Atry** -

Tuesdy 8th Fresh weather from S.W. – one Plough at Haugh and one Cart leading Turnips – two Carts to Stonehaven for Coals, Iron, Bread &c –

Wedy 9th Fresh weather & pretty brisk wind from Southward – 3 Ploughs at Haugh for Oats – Streighted for first time from the East side near to the River from S.E. End to North some of the ridges above 500 ells long – at Barns &c Jobs. Lapwings.

Thursdy 10th a Slight frost morning, day fine till night when Heavy rain & Gale of wind from S.W. – 2 ploughs at Floors & Haugh, and two Carts Leading Turnips – at Barns &c Jobs

Fridy 11th Fine Pleasant dry weather from S.W. – 3 Ploughs at Haugh for Oats & P.M. two Carts lead Home some turnips & two Cartful Hay – at Sundries &c – Observe Spring flowers out Viz Snowdrops, Primulas &c –

Saturdy 12th pretty strong frost this morning, but not to Impede Ploughing – 2 Ploughs at Haugh and two Carts Leading Turnips – at Barns and turning Dung in the Court Yard &c Jobs –

Mondy 14th Yesterday very pleasant weather from Southward – this day Thick Hazy weather till noon when wind from S.W. – 3 Ploughs at Haugh for Oats & P.M. at Barns Dighting Pease, and others spreading Dung on the Burns Ley – Observe some People Yoking twice a day –

Tuesdy 15th fresh wind variable from S.W. to West and some squalls of snow in the morning which however Cleared up & did not Lye – 2 ploughs at Haugh Award – and one Cart leading Turnips – P.M. people spreading Dung on Ley N.Burns – & others at the Turning Dunghill in the Dung Court.

Wedy 16th Strong wind from Westward & frost in the night day fair and a little before noon Came on a furious Gale from the N.W. fully as severe as any this Tempestous Season it abated by One, but Damaged the Houses very Considerably particularly the Tiled ones, 3 ploughs at Haugh for Oats and turning dunghill, laying Harrows.

Thursdy 17th Slight Skimering of snow which did not Ly – 2 ploughs nearly finished all Haugh for Oats and two Carts leading Turnips – at repairing of Houses, turning Dunghills – and Sundries repairing Harrows & laying the Tines &c Jobs – Sowed Garden Pease.

Fridy 18th Slight snow as yesterday – 2 ploughs finished all the Ploughing of E^t Haugh for Oats and one Cart to Stonehaven for Coals, Tiles & Sundries – P.M. at Barns dighting Pease, and turning Dung.

Saturdy 19th Showers of rain & Hail & rough weather from North – day D^o – 3 Carts A.M. driving out dung to finish the N.Burns Ley – and drove in dung turned Dunghill in Dung Court, on the Hirsty Part Head of Haugh to be sown with Oats & Grass seeds at rate of about 60 Cart load per Acre, mixed with lime & Clay – dighting of Rib wort or Plantain seed sold for Aberdereen &c.

Mondy 21st Yesterday fine pleasant weather but raw, this day some showers in the night, day dark moist weather from S.E. 2 ploughs at N.Burns ley and two Carts leading Turnips – at Barns & Sundries – Observed about Bervie some Beans sown in Drills – & Turnips flowred – Grain rising

Tuesdy 22d fair weather from South – and Mild Spring weather – three Ploughs at Nether Burns too small ridges – P.M. at Barns dighting seed Oats, and at Sundries. Corns & fodder sold on Saturdy last from 15/2 to 15/6.

Wedy 23d last night a very severe Gale of wind and strong storm from S.E. – rivers much out and fields Deluged with wet – one Plough at Nether Burns, and one with four strong Horses ploughed up the Road to Cottar Town Gathered for Graveling – no work without doors after Dinner – at Barns Dighting Oats – wind shifted at 5 P.M. to N.W.

Thursdy 24th In the morning frost & fair weather from Westward – Two ploughs very nearly finished the Nether Burns being all for Oats of season and one Cart leading Turnips – Gardner transplanting best Turnips Garden for seed. Dighting & lofting Oats for seed – to selves &c.

Fridy 25th Strong frost last night & this morning – Ice half Inch thick – day Calm & very warm Sun – One Plough finished all tillage for Oat seed the Earliest I ever had before – 3 Carts driving Dung from Cottar Town to Old Town Park – P.M. lofting seed Oats – Planted 5 large Pear Trees N side of Garden to the West Plott about 7 feet high &c.

Saturdy 26th Frost in the night pretty strong day fair & sun shine, one Cart to

Stonehaven for Coals, and Sundries and three D° Driving Dung and leading Home Turnips – lofting seed Oats P.M. – Sundries.

Mondy 28th Yesterday Moderate weather from N.W. and at night a little snow – this day mostly soft & the snow Chiefly gone in the low Country – Observed the Ploughs going in the Neighbourhood but not a good season, and Indeed not good either to Drive Dung which we did with four Carts to Auld Town Park – and Cleaning up from about Byres – and lead Home a Draught of Turnips – P.M. took in a 2^{d} Pease stack – dighting & lofting of seed Oats for self – Viscount Arbuthnot told me He had sold his Farms for 12/6 per Boll – Barley is worth about 14/ and Oat Meal & Oats giving 12/6 & 12/9 – seed Oats Have been sold for 14/6 & Fatt Cattle are rather more in Demand than I ever knew them & giving very High Prices – Beef 3$^{ds}\frac{1}{2}$ to 4^{ds} – Veal & Mutton about same.

Diary for the Month of March 1791

Tuesdy 1st But slight frost in the night & morning day very warm & pleasant – 4 Carts driving Dung to Old Town Park & Home some Turnips – P.M. Yoked Three Harrows & Begun Braking in Head Bank for Pease, Garden sowed Beans & Planted Potatoes &c.

Wedy 2d Strong frost in the night & white Rhime till 11 A.M. – after Moderate & slack – 4 Carts driving Dung for the Turnips Kavels done, & for West side Bank D° & P.M. 9 Harrows finished Braking in near 10 Acres N.E. side of Bank for Pease, at Garden two Men &c Jobs –

Thursdy 3d pretty strong frost in the night day fresh weather from Westward, with high wind P.M. – 2 Ploughs at the East side Bank seed fur for Pease, and One Cart Leading Turnips – at Barns – and turning Dung, D° Court.

Fridy 4th Fine Dry weather from S.W. & pleasant – 2 ploughs at Bank for Pease, and two Horse D° begun Ribbing acre of the North West side of Robs Hillock after Turnips very fine for Barley – P.M. Turning Dung, Dung Court & lead some Turnips. Planted Curleys &c.

Saturdy 5th Fine pleasant warm weather various with a Dagg A.M.- P.M. fair & Cloudy-2 ploughs at Bank for Pease-and a Cart to Stonehaven for Coals & Sundries- Begun setting Nursery & P.M. lead 5 Cartful Turnips – at Sundries –

Mondy 7th Yesterday fine dry weather from Westward – this day Cold & stormy from North with flirt of snow which did not Ly – 2 ploughs at Bank for Pease and two Horse D° steering Robs Hillock after Turnips for Barley – P.M. lead 4 Cartful Turnips – Barns &c

Tuesdy 8th Frost last night day fair & drought from S.W. – 2 ploughs at Bank – One D° Robs Hillock – and at Barns dighting Pease for seed, lead some Turnips, Planting Curleys, Cabbages, In Hay last of the Thornylee, Planting in Nursery &c Jobs –

Wedy 9th very fine dry weather from S.W. – Some People begun to sow Oats &

Pease – 2 Ploughs at Bank – and one at steering Robs Hillock for Barley – at Barns dighting Pease, & others at Nursery Planting – in Turnips.

Thursdy 10th Pleasant Season from S.W. – 2 ploughs morning yoking finished tillage near 10 Acres for Pease 3[d] Plough Robs Hillock – P.M. Yoked Nine Harrows to Harrow 7 Bolls Hastern Pease Sown this day – at Planting Nursery, Barns &c Jobs.

Fridy 11th Fine Pleasant dry weather from S.W. – 9 Harrows finished 8 Bolls some pecks Pease about $9\frac{1}{2}$ Acres Bank and at Planting of seedlings Nursery, Gardner & Several others – Severals sowing Oats &c –

Saturday 12th Soft weather from S.W. – 9 Harrows begun Harrowing first sown Oats 4B 3f Nether Burns Ley & P.M. sowed Over Burns with 3B 1f & Gave Covering to D° – Planting Nursery Flowering Shrubs – lead Home some Turnips – Brought Malt Home &c Jobs.

Mondy 14th Yesterday fine weather from S.W. – Evening a shower of rain slight – this day dry & fresh from S. – 9 Harrows finished Over & Nether Burns in fine order – One Cart leading of Turnips – at Barns, Dung Court & Sundries.

Tuesdy 15th Strong Gale from S.W. all day which Prevented Sowing of Oats on Bridge Shade after beginning said work & P.M. brought them Home again – the Ploughs steering at Robs Hillock after Turnips for Barley N.E. side with a Clean furr it being stronger & less fine – Harrowing a new Made road from Cottar Town as approach to the Mansion previous to more Ploughing Gathering & Graveling Dressing & Paving Garden Walks & Planting Flowring shrubs of various kinds in Gardens &c Jobs – Some gooseberries foliated Crocus, Snowdrops &c.

Wedy 16th In the morning Came on rain & stormy weather from N.E. 3 Ploughs at steering north side of Robs Hillock and driving Gravel to House approach, P.M. gave a second Gathering Ploughing to the road to y[e] House driving Gravel, at Barns dighting seed Oats, & Sundries.

Thursdy 17th About sunrise Came on rain from N.E. but light – almost whole People at Barn S dighting & lofting Oats for seed – 2 Carts driving Home Gravel for Garden &c Walks & some Turnips P.M. two Harrows at Intended road & Plough at the Old Dyke besouth Back Yards for Barley – Garden.

Fridy 18th This day fair & pleasant weather from Southward and drought – all People at Barns dighting & lofting Oats from South Barn – One Cart to Montrose for Sundries – and P.M. sowed 5 Bolls Oats on the Bridge shade, & partly Harrowed same; in Turnips

Saturday 19th Stormy weather & drought from S.W. veering to S.& SSE – Morning sowed 3B 3f 1p Oats which served all the Bridge Shade & nearly Harrowed the whole it being Tender, well Ploughed & pretty Clean – lofting Oats, and P.M. sacked some Oats for to sowe on Haugh –

Mondy 21st Yesterday Cold dry weather from S.W. some showers – this day very stormy from Westward & strong drought – Sowed & Harrowed Haugh north side House; D° foot of Shepherds Haugh & Harrowed the same also a Considerable Part of West side of Main Haugh from 9 to 10 Bolls Oats.

Tuesdy 22d Pretty strong frost in the night – day strong drought from S.W. – 8 Harrows all day at the Haugh for Oats Sowed P.M. 5B 2f – driving out Swines Dung to Cover Dunghill South side Barns for Turnips – Have now as much lead Out as serve 19 Acres of them,

Wedy 23d Moderate weather in the morning – day stormy with flirts of snow – 8 Harrows all day at Haugh, One Cart leading Turnips – & P.M. Gravel to Approach Road. Prices of Grain rising, & the Fodder will not be large.

Thursdy 24th The Best Spring day of this season from S.W. – sowed this day about 12 Bolls of Oats on the Haugh & partly Harrowed the same, leading Turnips and a Cart Graveling the Approach road to Mansion – am Informed all kinds of Grain & Cattle on the rise –

Fridy 25th Night and this day strong drought from Southward – A.M. finished all Oat seed by 30B 2f 2p – on whole of the Haugh Part – total Oats sown 47B 2f – or about 16 peck $\frac{1}{2}$ per Acre – & mostly Harrowed.

Saturdy 26th Early morning Came on rain from the Northward which lasted till Brakefast – Hills & Braes snow pretty deep – One Cart to Johnshaven for Coals – 2 driving Gravel to Approach Road – & leading in Turnips & Hay – sowed Colly Flower & some Salads on the New Hot Bed &c Jobs.

Mondy 28th Yesterday frosty morning – day fair – this day very Mild dry weather – 8 Harrows at Cross Harrowing of Haugh Oats & after D° at Braking in Back Waters. The Intended roup of Fatt Cattle did not take place as few Buyers only two attended – Sold However Privately about one Half the value of whole – Sundries –

Tuesdy 29th Very Pleasant dry weather from Southward – early 4 Men Begun sowing of East Side Haugh with Grass seeds, of which the finished 18 Acres before Compleat before 9 A.M. – Harrows Covered same by a strake & P.M. Begun Rolling Across D° from Head – Put sold Cattle on Hay & Water by Desire, and the others on Turnips for Water, & Hay – Sundries –

Wedy 30th Pleasant dry weather from South to SSE – One Plough at stirring Back Waters for Barley, and another at Robs Hillock D° for D° – a Roller whole day at Covering Grass seeds in very fine Indeed Condition equal to the Garden, at Chaffing Oats S.Barn. the vegetation forewarding but slowly &c Jobs Sowed onions, carrots, Beet &c in Broad land which W.Smith my Gardner Saw Practised last season about London to be Hand Houghed –

Thursdy 31st White frost in the Morning, day fair & pleasant spring weather – wind S to SSE Moderate – One Plough at Back Waters steering for Barley, Gathering the Hollow parts to Level it more, rolled out all the E^{t} side Haugh sown with Grass seeds in very fine Order – a two Horse Plough at Robs Hillock after Turnips, and a Cart lead Home whole of Turnip Crop, and driving Gravel to Approach Road – Prices of all kinds of grain rising, but the Prices not just fixed – the Oat seed not much done by many Neighbours – Barns & Sundries.

Diary for the Month of April 1791

Fridy 1st very Pleasant dry weather from S.E. – 2 Ploughs at steering for Barley nearly finished – P.M. sent three Carts to Polborn for three Cartful of shot Turnips – at Barns dighting & lofting Oats to Malt Barn &c Jobs. Capt. Leslie died.

Saturdy 2d This day fair & Cold drought from N.E. about noon Came on some slight rain of no Continuance – 2 Ploughs steering for Barley, Rolling at the Back Yards &c new sown Grass – Graveling of Approach road, dighting & Lofting Oats – took in Part of a stack of Barley S Barn for seed –

Mondy 4th Yesterday dark Cloudy weather from the Eastward – being at Johnston observed the Eclipse of the Sun very distinctly from 1 P.M. to 3 D° Apparently about $\frac{2}{3}^{ds}$, but little sensible diminution of daylight – this day fair & dry – 2 Ploughs at steering for Barley – rolling Grass ground – sent a Cart load of Lint to Mill of Drumtochty –

Tuesdy 5th last night a kind of white frosty Dew – this day Cool & Bleak weather from N.E. – 2 Carts graveling at Approach road all day, 3 Harrows Braking in Robs Hillock for Barley, and one Cart leading off stones D° – Put a Paling on the Strip Planting Shepherds Haugh Bank &c Jobs.

Wedy 6th a shower last night, day fair but Bleak Cold weather from N.E.-3 Carts driving Gravel, and three Harrows at Braking in Robs Hillock for Barley-but little Growth.

Thursdy 7th Thick Daggy weather & growing from N.E. – 3 Carts driving at Gravel to Approach road, well advanced. Harrows finished Braking in Robs Hillock Park for Barley – Garden &c

Fridy 8th In the morning after 8 came on Rain from Eastward – two Carts Graveling road, & three Harrows Braking in Black lands for Barley – a Cart to Montrose for Sundries – Observed to Montrose Oat seed not done and the Grass & Garden stuffs not Earlier than here.

Saturdy 9th Fine drought & spring weather from S.W. P.M.to N.W. Stormy – 3 Carts nearly finished the Approach Road from Publick road to Aberdeen to ye House. Two Harrows Braking in Black lands, Gardner sowing Annuals in Flower Garden. Barns &c –

Mondy 11th Yesterday fine dry weather from – this morning about 7 Came on rain & Sleet from the Eastw^{d} – one Cart to Stonehaven for Coals – all Harrows finished Braking in Black Lands, & P.M. the Floors – at Barns, Garden & sundry Jobs –

Tuesdy 12th This day fair but Cold & Bleak from Eastw^{d}. Hills very deep Covered with Snow, low Country very wet – at Barns A.M. – P.M. driving Dung to Floors Dyke for Potatoes, & at Sundries –

Wedy 13th Dark Cloudy weather from N.E. – 6 Harrows at Braking in Floors, and after at Backwaters for Barley – finished – & Brought up Harrows – at Barns and at variety of Jobs – put Cows on Grass – &c

Thursdy 14th Thick fog the whole day and growing weather – Three Ploughs begun Earing for Barley seed West side of Floors morning – Planting Potatoes on an Old Dyke leveled down North side Floors & old Road through the Haughs, before the one at Present Alongst the Brae head Thornylee was Made. Put new Calved Cows & some feeding Beasts on our Best Pasture grass for this two days – It is as yet but Midling good, but I have the best new grass, Backyards to be Met with.

Fridy 15th Thick foggy weather from Southw^{d} – three Ploughs at Floors at seed fur nearly finished, and People at Barns dighting Barley for seed, and at Planting Potatoes Dyke Floors – and at Sundries –

Saturdy 16th Closs Foggy weather and light Breese from Eastw^{d} – 3 Ploughs Part of day Others Harrowing, sow^{d} all Floors about 5 Acres with 3B 9 Pecks of Barley & finished Harrowing of same at the West side of Robs Hillock Park for Barley – in very fine Condition – at Barns dighting of seed Barley – and at Sundries – a very Considerable growth & tolerable Grass – There hath been a stagnation in the Sale of Fatt Cattle and a Demand for Grain of Kinds.

Mondy 18th Yesterday foggy Closs weather & a little rain – this day fair & growing weather from S.W. – sowed 2 pecks of Lintseed Haugh Head of Mill Dam Garvock & Harrowed it fine order – 3 Ploughs all day at Robs Hillock for Barley, and one Cart Driving Dung to N. side of D^{o} – at Barns Dighting seed Barley & at Sundry Jobs – sold 3 fatt Cattle for Arbroath.

Tuesdy 19th Fine Pleasant dry weather – various – 3 Ploughs at the Barley Ploughing Robs Hillock – One Cart driving dung all day to it – P.M. Sowed 4 Bolls of Barley West side of said field & partly Harrowed of D^{o} .

Wedy 20th a Rhime in the night but day very pleasant dry & warmer to the senses at least than for two summers by past – 2 Ploughs at Barley seed, and two Carts at driving of Dung to D^{o} – and two Harrows all day – Observe many People dunging their Land but few sowing – very Considerable growth of Grass Oats &c – Viscount Arbuthnot died unseen Aged 87.

Thursdy 21st a little frosty dew this morning, day fair & warm – 2 Ploughs at Barley seed fur – and two Carts at Driving Dung for D^{o} – at Auchinblae Markett Things but dull, a Considerable demand for grain but little in the way to sell, it being Scarce.

Fridy 22d Frosty Rhime, day Coolish & Bleak foggy weather N.E. – 3 Ploughs at Robs Hillock – & one Cart Driving of Dung to D^{o} – at Barns, spreading dung &c Jobs –

Saturdy 23d In the morning stormy Cold weather from N.E. and a good deal of rain, day more or less so with thick dagg – morning sent a Cart to Stonehaven for Coals, and Turning Dunghill South End Burns for Bank Turnips – and P.M. two Ploughs at N.E. Corner of Robs Hillock for Barley & One Cart driving dung to finish same &c.

Mondy 25th Yesterday Cold Bleak dry weather from N.E. – this day same from N.W. – three Ploughs A.M. finished all seed fur for Barley Robs Hillock – & P.M. sowed it out with 10B 1 firl of Barley & mostly Harrowed same – at Barns & at Sundry Jobs – Vegetation checkqued by the Cold –

Tuesdy 26th Bleak dry weather from the N.E. – 3 Ploughs begun West side of Black Lands Award for Barley & three Harrows prepared Robs Hillock for Grass seeds – Sold what Oats were lofted for Mill of Convey at 14/9 –

Wedy 27th Some small rain through the day, 3 Ploughs at the Black Lands, and P.M. drove up some dung to D°, on the Thin Parts of it spread same, at Barns & sundry Jobs Mixing Grass seeds for Robs Hillock.

Thursdy 28th Moderate weather – early sowed Robs Hillock park with Grass seeds Assorted to the different soils of it and Harrowed same, P.M. sowed three Bolls of Barley on West side of Black lands & Harrowed it – at Laurencekirk Tryst many Cattle of all kinds and prices being high little or no sale, Sold my Barley for Johnshaven – from 14/6 to 15/ per Boll –

Fridy 29th Cold Bleak weather from Northward & dry – Harrows & ploughs & dunging at Black Lands – P.M. sowed two Bolls of Barley in D° – Rolling Robs Hillock – at planting of Potatoes Floors leveled dyke, & at sundry Jobs in Barns – Vegetation much Checqued –

Saturdy 30th Cold & Bleak & dry weather from N.E. – Ploughs and Harrows & Roller at Barley seed Business and dighting Oats sold to deliver at the Mill of Convey – Observed towards Stonehaven that vegetation was more Backward than with us, and little or nothing done to the Bear seed, and farther to the Northward was Informed there had been much more rain than with us & the Lands too wet for Labouring – There is not such demand for fatt Cattle as Months ago, though there are few of them in it – Prices of Grain are rather on the rise, Oat Meal 11^{ds} & Bear 7^{ds} per peck – Manufactures pretty Brisk.

Diary for May 1791

Mondy 2d Yesterday and this very Cold stormy Bleak weather from N.E. but dry – This day Dunging at Old Town Planted Hillock park which nearly finished & Harrowing at Black lands – P.M. two Carts to Mill of Convey with 12B Oats sold &c.

Tuesdy 3d Cold Bleak dry weather from Northward – all Ploughs, some Carts & Harrows at Barley seed. Sowed One Boll Barley E^{t} side of above field & Harrowed it, at Barns &c Jobs. The Bear seed is not yet generaly begun –

Wedy 4th Cold Bleak dry weather – finished sowing all Blacklands with 8B 2f Barley in fine Conditiion – after at Potatoes ridge Back yards 2p $\frac{1}{2}$ Barley – P.M. 3 Ploughs at Back waters after Pease for the remainder of whole Barley seed & at Sundries –

Thursdy 5th Cold weather from N.E. – 3 Ploughs at Back Waters for Barley drove some dung to D° and at Braking in West side Bank for Turnips tender took in Part stack of Oats N. Barn, and at Sundries –

Fridy 6th last night a good deal of snow which did not ly till Brakefast time – 3 Ploughs at Back waters which finished whole for Barley seed – and Harrows at

Braking in the roughest of it before sowing – at Potatoes planting and variety Jobs about Barns –

Saturday 7th more Moderate weather – about noon wind veered to the Southward – 8 Harrows at Back Waters sowed it with 2B 10p Barley which finished the whole of this Season – P.M. Sowed same with Grass seeds & Harrowed the same – Vegetation much Checqued.

Mondy 9th Fair weather & Moderate from S.W. – One Plough at steering West side Bank for turnips and 2 Harrows Braking in Kavels for D° – One Cart to Johnshaven with 4B Barley & Brought out $\frac{1}{2}$ Boll Coals – at Barns dighting Barley & Jobs. viewed road to be Made or at least should be Made from Fordun benorth Castleton –

Tuesdy 10th day fair & very pleasant till about noon it turned Rough from Northward & then very Brisk rain at Intervals rest of day & warm – A.M. two Ploughs steering West side Bank & Harrows at Kavels and P.M. 3 Carts to Mill of Conveth 18B Oats &c Jobs.

Wedy 11th fair weather and Moderate drought from S.E. – 2 ploughs at steering for Turnip West side Bank & 2 harrows all day Braking in Kavels for Turnips – P.M. sent 15 Bolls Oats to Mill of Convey on 2 Carts &c Jobs. Sold Pease at 13/

Thursdy 12th last night very little rain – day fair wind S.E. – the Ploughs at Bank – finished Braking in Kavels – and P.M. Rolled the Back waters. Ploughs & Harrows – Sent remainder of Threshen Oats sold to Conveth –

Fridy 13th Fine pleasant growing weather from Eastward, 2 Ploughs at N.W. side Bank for Turnips, 2 Harrows at D° in very fine tilth already – Dighting & sacking of Pease sold for Johnshaven &c Jobs.

Saturdy 14th Coldish dry weather variable – 3 Carts to Johnshaven with 15B Pease & Brought home $1\frac{1}{2}$ Boll Scotch Coal at 8/4 per Boll in return, and at Planting Potatoes Floors, and at Sundries. The Best part of my Barley finely Breered, all my Hay done, & Pasture yet but Midling –

Mondy 16th Yesterday Coldish & rather stormy from the S.W. with some showers of rain – this day same but very strong drought – sent 3 Carts to Johnshaven with $14\frac{1}{2}$ Bolls Pease, and Brought Home $1\frac{1}{2}$ Bolls of Coals, the Roads to that place is as good as Bitter Bad Crowdie, three Harrows & a Roller at West side Bank for Turnips now in good Tilth – The Pasture in general very Indiferent, the Garden Produce very Backward, none of the Aples, Pears or Plumb & Cherrie Trees yet fully Blossomed Bear Seed yet very Backward in General every field Plant latter flowring than for years.

Tuesdy 17th Cold Stormy weather from Westward with showers of Hail and rain at Intervals, 2 ploughs at finishing West side Bank & P.M. at Head of Kavels cross ploughing for Turnips, at Barns dighting Barley sold for Johnshaven – and at Sundries –

Wedy 18th a very strong Gale from N.W. Braking down many of the Garden stuffs – 4 Carts to Johnshaven with 16 Bolls Barley to Mr Scott – 2 Harrows at Bank

– and others at Barns dighting Barley for above place. all manner of vegetation much Chequed & Jobs.

Thursdy 19th Early this morning a sharp shower of rain, and day a very strong Gale from N.W. with drought – 2 Ploughs at steering Kavels, & two Harrows at D^{o} – at Barns and at Sundry Jobs –

Fridy 20th Cold stormy weather from N.W. and strong drought – 4 Carts to Johnshaven with 16B Barley and Brought in return 2 Bolls Coals – Dighting Barley South Barn, 2 Harrows at Kavels for Turnips –

Saturdy 21st Cold dry weather from Westward – 4 Carts to Johnshaven with 16 Bolls Barley, and Harrows at Kavels, and flited the sheep Flakes, and turning Dunghill S Barn – a little before noon Came on a smart shower of rain of two Hours from S.W. – after Drought &c.

Mondy 23d Yesterday Cold Stormy weather from S.W. – this day Early a Considerable frost – and about $\frac{1}{2}$ past 11 Came on pretty Heavy rain from S.E. which lasted till near 6 P.M. – 3 Carts to Johnshaven with 14 Bolls Barley and Brought Back $1\frac{1}{2}$ Boll Coals – lofted new Threshed Barley 15B – Barns & Jobbs – 2 Wrights Ent^{d}.

Tuesdy 24th last night frosty weather & Cold, but Early morning a severe shower of rain – this day more Moderate & drought – 2 Ploughs at Kavels steering – and One cart to Balmanno for a Turin Chimney to Parlour and at Sundry jobs – Breer of Kinds good here –

Wedy 25th Frost white this morning – sent one Cart to Mathers for Lime shells for Parlour Chimney – new School House – 2 Ploughs at Kavels, and two Harrows at same – Miday at taking remainder of last stack of Oats but one, and at variety of other Jobs &c

Thursdy 26th White frost in the morning day a very strong Gale from S.W. & Hard drought – 2 Carts fliting the Cow Herd, and Ploughs & Harrows at the Kavels as before – Wrights & Masons pulling down the Parlour Partitions for a new dining Room – One Mason Enter^{d}.

Fridy 27th Strong drought from S.W. and Moderate warm weather – 2 Ploughs all day & Harrows P.M.- One Cart to Lauriestown for Stones to Dining room on West end of the House a new window – at Barns dighting Oats &c. am Informed Lean Cattle Sold very poorly at the late Glamiss & Forfar Marketts – the Pasture being so very Scanty & Season so Backward –

Saturdy 28th Strong drought from S.E. all day & Clear Sun Shine Ploughs and Harrows at Kavels for Turnips and a Cart to Stonehaven for Barley 6 Bolls the poor remainder of 40 Bolls sold to a Bankrupt at Doors in August last – and at Sundry Jobs –

Mondy 30th Yesterday warm dry weather from S.E. – this day the same – Ploughs & Harrows at Kavels, and a Cart to Laurieston for stones to new window West end House, and at Sundry Jobs – Observed Cattle Going South to Markett.

Tuesdy 31st Very strong drought from Southward – Two Ploughs at Kavels nearly finished and at Sundries the Fruit Trees not yet fully Blown, and upon the whole as far Back as in the year 1782 – the Grass very Indifferent in most places, and the Hay must soon get rain or it will be a very light Crop, Prices of Grain pretty High – Barley & Meal Oat about 15/ – Manufactures Brisk &c.

Diary for the Month of June 1791

June 1st Wedy A Constant drought Continues and Consequently a dull vegetation – Ploughs finished the 2[d] furr of the Kavels and P.M. three ploughs at 3[d] furr of the West side Bank tolerably fine, and at Sundries –

Thursdy 2d Thick fogg in the morning & through the day at Intervals – 3 Ploughs at West side Bank 3 fur – and two Harrows morning Kavels, at Barn N. Chaffing Oats and at Turning large Dunghill Golds Hillock &c Jobs.

Fridy 3d very strong drought from Southward all day – 3 Ploughs finished 3 furr Bank Turnips & two Harrows at Harrowing of same, at Chaffing Oats N Barn.

Saturdy 4th pretty Heavy dew in the night, day strong drought from S.E. – 3 Carts to Mathers for Lime, and 3 Harrows at Bank for Turnips – One Cart leading water and sand for the Lime, Turning Dung, flited sheep Flakes.

Mondy 6th very strong drought & warm weather – Thermometer P.M. at 67° – North room – Ploughs & Harrows & roller all day at West side Bank finished for Dunging – Vegetation Increasing however little here –

Tuesdy 7th drought from S.W. till past 10 A.M. when Came on a tolerable shower from that Quarter till past noon – Carts driving dung for Turnips on Bank at rate of 30 Cart Load per Acre, and set up part of the Cattle Flakes on red Craig Shade & at Sundries.

Wedy 8th Strong drought from Eastward, 4 Carts at driving Dung all day & 4 fillers of Carts – put up the Cattle Flakes and put the Bestials in for first time washed & sheared the few sheep – Grass growing – Hear there was little or no Sale at Cammock Muir.

Thursdy 9th Strong drought from S.W. – 4 Carts dunged out about 7 Acres West side of the Bank for Turnips – and P.M. Begun drilling of same – at Trinity Muir a great show of Cattle, but moderate sale, being so High – Bought 6 Cattle about 3/8 per stone – Sundries.

Fridy 10th Cold dry weather from Northward & Stormy, One Plough Drilling at Bank all day & Sowing Turnips – P.M. Rolled same in good order – others Men & Wewmen spreading dung – at Sundries &c.

Saturdy 11th Cold stormy weather with showers of Hail & Sleet at Intervals all day from northward – Plough from 4 A.M. to past 6 P.M. at Drilling Turnips Bank, and others at Rolling of D° – Spreading dung very near finished the whole of it near 7 Acres, Harrowing & Rolling at the East side of Kavels for Turnips & Sundries.

Mondy 13th Yesterday and this a very severe & Cruel gale of Northerly wind Blasting every kind of vegetable growth – 2 Ploughs at Kavels for 3[d] furr – and other people Turning Dunghill, Weeding lint &c.

Tuesday 14th very stormy weather from N.W. and showers of rain & hail at Intervals, towards Evening a good deal of rain. Ploughs and Harrows at Kavels for Turnips and at Turning the Dunghill for the Kavels Turnips – Put a new North stair window into the Dwelling House, at Garden & Sundry Jobs –

Wedy 15th last night and this morning a Considerable fall of rain from N.E., day fair & drought from same Quarter, Ploughs & Harrows all day Kavels, & P.M. finished the Head ridge of the Bank Turnips in good order – Observed towards Dava a Scanty Grass and the Oats & Pease scarcely making the ground as yet green – at a Roup of the Stocking of Mains Arbuthnott; Napier whose lease of 7 years was Purchased at as Many Hundred Pounds St[r] – everything sold high.

Thursdy 16th a very strong Gale from the Northward, Exceeding Cold – 4 Carts Driving dung all day at Kavels for Turnips – some Harrows at D°. Flited the sheep flakes – This season is the most Inclement I am Informed by others they ever remember off – a very great damage to Trees, flowers, vegetables &c and if the season do not Mend Miracoully every thing must suffer –

Fridy 17th Strong & severe stormy weather from Northward Blasting everything – Hear Duns Moor went much as the Trinity Market dull & dear – 4 Carts driving Dung all day – took whole of remainder of Oat Crop, and at sundries on the farm & Garden –

Saturday 18th Severe weather as yesterday – 4 Carts at Driving Dung to Kavels – and others Turning remainder of the large Dunghill, and at Sundries &c Jobbs – Have now 5 to 6 Acres Dunged for Drilling.

Mondy 20th Yesterday Coldish dry weather – this day same & some showers of Hail – Begun drilling E[t] side of the Kavels for Turnips, and Ploughs at the west side of what was not Dunged – at Sundries Dung spreading &c.

Tuesdy 21st a small shower from the Northward morning, day Bleak & dry – drilling, sowing & rolling of Turnips and Ploughs &c as yesterday – at Sundries &c.

Wedy 22d Dry weather as before untill about 1 P.M. Came on pretty Brisk showers from various till 6 P.M. drilling & sowing of Turnips, and Harrows and Roller on the roughest part of what is yet to be Dunged – Observe for some days past about Aberdeen and to near to Stonehaven everything quite stinted & Burnt or dryed up by the Cold & drought much more so than to the S.W. of Stonehaven – &c Jobs –

Thursdy 23d This day alternate showers of rain and fair growing weather – Carts driving dung for Turnips and P.M. Drilling for D° Others spreading Dung. there will now be a Considerable vegetation &c.

Fridy 24th This morning about 7 Came on a shower of rain which lasted till past 11 A.M. – 4 Carts and a Plough at Drilling P.M. Kavels for Turnips and at Barns dighting Oats, and at Sundries.

Saturday 25th This day fair & drought from S.W. – Ploughs, Harrows & roller at the Turnips & also the Dung Carts – at Barns dighting Oats – this day sold to the Mill Master at Conveth all my Oats & Barley on Hand at 15/6 per Boll –

Mondy 27th Yesterday a very Considerable rain & stormy weather from the Eastward – this day fair & strong drought from the Westward – two Carts to Johnshaven for Coals to the Smith, and others at dunging of Kavels, drilling & sowing D° – flited the Flakes – Observed the Crop about Johnston Excellent &c.

Tuesdy 28th Strong drought & high wind from S.W. till P.M. Came on pretty Brisk showers of rain which lasted more or less during the day – Carts driving of Dung, a Drill Plough & sowing & rolling P.M. at Barns dighting Oats sold, and at Sundries – The vegetation yet not very great – Have Had Cabbages & all kinds of sallad for some time – first Bee Hive swarmed.

Wedy 29th a Considerable fall of rain this morning the day drought & fresh Gales from S.W. – Carts & ploughs at Dunging & Drilling of Turnips not quite finished – P.M. sent to Mill of Conveth $12\frac{1}{2}$ Bolls of Oats sold to Andrew Young at 15/6 per B.

Thursdy 30th Strong drought from S.W. – and several showers of rain through the day – Ploughs and Carts finished the Turnip seed Business rather Tedious – about 18 to 19 Acres in good order – P.M. sent Bolls of Oats to Mill of Conveth, at fair of Megry Heared at Stonehaven that Many Cattle but not much sale, as there is little demand from the South – Prices of grain rather rising Oat Meal 1/ per peck, Bear 7^{ds} to $7^{ds}\frac{1}{2}$ and Barley about 15/6 to 16/ – and demand – the late rains Hath not raised either the Hay or Pasture as might have been Expected the sudden and Great Gales of wind succeeding preventing it –

Diary for the Month of July 1791

July 1st Friday Some small showers in the day wind Northerly – Sent a Cart to Conveth with Oats & then to Johnston for Wood Brought from Montrose – 3 Carts to Dava for stones to Build the Pidgeon House with, and at Barns dighting Oats and at Sundry Jobs on farm –

Saturdy 2d Fine growing weather from S.E. and fair weather – 2 Carts to Johnshaven for English Coal at 1/6 per Barrell and Two Carts driving stones from Dava, and other parts adjacent for Funds, Pinnings &c first plaister to Dining room. The Pasture is now growing much better – Field Pease Blooming.

Mondy 4th Yesterday some showers of rain from S.W., this blowing weather & rather stormy from D° – 2 Carts driving stones all day from Dava, a Plough for summer Fallow Kavels and P.M. leading sand & water to Masons for Pigeon House to be Built S. End New Barns, Sent $7\frac{1}{2}$ B Oats – Ent^{d} 3 Masons.

Tuesdy 5th very strong Gale & stormy dry weather from S.W. – One Plough at Fallow for Wheat West side Kavels and two Carts at Driving stones two draught from Dava, Houghing Potatoes Floors, and at Sundries –

Wedy 6th very strong drought from S.W. Checquing all Growth – One Plough at W side Kavels fallow & two Carts driving Home Stones from Dava, at Barns, Houghing very fine Potatoes on an old Dyke & old road Floors – the Wheat Floors good & tolerably shot – Sundries – Bees swarmed.

Thursdy 7th very strong drought from S.W. all day – being fast day no work except 2 Masons & 3 Wrights other Parishes – Evening some drops of rain but little appearance of it.

Fridy 8th This day some very light drops of rain, and one Plough at Fallow West side of Kavels & two Carts to Flemington for stones to Pigeon House – at Sundries Barns & other Jobs –

Saturdy 9th Moderate weather but strong drought from N.W. – One Plough finished the 3 fur Fallow, and sent $9\frac{1}{2}$ Bolls Oats to Mill of Conveth P.M. – the People at Church &c – all vegetation Checqued.

Mondy 11th Yesterday Cold drought this a perfect storm of Cold N.W. wind & drought Blasting every kind of vegetation – A.M. 2 Carts to Quarry for Stones & One Cart leading gravel to Aberdeen road – P.M. flited the Cattle & Sheep Flakes, and at Sundries –

Tuesdy 12th very stormy weather from N.W. and severe drought – 2 Carts driving stones from Dava, and two Carts driving Gravel to High road through this farm – a Man at spreading D° – Thistling Oats Haugh – Cut up the Turnip for Seed – and at Sundries –

Wedy 13th Uncommon severe drought from N.W. – 2 Carts driving stones from Dava, and 2 others at driving Gravel to Publick & Private roads – Some people Cutting Hay – Pease well Bloomed & Oats & Barley offering & Wheat shot – Some strawberries.

Thursdy 14th White frost Early – day warm & drought – Brought Home 8 Bolls Lime Shells, and two draught of stones from Dava, at Paldy fair near double of the number of Cattle usual, but the least sale of any year for above one Dozen years by past – Bought no Cattle though they are fallen 20 per C^{t}-

Fridy 15th Exceeding warm weather wind little from S.E. – People at Thistling & P.M. sent 4 Carts to Flemington for Stones to Pigeon House and two of them to Pitscandly for Slates –

Saturdy 16th Hazy weather in the night day very warm & strong sunshine, Hard on the Grass & Corns on the Thin ground, or what hath been much in Tillage – the 4 Carts did not return till 8 P.M. owing to the Heat &c – The Turnips are pretty Tolerable but in many places very unequal – owing to Sundry Causes – sowing, drought &c – the Pasture in general quite Exausted – I Have plenty here &c –

Mondy 18th Yesterday warm weather & very strong drought from the Eastward, this day the same from N.E. – sent a Cart to Finzean for Birch wood, and some Carts at taking a stamp of the North End of Corn Yard to level it & for the Dung Court, P.M. flited the Cattle and Sheep Flakes, Masons, Wrights &c as last week –

Tuesdy 19th This day foggy morning day warm & Misty Afternoon – 2 Carts to Laurieston for stones, Tatching & Peat stones for Pigeon House, and sent one to Flemington for Boles Pigeon House, begun Cutting the Hay B Yards a very Heavy Crop – the Crop failing in sundry places – Green Pease.

Wedy 20th Dark dry weather from the Eastward – Sent one Cart to Flemington for Pigeon Holes & two Carts driving stones from Dava, at hay Cutting & Sundries –

Thursdy 21st Warm weather from various – Carts driving stones, and other people at Hay Cutting – Weeding at the Nursery – and about 10 A.M. a shower of rain of short Continuance – and at Sundries –

Fridy 22d Strong drought but with some light skiffs of rain from the Southw^{d} of no Continuance – 2 Carts driving stones & one sand & water – One for Flemington & P.M. lead Home two Cartful of Turfs from Garvock Hill.

Saturdy 23d Some showers of rain & pretty rough weather from S.W. – 2 Carts driving stones from Quarry, & one leading Turfs from Garvock – Cut down most part of the Gibsons Park Hay, midling Crop, Masons 4 nearly finished the Pigeon House – Pasture Bettering &c –

Mondy 25th Yesterday fair weather & drought – this morning a frost & strong drought – Early begun Gathering & after Brakefast begun Coling Hay into pretty large Tramp Coles, 12 people Hired & own People – about 6 P.M. Came on flying showers from S.W. – Sundries.

Tuesdy 26th Some flying showers of rain from S.W. to prevent Coling of Hay, driving stones from Dava and some for Funding a Hen House S.E. side Barns finished Cutting up all Hay – P.M. dighting & sacking of Barley sold – and at Sundries –

Wedy 27th Showers of little note, Eleven or twelve people at Coling of Hay did out remainder of Back Yards and after all the Gibson's Park into Tramp Coles in good order – Masons at Pidgeon House.

Thursdy 28th Very strong drought from S.W. – People lead home 2 Cartful of stones from Dava morning – and others at weeding Corns – after Brakefast all at Garvock fair – not one Half of the usual Quantity of Cattle in the Markett & no Merch^{ts} – The Serv^{ts} however very dear, still a little on the rise Engaged mine.

Fridy 29th This morning a little before 6 o'Clock Came on a pretty strong Gale & Heavy rain from S.E. which lasted till near noon when faired – but some showers towards Evening – driving stones to Hen Houses and Earth from its found & Head Corn Yard into Dung Court

Saturdy 30th This day very strong drought from S.W. and rather stormy – 2 Carts to Montrose for wood – and others leading stones – flited the Cattle Flakes, and P.M. weeding of the Haugh Oats – and at Sundries –

Sundy 31st Strong drought & rather stormy – the Crop appears to be thin, short & many farms quite dirty except on good land, and in high order, it appears Impossible to be a weighty or Good Crop in any respects. The pasture is also more deficient than usual & in most places so to a degree – Prices of grain about 15/6 for Oat Meal, 7^{ds} $\frac{1}{2}$ per peck Bear D^{o} – the Hay Crop very light hear of no sale price – the Trade & Manufactures of all kinds Brisker than for several years by past – Gooseberries ripe

Diary for the Month of August 1791

Mondy 1st Mixed & various weather, but Moderate & Warm – Begun taking away the fur from the first sown Turnips Bank, and some people at Quarry – and others at leading stones, & sent for 2 Cart load of stones from Johnston – Masons at Hen House & Sundries –

Tuesdy 2d Moderate weather some slight skiffs from N.W. – One Plough at taking away first furr from Bank Turnips – and Eight people thining of D^{o} – lead Home some stones from Dava, and at Sundries –

Wedy 3d Moderate weather from S.W. and pleasant – 2 Carts to Johnshaven for English Coals, and one for Lime to Mathers, and Men & wewmen at thinning of Turnips Bank – and at Sundries –

Thursdy 4th very strong drought from Southward – People and Ploughs at Bank Turnips which nearly finished first dressing of – leading stones from the Quarry – the Drought Injuring the Barley & pease very much –

Fridy 5th Wind and weather but very Hott – finished out Horse and Hand Hoeing all Bank Turnips – P.M. at East side of Kavels D^{o} very Gravelly – about noon sent three Carts own & three of my Tennants to Flemington for Slates & Pigeon House Boles &c Jobs.

Saturdy 6th Very strong drought from S.W. – Carts returned at 4 P.M. with Slates & Boles – one Cart lead Home two Cart Load of Stones from Dava – finished Building the Hen House, Cutting Grass Water sides for Horses.

Mondy 8th Yesterday and this very strong drought from different quarters – sent 2 Carts to Mathers for Lime, and one Cart to Laurieston for a Lintel front Door & Sole window above it – nearly finished pigeon House – Horse & Hand Hoeing of Kavels Turnips a promising Crop – Sundries.

Tuesdy 9th In the night a small shower of rain, day fair and very pleasant warm weather from S.E. – Carts driving stone from Dava, sand & water for the Lime – finished Building Pigeon House. Tirred & rede out the rubbish of Old W^{t} Barn dressing by Horse & Hand Hough the Kavels Turnips &c

Wedy 10th Thick Hazy fogg all night, day fair & drought from the Eastw^{d} – One Plough all day at Turnips and several people Hand Howing D^{o} – brought Home the Remainder of stones from Dava – Sundries.

Thursdy 11th very strong drought from Eastward – People at Hand & Horse Hoeing of Kavels Turnips – and Masons at repairing Front House, and putting on the Roof on Pigeon House & Sundries –

Fridy 12th This morning a Moderate Shower of rain and after foggy dew till near dinner time – all people at finishing first dressing of all Turnips Amongst the best far or near – the Masons & Wrights at Sundries &c Jobs –

Saturdy 13th Strong drought & very warm weather from South^{d} Sent 4 Carts for Peats to Corsbitt Moss – P.M. flited the Flakes – Making a Pitt for Watering lint at West side of Floors Old road through Haugh – Sundries.

Mondy 15th Yesterday several showers of rain in the Evening – this day fair & pleasant from S.E. – Two Carts to the Corsebit Hill for remainder of Peats 6 Cartful in all at 10/ – After dinner 2 Carts at leading in Hay into a Rick 45 feet long by 12 Broad on Goulds Hillock a good situation & dry at all seasons – Making out a watering place for Lint – pulled & put some of the ripest of it into water – and at Sundries – Gooseberries & rasps in great Abundance

Tuesdy 16th In the morning a slight shower – Own people & 4 others begun leading in Hay – & Horse tramping of D° till 10 A.M. when Came on a Smart rain & stormy weather from the North till near 2 P.M. & stopped that Business – Covered over the sowe with 2 Cartful of loose straw – all people at Barns Making ropes For Thatching Hay – sorting up things within doors and Masons at Sundries within doors – Cutting P.M. of Rushes, Sprotts &c for Thatching of Hay &c Jobs.

Wedy 17th In the night a Considerable deal of rain from N.E. – this day several showers pretty smart but of no duration – P.M. strong Gale from N.W. and the Rivers much out Covering Gravel bank above the Distillery – lead Home 4 Cartful of stones from Dava Quarry – Masons & Wrights at Sundries &c.

Thursdy 18th Fine dry & pleasant weather from N.W. & P.M. S.W. – all our people and some Hired ones at pulling of Flax and after dinner put same into water & Covered it well. Sowed the spring Plants Garden, and lead Home 3 Cartful of stones from Dava being whole – one Slater begun.

Fridy 19th Fine pleasant weather from Southw^{d} – flited the Cattle Flakes, and after begun with 3 Carts leading in all Hay – which did & after Thatched it with Sprotts & Rushes, 18 to 19 H^{d} stones – at Sundries.

Sarturdy 20th Fine pleasant dry weather from Southward – Thatched & finished the Hay Sowe, and dighted up from two to three Bolls Rye Grass seed & lofted it – Harrowing the summer fallow West side Kavels & Sundries –

Mondy 22d Thick Hazy weather from the Eastward – 2 ploughs at West end of Kavels for Wheat – after at 2^{d} dressing of Turnips, and at Varieteys on the farm – light showers & fogg from 3 P.M. –

Tuesdy 23d fair weather and very warm from the Westward – 2 Ploughs at Fallow fourth furr – fine, and one Cart to Mathers for Lime to Slaters – after all Hands at 2^{d} Houghing of Turnips –

Wedy 24th Early morning a Heavy shower of rain from the Southward – day mostly fair with a few skiffs. People at Turnips Bank, good, Others at Sundries, preparing for Brewing and various Jobs –

Thursdy 25th Some showers in the morning – day fair & windy weather – People at Turnips – and at Laurence fair – many Cattle but little sale the Buyers at low prices & the sellers wanting high – at Sundries –

Fridy 26th Strong drought from N.W. and people at the Turnips Bank nearly done, after noon sent 3 Carts to Flemington for remainder of all the slates and at variety of Sundries.

Saturdy 27th Closs foggy weather from the Eastw^{d}, and from 9 to 11 A.M. a pretty Brisk shower of rain. One of the Carts for Slates Broke down at Inglismaldie & so obliged to put her load on the two others. Took out all the lint & spread it on the grass, it covers above double the ground it grew upon. Flited the Cattle Flakes P.M. & at Sundries.

Mondy 29th Yesterday Cold showery weather from the Northward – Went to see Lord Adam Gordons fine new Erecting seat at the Gannochy Bridge N.Esk – fine Thriving Plantations of wood – Observed the Crop to that place pretty good. Returned to Sir A Ramsays Fasque to Dine. This day Cold dry weather – begun Harvest by Cutting up Bank Pease a fine Crop, but stopped by rain by 5 P.M. – an ill beginning.

Tuesdy 30th a very strong Gale from the Northw^{d} with skiffs of rain – Early at Cutting up Pease which finished by noon, till dinner shearers at Cutting Rushes for Thatch to D^{o} in sundry places, Building up Turfs, Peats, &c – after Dinner Dismissed the Shearers Sine die –

Wedy 31st Cold weather from N.E. and dry – People at Horse and Hand Houghing of Turnips, and Tradesmen at sundries Finished Roofing of all Houses – a Welcome Business – Some very few begun shearing a little of their Earliest Bear – Hear of Hay being sold from Scotstown to Arbuthnot at 5^{ds} per stone out of the Cole – Oat Meal 1/ per peck – Bear D^{o} 7^{ds}- Spinning about 13^{ds} per spindle – the Cattle rather rising.

Diary for September 1791

Thursdy 1st Cold dry weather from N.W. – People at Horse and Hand Houghing of Turnips, Sundries.

Fridy 2d This morning a shower of rain, day moderate weather from S.E. till near 2 P.M. when Came on pretty smart rain & stormy from S.E. till 5 P.M. and stopped work without doors – at Barns after – Kavels Turnips.

Saturdy 3d Considerable Showers in the morning and at Intervals through the day from S.W. – all People at Hand & Horse Houghing of Turnips – finished the Horse on Bank & after at Kavels D^{o} – flited the Cattle Flakes to the Eastward of Cottar Yards &c Jobs.

Mondy 5th Moderate weather from S.E. to S.W. with drought midling D^{o} all day – all People at Turnips Kavels pretty nearly finished 2^{d} dressing, and two P.M. two Ploughs at D^{o} laying to 2 furr – & Sundries.

Tuesdy 6th Fine pleasant weather from Southward – at Turnips Hand & Horse Houghing at Kavels nearly done after Brakefast turned all the Pease Bank in good order, P.M. at Turnips again. Sundries.

Wedy 7th Strong Gale from S.W. with drought – before Brakefast at Turnips Horse & Hand Houghing. after D^{o} all Hands at West side of Kavels Cutting up Barley ripe & pretty good Crop –

Thursdy 8th Fair & very Pleasant weather from S.W. – flited the Flakes, wholly finished all Turnips, and after Brakefast shearing a little Barley and then all Hands Leading in 4 stacks of Bank Pease on wood in pretty good order – Sundries –

Fridy 9th Fine pleasant weather & drought from S.E. – after Brakefast yoked the Carts and lead Home all the Pease total 6 stacks & Thatched with Rushes 4 of them – took up & Built in Corn Yard two Cartful Lint and at Barns & variety of Jobs.

Saturdy 10th Dark weather but dry – Early all Hands at Cutting up Barley Old town park – after at Cutting up all the Wheat Floors & finishing Eddering Pease Stacks – Observed them leading Bear at Mundynes &c &c.

Mondy 12th Yesterday very sultry weather Thick fogg morning & Evening – This day same. after Brakefast Cutting up rushes and sprotts till past 11 A.M. when all shearers at Cutting up Floors Barley very unequally ripened, part too ripe & other but Half so & much Twined &c.

Tuesdy 13th Closs foggy weather till near dinner time, after fair – most people at Cutting rushes & sprotts, after at Cutting up Barley Robs Hillock – & Sundries –

Wedy 14th Foggy in the night & morning day fair & very warm slight Breeze S.W. – Shearers at Robs Hillock Barley, and others at Turning a stack of Pease which being put into too large a stack, & the uncommon warmth & fogginess of weather Hath Heated Considerably – if the weather Could Have been foreseen they would Have better on the field as yet but that is Impossible – flited the flakes &c A.M.

Thursdy 15th about 4 A.M. a pretty Brisk Breeze from the Northward which drove off the fogg and rendered the Barley ready to shear after sunrise – at Robs Hillock Barley, and at Turning Pease stack which must Have rotted owing to being put up too large at first – at Sundries –

Fridy 16th Early at Blacklands shearing Barley, West side not very Equally Ripened – Observe the Harvest now pretty General. It is the general Opinion at present that there will be a good deal of Grain, but not much fodder – the Turnips by no Means Growthy –

Saturdy 17th Fine pleasant Harvest weather – all shearers finishing Blacklands Barley – after at Floors D° now near Half Harvest – Glenfarquar little done –

Mondy 19th Yesterday a very severe Gale of wind from the N.W. – this day more Moderate from D° – Early lead Home the Wheat into a stack, and one of first Cut Barley West side of Robs Hillock Barley. Thatched the Wheat, & at Sundry Jobs on the farm.

Tuesdy 20th Very pleasant dry weather – all people at shearing of Oats Haugh, and at Sundries on the farm – Observe every body leading ours Barley not ready for it – Cattle & Horses Meat Enough –

Wedy 21st Cold Bleak weather from the Northward and appearance of rain – all people shearing till noon when Yoked 2 Carts & lead some Barley – after dinner Yoked 4 Carts to lead Barley Black lands in dry order but very Brokely – Wewmen shearing Oats Haugh –

Thursdy 22d Coldish frosty weather morning day fair, but some showers Evening – 4 Carts early & whole day lead Home all Black Lands & most part of Robs Hillock Park some of it but in Indifferent order, but the Major part; Wewmen shearing & after Gathering &c –

Fridy 23d Early at leading Barley from Robs Hillock and Floors which lead Home whole of in tolerable order before Dinner time – leading Rushes and after Thatching of Barley stacks – Flited the Cattle Flakes, and wewmen shearing Bridge Shade Oats &c.

Saturdy 24th Some showers of rain very light all people at Thatching & Making ropes &c till noon, after at Cutting up Bridge Shade Oats very rich –

Mondy 26th Yesterday and this very fine weather, frosty dew till Brakefast, after all Hands at Oats shearing Bridge Shade, and after at the Haugh, & at Sundries – Observed the Crop on the Coast about Hillton, Harvieston &c much latter than here, the Bear good, but Oats & Mashly Indifferent.

Tuesdy 27th This day fair weather from Southward – all people morning at Barns & afterwards at shearing Oats Haugh scrimply ripe – & at sundries.

Wedy 28th Frosty Rhime till near noon at Barns and afternoon at Backwaters Barley – the Plough at summer fallow Kavels & sundries.

Thursdy 29th Several showers of rain slight from the Eastward – House People at the Barns making ropes, and a Cart to Johnston for straw, after 11 A.M. all Hands shearing of Haugh Oats, and finished the whole of Back Yards Barley a tolerable Crop, and very Appearance of Young Grass amongst it –

Fridy 30th Early at Shearing of Oats Haugh, mixedly ripe, Observe the Crop very much Indifferently Conditioned – and that it is foreward in the Hollow of the Mearns by all Information & Eyesight before that of most parts adjacent – pulled up the onions and other Garden Stuffs As to the prices of grain there is no Markett yet in this Corner, but from the Haddington Sale prices in the news paper it seems not to Decline in value – There is little demand for Cattle except small store as Winterlings for the South & West Country – Manufactures & trade of all sorts Brisker than hitherto – Corns & fodder have been sold lately at 20/ Computed on the fact all Hazards Included –

Diary for the Month of October 1791

Saturdy 1st Fine pleasant weather from Southward – 4 Carts after Brakefast leading first Cut Oats Haugh and the wewmen shearing nearly whole of D° – got in nearly two stacks of them; flited ye flakes.

Mondy 3d Yesterday fine dry weather from S.W. this morning showery – 4 Carts lead home some Oats from the Bridge Shade & wewmen shearing Oats all day – finished all Haugh & Bridge Shade – after Dinner Leading Bridge Shade Oats – and at sundries –

Tuesdy 4th Closs foggy weather no drought – all people at the shearing of Oats, Nether Burns in Gayte by Dinner time fine Crop – after at Over D° fine Crop Heavy & ripe for the first one ever so on it – P.M. Sorting up some stacks of Oats Barn Yard – rain Evening.

Wedy 5th Weather as yesterday, Evening some rain but light, the people at shearing Oats Burns, and sorting up of the Corn Yard – by noon finished whole shearing – The Cows & other Cattle now pretty well foddered – 26th night.

Thursdy 6th last night a strong Gale with showers of rain from the Southward, this day strong drought from S.W. – all People at Corn Yard, Threshing Wheat for seed & Thatch & P.M. lead Home some Oats in Gayte and Hauded out some stacks with it – Sundry –

Fridy 7th Fine dry night but frosty – Early 4 Carts at driving in Oats from the Bridge shade & Nether Burns in but very midling order till after Brakefast after that in fine Condition – Wewmen & Men at Binding up Gayte, Burns & Haugh, Built 5 stacks of Oats, & left but two stacks to lead –

Saturdy 8th Very fine droughty weather from the West & Coldish – Carts leading at Haugh Oats, & Barley in fine order, and finished the whole Ingathering in the Afternoon, 29 stacks & One Hutt – Cleared up the Corn Yards and made everything snug –

Mondy 10th Gave the Feast. Yesterday uncommonly pleasant & warm weather but the Barometer for days preceeding showed a Change of weather which accordingly Came to pass – last night about 11 Came on rain & stormy weather from North, which Brought down the Rivers much & overflowed a good deal of the low Grounds, began to subside a little after One P.M. – People Employed at Barns in dighting Wheat for seed & drawing Thatch & Making ropes for D° – & at variety of Jobs &c.

Tuesdy 11th last night & this morning pretty strong wind Northerly with flying showers, but slight – all people at Thatching of whole stacks which finished before the Evening – The Grounds pretty wett –

Wedy 12th Fine warm & pleasant day – Ploughs & Harrows all day at West side of Kavels for Wheat, now ready for Dunging – Others Eddering out the stacks and P.M. flited the Cattle Flakes.

Thursdy 13th This day fair & pleasant weather variable – Carts driving dung for Kavels Wheat, a good dunging. at Drumlithie Markett a good Many Cattle, but very little sale – Sold one stott pretty high near 5/ per stone – Observed Crop very late there –

Fridy 14th This morning and through the day showers of rain, but P.M. pretty Constant to Stop work of Horses &c driving dung to Wheat A.M. afternoon put about 5 sacks fine Potatoes into Pitts Cornyard 2 feet deep & Covered them above the Earth with Rushes thick – Sundries –

Saturdy 15th last night & this morning a good deal of rain & very wett weather from the N.E. – with rather growing weather – Yoked two ploughs to clean plough the South side Bank after pease and Across for Barley – twice in the day – Digged up whole of Potatoes about 5 Bolls and sorted them for seed & at ye first – My Corn Yard is Keeping well, and fortunate with the now Bad weather – Stopped from Dunging for the Wheat for Spoiling the Ground – many are Backward.

Mondy 17th Saturday night, all Sunday, and this day Constant & Heavy rain &

high wind from N.E. Rivers much out and whole fields over flowed with water – The Cattle Have Had three very severe nights of it – which must Injure them Considerably – Could not prevent it.

Tuesdy 18th Rain & stormy weather in the night & more or less so whole day – from East – Morning Yoked two ploughs at Bank Cross Earing till Brakefast – after flited the Flakes – and after dinner at Plough again – Housed the Cattle 19 for stall feeding and 33 other Cattle Young & Old &c.

Wedy 19th last night and this morning Excessive Heavy rain which brought down the rivers higher than for a long time by past – morning Ploughs at Bank Head – & P.M. sent to L.kirk to Jo Scott Smith 440 slates formerly Borrowed – on 3 Carts – the Rivers falling in & fair sunny day –

Thursdy 20th Early this morning Came on very Heavy rain from the East with a stronger Gale than before which Increased the Deluge, and Penetrated Houses &c not hurt hitherto – Yoked the ploughs but Obliged to stop soon – Housed all Cattle again.

Fridy 21st moderate weather all day – 2 ploughs at Bank from Brakefast to Dinner, a Cart driving stones to a new sunk fence for an Intended Belt of Planting N.W. side of Garden. Kept Cattle for stall feeding in House all day &c.

Saturdy 22d Stormy weather from the North all day & showers in the Hills, 2 Ploughs at Bank, and One Cart leading Turnips, & after leading of stones to build the Dyke, at Barns & Sundries.

Mondy 24th Yesterday Cold Bleak weather and some snow on the Hills, this day frosty morning and day fair & droughty from N.W. – 2 ploughs at Bank & leading Turnips – took in some Barley first shorn for Malt to family &c Jobs

Tuesdy 25th Blowing dry weather from the Southward all night, betwixt 9 & 10 A.M. Came on very Heavy rain & stormy weather from the Southward which stopped all work without doors – after at a variety of sundry small Jobs about ye Place. Begun Threshing by Lott.

Wedy 26th The rain I believe Continued most of the night & hail Morning, day showers at Intervals, 2 Carts driving dung on Ley Cottar Shade Bank, and one Plough Broke up piece of old Sward ground on the south End of it, to Fallow, & Lime & add to same P.M. – Spreading the Dung on ley & Sundries –

Thursdy 27th Fair weather from Eastward, One Plough finished the Ley sward South End Cottar Shade Bank and two Carts driving dung to above field & spreading D^{o} – fliting sheep flakes & taking in a stack of Oats into South New Barn.

Fridy 28th Fair & pleasant weather from S.E. – One plough at the Bank & two Carts driving to Cottar Shade Ley , and Brought Home Cattle Flakes Housed them and dighting Barley P.M.

Saturdy 29th Fine pleasant dry weather – a Cart for Lime Shells & rigging stone, leading Turnips, and a plough at Pease land – Spreading dung on ley &c.

Mondy 31st Yesterday fair weather from Eastward, and this day the same, three

Carts driving dung to West side of Kavels for Wheat, rather wett. Observed from Stonehaven and Indeed from Little Fiddes a great deal of the Crop out & much to shear all the way to Bridge of Dee, and in ill order – It is said there is a good deal of Demand for Grain, and Cattle very dull –

Diary for the Month of November 1791

Tuesdy 1st Fine pleasant dry weather from N.E. – Carts at Dunging for Wheat and at Sundries –

Wedy 2d Weather as yesterday,Carts at dunging as before,leading turnips, at Barns dighting Barley for Malt to family –

Thursdy 3d Fine dry weather from the Northward – Carts at Dunging for Wheat, filling up a sewer N.W. Corner of Kavels with stones & rushes to drain it – and at Sundries – Observed much Oats to Cut from Aberdeen to Stonehaven.

Fridy 4th Cold Bleak weather with skiffs of sleet from the Northw[d] – Carts finished the Dunging for Wheat and filled up the sewer & Covered it – leading Turnips & Begun ploughing seed furr for y[e] Wheat –

Saturdy 5th Cool weather & some slight showers of Hail from the North – 3 Ploughs at seed fur for Wheat in Midling Condition – leading Turnips & Sundry.

Mondy 7th Yesterday a strong frost, and snow lying on the fields a Mile to the Northward till near noon, this day frost & fair weather – 3 ploughs nearly finished all the Wheat in good order – at Barns and at Sundries, dressing Garden Hedge &c.

Tuesdy 8th Fine pleasant weather from the Westward – Ploughed all the land for Wheat about $3\frac{1}{4}$ Acres & after Dinner Sowed & Harrowed same with $2\frac{1}{2}$ Bolls Wheat, properly prepared in good Condition – put up to feed 4 Cows –

Wedy 9th Fair & pleasant dry & fresh weather from the Westward – finished Harrowing, and water furring of the Wheat before Dinner, 2 ploughs at the Bank nearly done, and P.M. took in a stack Oats –

Thursdy 10th Fair weather and Cold drought from S.W. – Ploughs at finishing Bank after Pease, begun ribbing of West side of E[t] Haugh for Turnips – P.M. at Sundries on the farm.

Fridy 11th Blowing weather with some showers of rain from S.W. – 3 Ploughs at Haugh ribbing for Turnips very Clean & tender, at Auchinblae Crafts about 2 score Cattle – One Half sold.

Saturdy 12th Rough Stormy weather from S.E. and P.M. some rain – three ploughs at West side of the Haugh for Turnips, leading Turnips, and at South Barn dighting Oats &c Sundries.

Mondy 14th Yesterday Closs dark weather and P.M. rain – this day morning same, 3 ploughs nearly ribbed all the West side of Haugh, at Barns & Sundries – am Informed that Barley & Bear is high Priced Barley to 17/ per Boll Montrose Measure, the price of Cattle dull & low –

Tuesday 15th frost in the morning day very pleasant weather – 3 ploughs ribbed out all the Haugh – & after part of the Black lands for Turnips a very fine Tillage, flited the sheep Flakes, and lead Home some Turnips-now Increasing much in size and by no means an Indifferent Crop

Wedy 16th last night and this morning a very heavy fall of rain & Closs Moist weather – the rivers out Much, lead some Turnips, and three Ploughs at ribbing Black lands, a Cart Gathering Gutturs & Cleaning about the Houses, P.M. at Barns dighting Oats for the Mill – Hear that there is much demand for Barley from 16 to 17/ – Meal fallen to 13/4 – Butcher Meat from $2^{ds}\frac{1}{2}$ to $3^{ds}\frac{1}{2}$ per lb – Trade & Manufactures very Brisk – dressing Garden Hedges &c.

Thursdy 17th fair weather from the Westward – lead some Turnips and three Ploughs at Ribbing of Black lands – at Barns dighting of Oats for the Mill – Garden pruning Trees –

Fridy 18th pretty Early this morning Came on rain from the Northward & from 10 to 11 A.M. stormy, Ploughs stopped – One Cart to Johnshaven for Coals, and lead 4 Cartful Turnips, dighting and sacking of Oats for Meal to family –

Saturdy 19th Strong wind from the N.W. – lead some Turnips and after three Ploughs nearly finished the Ribbing of Black lands, being all of it for Season, at Barns and other Jobbs –

Mondy 21st Yesterday frost and a shimering of snow lay most of the day – this day fresh weather and all snow & frost gone – 2 ploughs finish all Ribbing for season Black lands – sent Oats to the Mill of Kair for family, lead some Turnips, and Graveled roads &c –

Tuesdy 22d Fair weather & some drought from the Eastw^{d} till about 2 P.M. Came on rain from S.E.- 2 ploughs water furred the Bridge shade out and after begun seed furr for Oats Upper Burns – at Barns dighting Oats 27B to Mill – 5B lofted.

Wedy 23d Slight frost in the night day fair but little drought, 3 ploughs at the Over & Nether Burns, lead some Turnips – and at Barns – flited sheep flakes &c.

Thursdy 24th last night about 6 P.M. Came on a Severe Gale of wind from S.E. with Heavy rain – and this morning the rivers much out and fields deluged – lead Home 4 Cartful turnips – after took in a stack of Pease Old West Barn, at dighting Oats new N. Barn & Sundries.

Fridy 25th Fine pleasant day from various, 3 ploughs at the Ley Old Cottar Land, and at Barns &c Sundry Jobs – Hear the prices of Barley high & much demand from all Quarters.

Saturdy 26th Frost in the night and dry pleasant weather – 2 ploughs at Ley as yesterday, one Cart to Johnshaven for Coals, and other people at sundries about Barns &c. Jobs –

Mondy 28th Yesterday Bleak weather – this day strong drought from the Westw^{d} – 2 Carts to Johnshaven for Coals and one Plough at Ley – people at the Mill and P.M. lead Home some Turnips &c Jobs – 23B 2p Meal.

Tuesdy 29th very strong Gale from N.W. with flirts of sleet &c at Intervals –

Carts at the Mill of Kair for 2 Bolls of Meal, for Malt at Bridge of Leppie – and driving of Turnips one Plough at ley – and at sundries –

Wedy 30th pretty Brisk frost in the night day Moderate but Came on a Great fall of snow about 11 A.M. – and lasted till past 1 P.M. when fresh – two Plows ribbing Bridge Shade for Turnips – and four Horse Plough at the Nether Burns for Oats – P.M. Girnelling the Meal first of Crop from the Mill of Kair – Prices of Bear about 15/ per B, Barley 16/6 to 17/, Wheat 20/ – Oat Meal from 14/ to 14/6 – Hay $5^{ds}\frac{1}{2}$ per stone – Servants very scarce & dear – Cattle very dull Beef $2^{ds}\frac{1}{2}$ to 3^{ds} per lb – Spinning 15^{ds} per spindle – Trade & Manufactures Brisk.

Diary for the Month of December 1791

Thursdy 1st In the night & morning some snow & frost pretty strong – three ploughs at Bridge shade, and lead Home 2 Cartful of Turnips – Barns dighting Oats for the Mill and at Sundries –

Fridy 2d last night a good deal of rain from N.W. and morning Frost – 2 ploughs at Bridge Shade, one Cart at the Turnips & P.M. Barns – Sold for Montrose 100B Barley at 17/ per Boll –

Saturdy 3d very strong frost last night, and this day the same, Could Have Ploughed, but dighted up Oats for the Mill, and taking in P.M. a stack of Barley N. Barn – Horses at Smithy &c.

Mondy 5th Yesterday a very strong frost, and this day the same – 2 Carts Carried to the Mill of Kair 2 Bolls of Oats for family Meal, and One Cart leading Turnips & One Man Digging of same – P.M. at Barns Chaffing of Barley &c.

Tuesdy 6th Fresh weather from S.W. – all people at Barns dighting Barley for Montrose, lead some Turnips – and at variety's – Some additional Hands at Barns – The Thaw made not much relax of the Frost – river frozen over.

Wedy 7th Fresh dry weather from S.W. – 3 Carts to Montrose with 12 Bolls Barley – and one Cart driving Turnips, Brought out 10 Barrell of English Coal, Iron &c Jobbs –

Thursdy 8th pretty strong frost in the night day very variable, P.M. Stormy – One Cart lead Home some Turnips, all People dighting Barley –

Fridy 9th Cold stormy weather from North W – lead Home some Turnips, and at Barns dighting Barley – and also at Pease Barn – Sundries –

Saturdy 10th a small fall of snow and very hard frost, 3 Carts to Montrose with 12B Barley – and One Cart leading Turnips – at Sundrys, Brought out 4 Barrell English Coals &c –

Mondy 12th Yesterday Excessive Hard frost Th^{r} at 11 A.M. open Air at 21 degrees – this day Came on at a little after 10 A.M. a most severe storm of snow & drift with a strong gale of wind from the Southward – in the Evening it Abated with sleet & rain – lead some Turnips, Barns &c Jobs.

Tuesdy 13th Frosty in the night, about 9 A.M. Came on a strong Gale from the N.W. but fair, led Home three Cartful of Turnips from Kavels and at Barns dighting Barley for Montrose.

Wedy 14th Strong frost this day & roads a sheet of Ice – Digging & leading Turnips – Horses at Smiths and morning took in a stack of Burns Oats Old West Barn – dighting & sacking Barley for Montrose – and at Sundries Garden Covered of late, Strawberries, Artichokes, rasps, all Fruit Trees at their roots with Pease Holme & Dung – dressed Gooseberry Bushes.

Thursdy 15th very strong frost Continued Digging at Turnips and leading of them, Horses at Smithy Shoeing, and at Barns, Sundries.

Fridy 16th very Hard frost Continues, not much snow on the Ground, the Turnips very ill to Digg, 2 Men at that Business and One Cart leading them, 2 Carts to Mill for the Meal 22B $5\frac{1}{2}$p. P.M. at Dighting of Barley North N. Barn. There is the greatest demand for Bear I ever remember of – Barley 17, Chester 15/6, Oat Meal 15/3, Hay $17^{ds}\frac{1}{2}$, Corn & fodder 18/ – &c Jobbs.

Saturdy 17th This day appearance of Fresh weather from the Southw^{d} – 3 Carts to Montrose with Barley – and one Cart leading 4 draughts of Turnips, two Men digging them – Sundries.

Mondy 19th Yesterday pleasant weather & fresh from S.W., this day pretty hard frost & Cold North wind – One Cart leading Turnips & severals digging D^{o} – P.M. took in a stack of Barley N.Barn &c. the Gardner driving by Barrow Dung to Garden.

Tuesdy 20th Frost Continues some little snow, One Cart leading Turnips, and all other People at Chaffing & dighting Barley for Montrose.

Wedy 21st Strong frost & fair weather – 2 Carts to Woodstock with Hay for a Neighbour, and one Cart leading & 2 Men Holing Turnips, if this weather Continues it will Shorten the time of Expending them, at Barns dighting Barley.

Thursdy 22d a strong Gale from the Southward last night, this day Extreamly bitter Cold from D^{o} – P.M. rain & very stormy – lead some Turnips and Men Turning the Dung Court dung into stacks to rot – at Barns &c Jobs –

Fridy 23d Moderate weather but roads very bad – One Cart leading Turnips, and others turning the Dung into Stacks – at Barns & Sundries.

Saturdy 24th a very strong gale from the N.W. and Hard frost – early sent 3 Carts to Montrose with 12B Barley – lead some Turnips – and at turning Dung in D^{o} Court.

Mondy 26th Yesterday Clear & fair weather but frosty – this day the same but Harder, One Cart leading Turnips, One Man digging at D^{o} – Others at Turning Dung in Dung Court & at Sundries.

Tuesdy 27th This day morning took in a stack of Robs Hillock Barley S. Barn, after 10 A.M. Came on Heavy Snow Melting from Southward which lasted till past 3 P.M. – dighted & sacked $3\frac{1}{2}$ Bolls Oats – lead some Turnips.

Wedy 28th last night and most of this day a very strong Gale from N.W. and rather fresh – all people at Barns Chaffing Barley, and some Oats, P.M. lead some Turnips – this weather will much Hurt them & the sooner done than wished for.

Thursdy 29th very strong frost & very slippery roads – One Cart leading Turnips, and two Carts driving stones to the Dyke Intended West side Garden strip of Planting, at Barns dighting Barley and at sundry Jobs.

Fridy 30th rather fresh this day and after Brakefast came on sleet & rain from Southward – dighting & sacking of Barley for Montrose and at Barns, Smithy frosting Horses & Jobs. a Young stott 6 Months old Died unexpected of the Quarter Ill vulgo a pretty general Obstruction & Extravasation of Blood in the different parts of the Body –

Saturday 31st Black frost in the night, about 8 A.M. Came on rain which lasted more or less all day – 3 Carts to Montrose with Barley – and one at Smithy & leading Turnips – the Ice on the River Broke a little after 4 P.M. with a flood which Carried off the Timber Bridge at Hercules Haugh to Garvock & our small D° – soon subsided.

Diary for the Month of January 1792

Mondy 2d Yesterday Bleak weather from the Northward, this day Moderate & the roads much better 2 Carts leading whole of Bank, and part of the Kavels Turnips, at Barns &c.

Tuesdy 3d Fresh weather in the night, day fair & pleasant – Men Employed at Barns dighting & sacking Barley for Montrose, and at Sundries –

Wedy 4th Slight frost & foggy weather all day, 3 Carts to Montrose with 12B Barley – and others at Sundries – Brought out $1\frac{1}{2}$ Bolls Coals –

Thursdy 5th Strong Hoar frost in the night, and day moderate & good winter weather – all the people at their several Christmass's – Hard frost Evening.

Fridy 6th Weather as yesterday – this day 2 Carts and &c – the other people Digging & leading of Turnips, at North Old Barn Chaffing of Oats –

Saturday 7th Fresh in the night day various, frost & stormy from N.W. – people Employed at taking in Pease Old Barn, & Barley N. North Barn dighting & lofting Oats for family use, & Sundries.

Mondy 9th Yesterday Constant frost and a skimmering of snow lay all the day – this day fresh & A.M. some showers of rain – river some out – lead some 3 Cartful Turnips – at Barns dighting of Barley for Montrose and Sundries.

Tuesdy 10th last night and this strong frost and Heavy Gusts of Snow from N.W. – lead Home 2 Cartful of Turnips, and sent $8\frac{1}{2}$ Bolls to Mill of Kair, at Dighting & sacking of Barley for Montrose &c.

Wedy 11th a very severe storm, fall of snow, and Hard frost, People at filling of Barns, begun to take in first of Hay for Cattle. Malt Barn loft, at Smithy Frosting Horses & Jobs –

Thursdy 12th Keen frost from N.W. & very Cold Th^{r} 11 A.M. 24 degrees – 3 Carts to Montrose with Barley One D° Leading Turnips, and at Sundries –

Fridy 13th very strong frost – wind Westerly – one Cart leading Turnips & some people Digging D° – at Barns dighting pease lofted for Horses – dighted & sacked $8\frac{1}{2}$ B Oats for the Mill – took in remainder of stack pease West Barn.

Saturdy 14th Thick weather and for two Hours about noon some Snow, but Moderate & Mild – 2 Carts to Mill of Kair with Oats for family Meal & then brought Home 2 Cartful of Turnips – at Barns dighting of Barley for Montrose and lead P.M. in a 2^d Cartload of Hay.

Mondy 16th Yesterday a kind of Fresh weather and small rain from S.E. – this day very little of the snow gone, and a Fresh Continues but Moderate – lead Home 2 Cartful of Turnips – and other People at Dighting Barley North Barn and sacking for Montrose. There is a stoppage of the sale of the Bear owing to the new Corn Bill & other Causes – Corns & fodder selling 21/ 22/ per Boll, Hay at 6^ds- Sale Yarn & spinning high

Tuesdy 17th last night a strong Fresh with rain but in the morning Hard Frost again the River Broke, people at Barns dighting Barley and took in remainder Crooked Oat stack N.N.Barn Cold Stormy wind Northerly &c.

Wedy 18th Strong Black Frost & high wind from N.W. roads very slippery – lead Home 3 Cartful of Turnips, and others at Barns dighting and sacking Barley – and at Sundries Smithy.

Thursdy 19th Moderate weather but still the frost not remitting – 4 Carts to Montrose with Barley, Men at Making up Dunghills Dung Courts and at Sundries – Oat fodder Setting ye Cattle.

Fridy 20th Slight frost & Moderate weather from Eastward – Carts leading Turnips & people digging at D° – lead in a Cartful of Hay – at Barns dighting and reding up of N South One for Oats & Sundries.

Saturdy 21st A kind of Neutral weather neither fresh nor frost wind East^y – took in a stack of Oats S. Barn, lead 2 Cartful Turnips, Making up Stacks of Dung D° Court, & Sundries.

Mondy 23d Yesterday a Gentle Thaw, and thick Hazy weather with small rain – the River began to Brake about 10 A.M. & supposes was open by Midnight – this day same weather – sent a large Cartful of Lint to Pitnamoon, and leading some Turnips – Dighting Oats for ye Mill.

Tuesdy 24th Dark Gloomy weather & fresh, frost little out, 2 Carts driving Dung from Cottar Town to the Bridge Shade, and two Carts to Mill with Eleven Bolls of Oats, took in a Cartful Hay Old Barn.

Wedy 25th last night Came on a Constant Closs rain from the Eastward, with High wind, which Brought down the Rivers very high & Sweeped off all the shoals of Ice of the Season-Dighted & lofted One Boll rye Grass seed shaken, and P.M. lead home 2 Cartful of Turnips – &c Jobbs.

Thursdy 26th Fair weather for most part & fresh Frost not near out 4 Carts driving dung to Bridge Shade and P.M. at Dung Court, and at Sundries.

Fridy 27th Frost & fair weather from N.W. 4 Carts driving dung to Bridge Shade, and Home some turnips, at Smithy Making Iron Rack Bridles for Horses and at Dung Court – sundries – &c.

Saturdy 28th very hard frost, day wind Easterly – about noon Came on Hail & P.M. a good deal of snow – 4 Carts driving dung to Bridge Shade, and leading Home Turnips, & one Cartful of Hay –

Mondy 30th Yesterday thick foggy fresh weather & some showers of rain – this day the same – 4 Carts driving dung to Bridge Shade for Turnips – and morning took in a stack of Oats N.N.Barn – P.M. at Barns Chaffing Oats, and at Sundries –

Tuesdy 31st last night a strong Gale from Southward and a very Considerable deal of rain and this day fresh wind from S.W. – 4 Carts driving dung to Bridge Shade and Brought Home 4 Cartful of Turnips, at Sundries – A second of our last years Calves died of the Quarter Ill, must Bleed all the rest without further delay – Hear of no Farms of Estate being sold as yet Except Glenbervie seed at 15/ per B. The Frost is not yet near out having a foot In.

Diary for the Month of February 1792

Wedy 1st This day mild soft foggy weather – all People at the Taunton fair – Hired the shearers and most of the other Hands – they are very High – Brought Home from Kair 26B 6p Oat Meal.

Thursdy 2d In the night High wind from S.W. and a Considerable fall of rain which brought down the rivers, and through the day fresh but the frost not one Half of its depth removed, at Barns Chaffing Oats, & putting Meal into Girnells &c – Bled all the Calves for fear of more Ill Consequences – Crocus & snowdrops springing.

Fridy 3d Stormy Blowing weather from S.W. – People at Barns Chaffing & dighting Oats South Barn, Smithy lead Home 3 Cartfull Turnips – Observed several Neighbours Ploughing Ley, but surely not very deep, as I assured by the Gardner frost had been about 15 Inches deepth & remaining about 4. flited the sheep flakes –

Saturdy 4th very strong wind from Westward – this day yoked 3 Ploughs on long Burns & Cottar Shade Ley – but the frost Scarcely out Enough for the Plough, lead Home some Hay, and at Sundries.

Mondy 6th Yesterday a pretty strong frost – this day Milder and 2 ploughs at the Cottar Bank ley, and one 4 Horse D° at the Long Burns, the frost not out, lead a Cartful of Turnips & P.M. took in a stack of Oats Haugh N S Barn, dighting & sacking Oats for the Mill – and at variety of other Jobbs –

Tuesdy 7th Some light rain from S.W. – 2 ploughs at the Ley – and one at Long Burns – P.M. at dighting rye Grass seed 6 firl: lofted and &c.

Wedy 8th Some small rain A.M. and Cold weather from N.E. – One Cart to Stonehaven for Coals – One Plough on Burns & One at ley – took in a part of a stack of Pease West Old Barn, &c Jobbs.

Thursdy 9th Slight frost in the night, day fair weather – 3 Ploughs at Burns, and

Cottar Shade Ley – People at Barns, Trenching a strip of Ground West side Garden for Planting a Belt of a variety of Trees as a Screen to it.

Fridy 10th Strong wind Westerly last night – this day fresh, 3 ploughs at Burns & Floors, and at Cottar Shade Ley – P.M. lead Home 2 Cartfull of Turnips, and sent 3 Carts with the $21\frac{1}{2}$ B of Oats to the Mill of Kair &c –

Saturday 11th Pleasant fresh weather from the Westward – Ploughs as before, lead some Turnips &c – Observed in my way to Montrose not much ploughed and the roads & fields wett – At Sundries –

Mondy 13th Yesterday very pleasant fresh weather from W. This day Calm & warm sunshine all day – 2 ploughs at Ley, and one at Floors, a Cart leading Turnips, Hay, & Stones Barns &c Observe some Early flowers in Blossom. Snowdrops –

Tuesdy 14th Slight frost in the night day warm & pleasant – 2 ploughs ribbed out Bridge Shade for Turnips and one at Floors for Oats, at Barns &c Jobs –

Wedy 15th fine pleasant weather & warm foggy – 3 ploughs as yesterday – at Barns leading Turnips & some stones for Garden wall, seed Oats at 16/- Farms of Glenbervie sold to Trade of Montrose at 15/6 payable at Whitsunday – &c

Thursdy 16th Thick foggy, sharpy weather from the Eastward – 3 ploughs as yesterday – P.M. at Barns dighting of Oats, Gardner Trenching &c Jobs.

Fridy 17th very pleasant weather slight frost in the night – 3 Ploughs as yesterday – Trenching, taking in Oats, lofting Oats, dighting Pease &c Jobs.

Saturday 18th Cold weather & some snow on the Hills from N.E. – 2 ploughs on Cottar ley Bank, and one at the Floors, took in a stack of Wheat & Oats, the first N N Barn, the other N S D°- dighted & lofted seed Pease, In some Hay-

Mondy 20th Yesterday morning a good deal of Snow from the Eastw^{d} – this morning the same and also through the day – 2 Ploughs at Ley and one on stuble ground, Trenching for a Belt and taking out Ash Trees from it – at sundries.

Tuesdy 21st a Considerable fall of snow in the night some two or 3 Inches – day fair wind N.W.- three Ploughs as before – Transplanted several Ash Trees as Gate Posts – to Bridge Shade, red Hall park and Robs Hillock &c – P.M. Barns Chaffing Oats –

Wedy 22d Showers of Snow night & in the day-3 Ploughs at Cottar Shade ley, & Begun red Hall Park ley – P.M. at making up Dunghill and at Trenching for Belt &c at ye Mill.

Thursdy 23d pretty strong frost in the night day fair & fresh from S.W. – three ploughs at ley Cottar Shade and red Hall park ley – lead some Turnips &c at Smithy sorting & Ironing two Ploughs &c Jobs.

Fridy 24th very Considerable frost in the night, day very Chilly Cold & Surly – about 4 P.M. Came on Sleat & Snow & stormy weather from S.E. – 2 ploughs at Red Hall park ley & 2 Carts driving Dung to Bridge Shade &c. People at Barns dighting Oats –

Saturday 25th Stormy weather & Hail, Sleet &c, but this day mild, misty & Cloudy weather from S.E. – 3 ploughs at Red Hall park – and one Cart at leading in Hay, Turnips and sundries – P.M. Sundry.

Mondy 27th very Stormy weather from the S.E. – 3 ploughs at red Hall Park ley – One Cart to Stonehaven for Coals – at Barns &c. This day Bought the lands of Easter & Wester lair from Mr Smart Polburn for Two Thousand p^{ds} S^{tr} –

Tuesdy 28th Fine pleasant dry but dark weather from the S.E. – 3 Ploughs at Red Hall Park & one Cart leading Turnips – Planting Ash Trees Back Yards, trenching Belt West side Garden of D^{o} – P.M. at Barns dighting Oats & sacking of D^{o} – at Sundries –

Wedy 29th This morning Came on rain & stormy weather from S.E. which lasted the whole day & prevented all work without doors – people at Barns dighting Oats, Pease, & Wheat and lofting part of D^{o} – The victual prices rather duller of late, and hear of no farms sold lately – Seed Oats is sold for 16/ & Seed Pease for 14/ per Boll – Oat Meal 14/8 and the Bear 15/ Hay 6^{ds} per stone & the Purchaser to remove it at their own Expence. Corns & fodder rather fallen – Manufactures & Trade very Brisk in General, and land selling & setting at Double the late prices – Gardner & assistant trenching a strip of fine Ground for planting West side of Garden – Exchanged Missives here for the Above Estate.

Diary for the Month of March 1792

Thursdy 1st Closs foggy weather from S.E. & grounds very wett, took in some Oats & Pease and after 10 A.M. 4 Ploughs at Red Hall park – P.M. lead some Turnips, & dighting of Wheat &c –

Fridy 2d This morning Came on heavy rain from the Eastw^{d} which Continued till past 10 A.M. – Yoked 3 Carts driving Earth to the sand Hole by Hay Sowe from a Dyke full of Quicken &c bewest the House – Gardners at Trenching &c Jobbs –

Saturdy 3d This day fair weather from S.W. – all Carts at driving away Earth from West side House to sand Hole and part on Goulds Hillock in Heaps, taking in Hay – lead some turnips, and Trenching the Ground for Belt.

Mondy 5th Yesterday a Considerable deal of rain from N.E. this Bleak weather with little rain – 3 Ploughs at the Red Hall park Ley – and after Rouped & Sold 16 of twenty fatt Cattle at but Moderate prices –

Tuesdy 6th This morning some snow, this day fair but Cold Bleak weather from N.E. – 3 Ploughs at the red Hall Park – One Cart driving Earth to Golds Hillock ley – delivered 2 fatt Cattle for Montrose – at Barns &c.

Wedy 7th This morning Came on a Severe Storm of Snow & drift from the Eastward, which Continued all day – People Employed within doors, at dighting Wheat, Pease, & Rye Grass – P.M. frost Conjoined – &c Jobs

Thursdy 8th a great deal more snow from N.E. and now very deep at least in wreaths – 4 Carts driving dung and Earth till dinner time, after at sundries Gardner Trenched the Ground for Belt.

Fridy 9th very strong frost & snow lying deep in many places. 2 Carts to Montrose with 8B 5p Wheat and 2 Carts driving & Earth to Bridge Shade & Golds Hillock – at taking in Hay, leading Turnips &c.

Saturdy 10th This day Rather fresh weather and some Gusts of Snow from the Northw^{d} – 4 Carts driving dung to Bridge shade & Brought Home 4 Cartful of Turnips – P.M. took in a stack, stathel Oats for seed.

Mondy 12th Yesterday very pleasant Clear Sun Shine weather & warm, this day pretty strong frost from the Southw^{d} – 3 Carts to Johnshaven for Coals, and one D^{o} driving Dung &c lead Home some Turnips, at Barns &c Jobs.

Tuesdy 13th Cold & stormy weather from Eastw^{d} & most of the day Snow & Sleet – Stopped Carts driving dung by 11 A.M. – after at Barns Chaffing & dighting of Oats for seed, and at sundries &c.

Wedy 14th last night & this day strong fresh S.W. wind and some rain – took in a part of a stack of Pease, and dighting seed Oats, P.M. Yoked 4 ploughs finished Red Hall Park, Cottar Bank ley & begun the Haugh above the Distillery.

Thursdy 15th Strong N.W. drought all day – 4 ploughs at Haugh ley & Loup Yards D^{o} above the Distillery, and at Barns lofting Oats 17B 2fi – Gardners at Planting some fruit Trees in ye Garden, & some firs Sheperds Haugh, put a Handsome Paling on the House side of Belt of Plantings: and at sundries &c.

Fridy 16th pretty strong frost in the night, day fair & Moderate drought from Southw^{d} – 4 ploughs finished all ley Tillage of Season, leveled down a Brae foot Cottar Bank by the Plough never before laboured with Difficulty – after at Barns Sacking seed Oats, Mason at dyke West side of Garden, and at sundries, sowed Garden pease

Saturdy 17th very strong Gale from the Westward, and fresh weather – 3 Ploughs finished out the Floors and two Carts driving out Earth to ley and leading Home Turnips – Planted Garden Beans – P.M. rain.

Mondy 19th White frost in the night day fair & Cold from S.W. – 4 ploughs at Steering West side Bank after Turnips for Barley – and others at sundries, sent for the Lint Crop 6St.4lb after 2 pecks.

Tuesdy 20th The first part of this day Cold Stormy weather from the Southw^{d} – about 3 P.M. Came on a Severe Storm of wind & rain from the S.E. which stopped all work without doors, at Barns &c – before Dinner 4 Ploughs at Bank steering for Barley.

Wedy 21st This day fair weather and strong drought from the Westw^{d} – 4 ploughs at Bank steering and after at Kavels Ribbing after the Turnips for Barley – at Sundries &c Jobs –

Thursdy 22d This day cold dry weather & much drought from S.W. – 3 Ploughs at Kavels ribbing, and one Cart for Lime, but obliged to Bring Coals from Johnshaven there being no lime shells Burnt – at Grinding Malt – 3 Masons at Stone Dyke & Sundrys

Fridy 23d Yesternight & this morning a good deal of rain and stormy weather from Southw^{d} – after Brakefast yoked 3 ploughs on Kavels ribbing – and one Cart Brought a Cartful of Lime soured from Mill of Garvock for Garden Dyke – P.M. took in a stack of Barley Floors & R.Hill^{ck} S.Barn.

Saturdy 24th Cold weather & rather stormy from Southward, 3 ploughs at Kavels steering for Barley – One Cart driving stones & Earth from West side of the House and slopping & dighting pease West Barn.

Mondy 26th Yesterday tolerably dry weather, this day more so – 3 Ploughs at South side Kavels, morning and P.M. yoked 9 Harrows & Begun sowing Oats on Loup Yards, & long Haugh red Craig sowed & partly Harrowed 3B 1f Bridge Shade Oats

Tuesdy 27th a severe shower of rain & wind about 7 A.M. which stopped sowing & Harrowing when begun, 3 Ploughs finished the steering of Kavels – and P.M. three Harrows at Oats Harrowing – &c Jobs. One Cart for Lime.

Wedy 28th Fair & strong drought from S.W. till 4 P.M.Came on rain which Increased to be Heavy & stormy & obliged to stop Harrowing by $\frac{1}{2}$ past 5 – Sowed 5B 1f more Oats South side Cottar Bank – driving Earth to Ley from Intended Belt – and at Barns dighting & Chaffing Oats &c Jobbs.

Thursdy 29th The storm of rain & wind Increased & was severe all the night & most of this day – 2 Carts at driving Earth A.M. & 2 P.M. Dung to the Bridge Shade, while no Neighbours were working – at Barns & sundry Jobbs.

Fridy 30th Friday fair weather & 3 ploughs at Upper Burns Wet and not good work – at Barns and sundries – Observed to Aberdeen Scarcely any Oats Sown, but the grass & Greens much better than in the Hollow of the Mearns –

Saturdy 31st Fine pleasant drought from the S.W. – the Harrows at the Cottar Bank ley before Sowing – Broke in part of Black lands for Pease very wett – finished Turnips – at Garden sowing small seeds of kinds – Observed the Grass & the other vegetables not growing much. Most of the Fatt Cattle now sold in ye Country.

Diary for the Month of April 1792

Mondy 2d Yesterday drought & Blowing weather from the S.W. Evening rain – this day generally fair the wind Westerly – but showers – taking down a Dyke Haugh for a Midding layer for Turnips – & P.M. 4 Carts driving Dung to it & leading stones to Garden strip.

Tuesdy 3d Fair weather and rather drought from Westw^{d} – 2 Ploughs at Black lands for Pease, and two Carts driving dung to Haugh, & P.M. at driving Earth to Goulds Hillock ley – at Sundries –

Wedy 4th Showery weather & wind Easterly – 3 Ploughs at Black lands for Pease, and three Harrows Braking in for D^{o} – the Ground by no means in Order – Planting at the Belt West side House &c Jobs –

Thursdy 5th Cold stormy weather from the N.E. with Rain in small Quantities, 4

Ploughs at Black lands for Pease, in not good Condition – took in first Barley stack S Barn – Sold remainder of fatt Cattle 4 stotts at 40£ for Montrose – Planting Belt.

Fridy 6th fine dry weather from Northw^{d} – and about noon Became pretty warm – 4 Ploughs at the Blacklands for Pease, and dighting Oats middle of the day – Planting Belt & Building Sunk fence.

Saturdy 7th Fair weather and drought from S.W. but no Sun Shine, 9 Harrows at Cottar Bank Ley – Sowed 10 Bolls – Oats Making in all 18$\frac{1}{2}$ total on Cottar Shade Bank – & Sundries –

Mondy 9th Yesterday fine dry weather from S.W. – this day very pleasant & warm weather from Southw^{d} – 9 Harrows finishing the Cottar Bank Shade, and a two Horse Plough finished all Tillage for Oats – sowed onions carrots.

Tuesdy 10th very pleasant & warm weather – 9 Harrows all day at the Over & Nether Burns Award – Sowed 8 Bolls – Oats & partly Harrowed the same, took in remainder Barley stack & dighting Oats.

Wedy 11th Remarkable pleasant & warm weather from the S.E. – 9 Harrows all day at the Burns which finished and after at the Floors Sowed with 5B 1f Oats & partly Harrowed same.

Thursdy 12th Weather as yesterday rather Hotter – 10 Harrows all day at Floors, & red Hall park sown with 11 Bolls Oats & partly Harrowed which Finishes all for Oats 42B 3f 17 to 18 pecks on 38 to 39 Acres.

Fridy 13th rather Chilly weather & Hail from N.E. – 4 ploughs at Blacklands for Pease which finished in the Evening – 4 Harrows at the red Hall Park – Sacking at Pease noon &c Sundries –

Saturdy 14th Bleak kind of weather from the N.E. but fair – 9 Harrows at red Hall park Oats and after Harrowing at Black lands sown with 9 Bolls of Pease & Covered the same.

Mondy 16th Yesterday dry & Cold Bleak weather from the S.E.- this day the same – 9 Harrows finished out the Harrowing of the red Hall park for Oats, and at Barns in the Middle of the day – Observe that the Oat seed is not above Half done in the General and the Pease same – the pasture but very middling & the sown Grasses Backward – Fodder scarce & scrimp. The price of Grain rather lower

Tuesdy 17th Cold Bleak weather from the N.E. and Harrows at Upper Burns, and after at Braking in the Bank for Barley very foul, &c Covering the Grass seeds Floors: P.M. Rolled D^{o} – and at Bank Braking in &c Jobbs –

Wedy 18th Weather as yesterday – Two Carts driving dung to the Coldest part of West side of Bank after Turnips for Barley – and six Harrows at Braking in Bank & P.M. at same on the Kavels after D^{o} for Barley lofted 10 Bolls of Oats N N Barn &c Jobbs –

Thursdy 19th Cold Stormy weather from the N.E. and some light showers of sleet & rain morning – 6 Harrows at Kavels, and two Carts dunging at Head of West side Bank after Turnips – the Ground rather Cold to secure the grass Seeds –

Fridy 20th Strong Black frost last night & this morning – 2 Carts to Dava for Stones to a new Kitchen twice this day – 2 ploughs at West side of Bank Steering it out for Barley, & Sundries.

Saturdy 21st Strong wind from S.W. and skiffs rain in general dry – 2 Carts leading stones & Ploughs & Harrows at West side of Bank now in fine order – at Sundries – took in a part of a stack of Oats N N Barn &c.

Mondy 23d Yesterday stormy weather but dry till Evening a pretty good shower of rain from ye S.W. this day same – 3 Ploughs at steering the Late Pease ground Bank & two Harrows at D^{o} – a Cart to Johnshaven for Coals – Middle of day & wewmen P.M. Chaffing & dighting of seed Barley & sundry Jobbs –

Tuesdy 24th Strong drought from Southw^{d} – 4 Ploughs at the Bank, and two Harrows all day – at Barns dighting Barley sold to Jas Croll Polburn for seed – and at Garden & other Jobbs –

Wedy 25th Strong drought from S.W. Continues – 4 ploughs at Bank A.M. & two Harrows D^{o} – P.M. 3 Carts driving Dung to Old Town Park for Potatoes, One Plough & two Harrows at Bank &c. Observe most of the Neighbours preparing for Bear.

Thursdy 26th Strong drought from S.W. till near noon when skiffs of rain – 4 Ploughs at Bank P.M. & one on D^{o} – Planting the Potatoes on the Old Town park Every other Furrow by the Plough, at Laurencekirk Tryst a great many Cattle and Considerable sale & prices &c Sundries.

Fridy 27th last night a shower & after & this day much drought & high Boisterous wind from the S.W. – 2 Carts driving stones for the new Servants House & new Kitchen – One Cart to Montrose for sundries, and one Plough, Harrows &c &c

Saturdy 28th very strong Gales Continue from S.W. & drought, 2 Carts driving stones from Dava, and Plough & Harrows at Bank – a Cart to Mathers for Lime 4 B – 2 Masons Henghing Stones – Soured ye Lime.

Mondy 30th Yesterday stormy weather and this day pleasant weather from S.W.- 4 Ploughs at West side of the Bank for Barley – Observed to Stonehaven that the Grass was very poor in many places, and fodder mostly gone and that Hay was up to 7^{ds} & 8^{ds} per stone – Several people sowing Bear & Barley and in General the Oat seed finished except at Arbuthnott – Sundries –

Diary for the Month of May 1792

Tuesdy 1st Cold stormy weather from N.W. with flirts of snow which lay a little on the Hills – 4 ploughs, and some Harrows at West side Bank, begun the sowing of Barley about 4B in fine order, Barns &c.

Wedy 2d Remarkably stormy & Cold weather & severe Gusts of Snow from the Northw^{d} whole day – 4 ploughs at Kavels for Barley, and at Sundry other Jobs. Masons & Wrights at new Servants House –

Thursdy 3d Excessive strong Gale this morning Early, day tolerable good & about

11 A.M. Came on a good shower of rain from S.W. which lasted more or less all P.M. – Sowed West side of the Bank with 3 Bolls R Grass – 40 lb ams red Clover & 12 lb White D°

Fridy 4th Fair weather from the Northward & Cold – Sowed 2B Barley in Kavels & partly Harrowed it – but the Ground too wett for Harrowing it, or ye Bank – 4 ploughs at Kavels dunged some of it again &c Jobs.

Saturdy 5th Strong drought from the Northward & Cold weather – Sowed out the Kavels with total on it of Barley – 7B Scrimp, Ploughs & Harrows at Kavels harrowed the Grass seeds Bank &c Jobs –

Mondy 7th Yesterday Moderate weather from N.E. – this day Milder from E^t^- A.M. sowed Kavels with 7 Bolls Rye Grass, 100 lb Ams^td^ red & 20 lb white Clover all mixed together & Harrowed in same – 2 Ploughs at foot Bank steering for Barley –

Tuesdy 8th Moderate weather from the various points – 4 ploughs at steering 3^d^ furr after Pease Bank – & a Roller on West side Bank sown with Grass seeds Water furring of Bank & Kavels before Rolling,

Wedy 9th Cold weather and Stormy but dry – 4 Ploughs at the Bank steering and a Roller on Kavels – the same P.M. – Bought 6 Ewes & lambs N Hill Halkerton at 9/7 each – Observe the Gean, Strawberries & some Plumbs & Peers in Flower – Dr Stuart Redmyre Died 81 y^rs^.

Thursdy 10th Cold Stormy dry weather from N.W. – 2 ploughs finished steering of Bank for Barley & 5 Harrows on D° – a Roller on Kavels nearly done – Gathering &c Couch Grass on Bank – 2 Masons dressing stones Quarry for new Kitchen.

Fridy 11th Snow on the Hills White, Cold & Stormy weather from N.W. – 4 Carts driving dung to Bank – Rolled out the Kavels, and Burnt all the Quicken on Bank, at Sundries – rain Came on 7 P.M. –

Saturdy 12th Strong frost & some snow on the Hills & even on the Laighs – 4 Carts driving dung all day and Hired people filling D° – Spreading at D° Bank and at sundries – No Growth –

Mondy 14th Yesterday most of it more or less rain – at 6 o'Clock yesterday, this day fair but Cold weather from the East, 3 ploughs at the Bank for Barley, and one cart leading stones from Dava Quarry, at Barns spreading dung the Barn Men & Wewmen, & sundry Jobs –

Tuesdy 15th a Considerable deal of rain this morning but Cleared up before noon – People at the Barns dighting & sacking Barley – at 1 P.M. yoked 4 ploughs on Bank tolerably warm & vegetation Increasing –

Wedy 16th Fine pleasant weather from Southward – Morning sowed Bolls Barley & Harrowed Partly the same – 2 Carts to Quarry – P.M. 4 Ploughs at Bank but did not Finish it out. Evening rain from the S.E. and appearance of more, dighting of rye Grass – Observe a Growth – several Polburns &c done & other Farmers not well begun – Fruit Flower & strawberries Blossomed.

Thursdy 17th a new Markett Bervie last night a Considerable fall of rain from the S.E. which was this day however fair – 4 ploughs at Bank A.M. & P.M. 3 D° sowed all the Bank with Barley and after dunging the two strips of Haugh North side of the river Ploughed & Sowed – which finishes Barley seed.

Fridy 18th Pretty strong drought from the Westward all day – finished out Harrowing of the Bank, and two Carts to Dava Quarry for stones to new Kitchen, P.M. Broke in the Bridge Shade for Turnips – Dighted up the remainder of 2 stacks Barley S Barn – and took the whole of the Oats remaining S Barn &c Jobs –

Saturdy 19th Strong Gale from the Westward and Drought. 2 Carts to Gurdon for Coals, and one to Mathers for Lime, but Missing it Brought stones for the Chimney Head from Laurieston dressed, Covered the Grass seeds Bank & Braking in Haugh for Turnips.

Mondy 21st Yesterday stormy weather from the Westward – this day frosty weather & some skiffs of rain – a Cart for Lime to Stonehaven Bad, Harrows Broke in the Haugh for Turnips, P.M. 2 Ploughs begun giving 2^d^ furr to lower End Bridge Shade for Turnips – Rolling Bank Barley &c. Horses to Grass yoked at 5 A.M.

Tuesdy 22d Moderate weather from the S.W. till P.M. Heavy showers of Hail & rain from Northward and Cold weather – 2 ploughs at Bridge Shade, Rolled out the Bank, Chaffing Oats N N Barn & Sundries –

Wedy 23d This day fair & pleasant weather – Carts driving some stones, Ploughs & Harrows at Bridge shade, and Roller at Haugh Grass &c. a Ridge of Blacklands being Missed in the sowing – Dunged & planted it with Potatoes by Plough –

Thursdy 24th White Frost in the morning – day variable but strong drought from Southward – 4 ploughs at Bridge Shade all day – Chaffing Oats N B.

Fridy 25th Cold Bleak weather from the S.E. and P.M. some showers of rain, 2 Carts drove 6 Cart load of stones from Dava, One Cart to Mathers for Lime, 3 Harrows at Bridge Shade, Chaffing Oats &c Jobs.

Saturdy 26th Growing weather – 3 ploughs at Bridge Shade A.M. – P.M. four Horse Plough at D° very stiff – Observed to Montrose to Westward of Laur:kirk every article of vegetation much behind with us &c

Mondy 28th Yesterday some showers of rain, this day for the most part Heavy rain from the East, but particularly from noon to near 5 P.M. – Sent to Mathers for a Pair new wheels Iron Axeltree and Brought two Cart load dressed stones from Laurieston for West chimney of House &c Jobs.

Tuesdy 29th last night a great deal of rain & stormy from the Eastward, this day at Intervals much the same, One plough did out 2^d^ fur Bridge Shade and two Carts driving stones from Dava – Sundries.

Wedy 30th fair weather and drought from the Southw^d^ – 4 Carts all day driving dung to Haugh for Turnips on a layer Old Dyke – but too late in doing –

Thursdy 31st Good weather & Moderate from Southward – 3 Carts driving dung all day to Haugh and One Cart leading 3 draught of stones from Dava for the

Chimney Kitchen a perfect Glutton – People at sundries – Hear Glammiss Cattle Markett went very dull little sale & much loss to the Sellers – Prices of Grain falling owing to the new Corn laws not being understood by the Merchants – Potatoes & standard Apples Flowring.

Diary for the Month of June 1792

Fridy 1st Some showers of rain & rather stormy from S.E., but Cleared up before noon – Carts driving dung all day & some stones from Dava – Strong Drought P.M. –

Saturdy 2d Fine Pleasant dry weather from the Westw^{d} – 2 Carts to Johnshaven for Coals to Smith, 2 D^{o} Driving Dung to Haugh and Cleaned out Pigdegeon & Hen Houses, at Barns &c Jobs. Sold all the Barley at 17/ to Mr David Scott Johnshaven &c Jobs –

Mondy 4th Growing weather & rather windy N.W. morning, day very warm & strong sunshine – 4 Carts to Mathers for 16B Shells for Haugh Turnips, and two Horses mostly to Aberdeen with James – put up the Cattle Flakes Thornylee & lodged them first night – begun Threshing last stack of Barley S Barn &c Jobs –

Tuesdy 5th Cold & stormy weather from the North – 3 Carts to Mathers for Lime but Missed it Brought 2 Bolls Coals – One plough at 3^{d} fur Bridge Shade, & P.M. driving dung to Haugh – Hired Men turning dunghill –

Wedy 6th very stormy weather & Cold from the North by W – 4 Ploughs at Bridge Shade & 2 Harrows all day – at Barns and sundries about the Farm, turning dunghill &c.

Thursdy 7th Cold & Stormy weather with some rain from N.E. – 4 Ploughs all day Bridge Shade, & Harrows A.M. Dighting & sacking Barley sold for Johnshaven – Ditchers Inclosing Bog Atry for Planting &c Jobs. Stricking out a passage from new Kitchen E^{t} Gavel House. Pease garden Blooming.

Fridy 8th Cold Stormy weather betwixt 7 & 8 a severe shower of rain & high wind from N.E.- 4 ploughs & Harrows at Bridge Shade – afternoon D^{o} – dighting Barley for Johnshaven, Finished turning Haugh Dunghill.

Saturdy 9th pretty Early this morning Came on very Heavy rain and severe Gale from the North by W – Sent 4 Carts to Johnshaven with 12B Barley & Brot back 16 Bolls of Lime Shells from Mathers, at Barns & Sundries –

Mondy 11th Yesterday very Heavy rain from S.W. – this day Closs & rainy most of the day – 3 Carts to Mathers for Lime for Haugh, and taking in last stack of Barley – Indeed last of the Crop S Barn, at fliting the Flakes and at sundries about repairs of Inside House –

Tuesdy 12th This day fair weather till 10 A.M. when almost Constant & P.M. Heavy rain – 2 Carts to Mathers for Lime & one four Horse Plough Haugh – one Cart leading stones for Paving new Kitchen, & at sundries on the Farm &c – at road Making –

Wedy 13th Mixed various weather with showers & Growing – two Carts to Mathers for Lime to Haugh, and One Plough at D° Cross ploughing very deep to 18 Inches – at Barns dighting Oats – Masons at New Kitchen and at new fitting up of Cellars & Milk House &c Jobs – Observe the Caterpillar amongst Gooseberries picked of the damaged leaves & Inside & Cut off all new Shoots –

Thursdy 14th rather frosty in the night day generally fair but some showers, 2 Carts to Mathers for Lime to Haugh, and One 4 Horse Plough Haugh, & two P.M. 2 Harrows at D° – a Power of People at the Cellars, West Chimney Head taking Down &c – planted Colly flowers for the larger Crop – Heard that Trinity Muir Markett went well in Sale of Good Cattle & good Yeld Cows &c.

Fridy 15th Th^r 50° rather Frosty dew – fresh Breeze & fair weather from the S.W. – One Cart to Laurieston for dressed stones to Chimney Head, and One Cart for Lime to Land – One Plough & some Harrows in the Haugh – the Ground now drying & warmer, but no great vegetation; 5 Ditchers Inclosing Bog Atry & Jobs.

Saturday 16th Pleasant dry weather variable chiefly Easterly – 2 Carts to Mathers for Lime to Haugh, and Six Harrows at Bridge shade for Turnips – and at N N Barn dighting Oats to Loft, and at Garden Planting Curleys, Weeding & dressing Borders &c – Oat Meal from 14/8 to 15/ – Bear 15/ to 15/6 – Beef 4^{ds}to 4$^{ds}\frac{1}{2}$.

Mondy 18th Yesterday a severe Gale of Wind from S.E. with Heavy rain & water much out – This day fair till past 9 A.M. when a severe storm of wind & excessive heavy rain from the N.E. which Deluged the Country Immensely, Self to Stonehaven on a Corn Jury Bill Compleatly Drenched. rivers very high.

Tuesdy 19th very stormy & strong Gale from North with fair weather – 2 Carts to Laurieston for stones to Mansion Chimney Head, to the Ornamentals; One Plough at Haugh and at sundries, Masons could not work out &c.

Wedy 20th Pleasant weather and Drought from Northward – P.M. Begun 4 Carts to Dung West side Bridge Shade for Turnips – Observed to Netherley Mr Silver the Crop Midling, and no Turnips sown all the way to Aberdeen – the Cutting Grass Better.

Thursdy 21st Fair weather from the N.E. – 4 Carts all day the Dunging Bridge Shade at rate 24 Cart load per Acre – Gardner & Wewmen Hand Houghing the strip of Planting West side Garden – Observed the Crop of Grain & Grass very fine on the Farm of Scotstown 4 Miles North Aberdeen.

Fridy 22d In the morning Came on rain before 10 A.M. which Continued two Hours, at Dunging nearly finished Bridge Shade & Wewmen P.M. spreading D° & begun drilling with two Horses a trip double Bouts very raw

Sasturdy 23d Thick Closs & Growing weather and after 8 A.M. a very Heavy shower stopped Drilling. Ploughs at Haugh field for Turnips and People Chaffing & Dighting Barley sold.

Mondy 25th Fair weather and some drought from the Eastw^d – 3 Carts to Johnshaven with Barley 10B & Brought Back six firlots of Scotch Coals – One Plough at Haugh for Turnips & Sundries.

Tuesdy 26th Fair weather from S.E. but little drought & foggy – One Plough drilling double Bout Bridge Shade, two yoking, others spreading dung, Plough & Harrows at Haugh and lead some stones from Dava and at sundries – Bought up the Tack of Arthurs & of this Estate but failed afterwards on differences –

Wedy 27th Strong drought & warm weather from the S.W. – One Plough at drilling Turnips Bridge Shade three yoking – One Plough finished fur Haugh for Turnips, and other people at spreading of Dung, opening and Covering of D° – and at Sundries –

Thursdy 28th pleasant dry weather from S.W. – One plough all day at drilling Bridge shade for Turnips – people spreading Covering on D° – lead two Cartful stones for steps for the new Kitchen to lower stair & P.M. Harrows Haugh – Observe towards Johnston that few Turnips Sown and those in very Indifferent Condition –

Fridy 29th Fair dry weather and drought from S.W. – One Plough all day drilling of Turnips Bridge Shade & people spreading of Dung & rolled most of what was sown in but tolerable order, Harrowing at the Haugh for Turnips not yet fine Enough, and Sundries –

Saturdy 30th Cold drought from the N.W. & some very slight showers, drilling Turnips all day – Harrowed out the Haugh for Turnips & people at sundries. The Cattle Markets owing chiefly to the want of Grass quite sunk and no demand – The Crops on all wet land looking very bad and not very well on the best – The Wheat Crop failed in the South of Scotland, and the Hay there & here very light Indeed – The Turnips must be a very Bad Crop in General prices of Grain are rising in the South & must soon do so here in all probability – Fruits very Bad –

Diary for the Month of July 1792

Mondy 2d Yesterday Cold Bleak dry weather from the Northw[d] this day drought from the Eastward & little Groath – One plough drilling Bridge shade turnips – & others steering & rolling D° – P.M. two Carts begun dunging of Haugh for Turnips from 24 to 30 Cartload, at Barns &c.

Tuesdy 3d Fair weather and drought from S.E. – 3 Carts driving dung to Haugh field – One plough drilling the Bridge shade all day – Men & Wewmen at spreading of Dung, filling Carts & sundries, all the Masons Dismissed but one at repairing old Kitchen.

Wedy 4th Fair & drought all day Carts driving dung at Haugh, and finished the drilling & rolling of the Bridge Shade a tedious work, at sundries &c Jobs.

Thursdy 5th a good deal of rain afternoon & Growing weather – being the Fast day people at Church – very little work –

Fridy 6th Moderate & Growing weather with some light showers – 3 Carts driving dung all day to Haugh for Turnips and one Cart to Mathers for Lime shells to Haugh, Gardner & Wewmen Hand Houghing Potatoes &c Jobs.

Saturdy 7th Th[r] 63° Moderate showers last night, this day warm from S.W.-

finished Dunging all of the Haugh Intended for Turnips about 9 Acres well dunged, on y[e] Hirsts particularly – Houghing at Potatoes Old Town Park and at Sundries on the farm – vegetation better –

Mondy 9th Yesterday fair & pleasant weather from S.W. – this day the same from S.E. – spreading lime on Haugh & Dung on D[o] – also Houghing Potatoes, People at Church. Observe the Crop to Bervie very Indifferent –

Tuesdy 10th Frosty dew – day fair & pleasant weather from N.E. – sent two Carts to Johnshaven 4 Bolls Barley & Brought home 8 Bolls Lime from Mathers – drilling & sowing Turnips, spreading lime, at Barns Miday & Sundries –

Wedy 11th Coldish Bleak weather from the North – One plough drilling at Turnips Haugh, Others spreading dung & lime, and One Cart with Barley to Johnshaven 3B – Houghing of Potatoes, and Gardner at D[o] the Nursery –

Thursdy 12th Cold stormy weather from North – 3 Carts to Mathers for Lime shells Haugh, sent for a Roller Johnston – People Houghing Potatoes out & some Early Turnips Cottar Yards, dressing at Nursery & variety sundries –

Fridy 13th Dark weather & very warm from Northward – sent 3B Barley to Johnshaven being whole of Crop, Brought Home with Mr Leslies Cart 8B lime shells – Drilling sowing, rolling Turnips, spreading Dung Evening about 5 P.M. Came on rain – dish of strawberries –

Saturdy 14th a Considerable deal of rain from the N.E. & Closs Growing weather – faired up about 2 P.M. – One Plough the whole day at drilling Turnips which finished all of about 17 Acres of Season, a Bad one to Begin with, and the latter sowing by far too late owing in part to the driving of lime for them, and obliged to sow the last of them without lime & to spread on them as soon after as possible –

Mondy 16th Early morning a fine shower of rain and fine growing weather from the S.E. – One Plough at Summer fallowing the Haugh West side of Haugh Turnips – two Carts to Mathers for Lime but like many others Dissapointed and obliged to leave the Carts till tomorrow at the Quarry- Spreading lime on the South drills with Turnips not being got ready before, some Stotts sold in the Neighbourhood about 4/6 per stone –

Tuesdy 17th last night Heavy rain for several Hours, this day rain Evening – 2 Ploughs summer fallow Haugh and 2 Carts Early for lime – and at ridding up about the House Doors &c.

Wedy 18th Rather a frosty dew in the night day fair and Growing weather – Ploughs at West End of Dickies Haugh for Summer Fallow, and Harrows at Braking in part of South End of Cottar Shade Bank formerly waste ground & Riskue to summer fallow, lime & dress about 2 Acres of it. Hear the sheep at Paldy fair very low priced and that Linnens 30 or 40 per cent above former prices. Vegetation & pasture much Increasing – Course Linnen at 1/ to 1/5 per y[d]

Thursdy 19th This day Moderate weather from S.W. – Ploughs at finishing 2[d] fur Fallow Haughs, and at foot Cottar Bank – Thistling Oats & pease – at Paldy fair a great show of Cattle & numerable purchasers, Cattle of sizes large above 36 stones &

large stotts very high 4/ to 4/6 – Smaller Cattle old or wrought, about 32 stones not above 3/ per stone – Small Cattle Cheap – Green Pease.

Fridy 20th Very pleasant dry weather, sent 4 Carts to Mathers for Lime – flited the Cattle Flakes, and at Sundries – Hear Horses at Paldy no Sale – flited ye Flakes

Saturdy 21st Cold stormy weather from N.W. – 3 Carts to Mathers for Lime to Haugh Turnips – P.M. giving a second fur to south side Cottar Bank fallow – Two Cattle strayed from ye Markett found at Chapelton.

Mondy 23d Yesterday Bleak dry weather from the N.W. – this day same from the Eastward, two Horses Drag Harrow at South End Cottar Shade Fallow very Tough – People at Houghing of Potatoes Old Town Park.

Tuesdy 24th This day fair & strong drought from S.E. – and sent 3 Carts lying Bodies to Finzean for Birch & fir wood sawed out there some Months ago, and two Harrows at Fallow Cottar Bank – & P.M. at Haugh D^{o} – Sundries. Observe the Bear in general very poor – Sowed all the Garden Spring Seeds.

Wedy 25th Foggy weather from the S.E.- and about One P.M. Came on a pretty smart shower with more or less during the remainder of the day – the 3 Carts with Birch Returned by a little after 9 A.M. – two Harrows on the Haugh fallow – P.M. flited the Cattle Flakes &c Jobs.

Thursdy 26th last night a Considerable deal of rain till 11 A.M. – fair till near 3 P.M. a very Heavy fall of rain all the reminder of the day – Two Carts driving Gravel to the Aberdeen road.

Fridy 27th This day fair till past 3 P.M. Came on more or less rain – 3 Carts to Finzean for sawn fir wood, and one Cart to Montrose for sundries for family use, at repairing the Well.

Saturdy 28th Very pleasant fair weather from the Southw^{d} – Carts returned from Finzean after 3 P.M. – Other people at Fliting the Flakes P.M. – the Growth Considerable & very fine pasture – Bees swarmed first time.

Mondy 30th Yesterday very warm weather & P.M. some very Heavy showers of rain – this day warm & fair, 2 Carts driving Gravel to the publick road on North side of this farm – and one Plough at Giving first fur to a piece of Fallow ground North side of Bogatry – at Smithy &c Jobs – &c Yesterday a swarm of Bees smothered by Missmanagement.

Tuesdy 31st Droughty weather from the S.W. and fair weather – 3 Carts to Finzean for remainder of fir wood, and three Harrows at summer fallow Haugh few Enough – Bought 10 Ewes at Garvock fair at 4/ each about 8 lb per leg. Observe the Bear Crop in the General very thin & foul, on the whole in many places promises ill – here very Good –

Diary for the Month of August 1792

Wedy 1st Th^{r} 71° perfect shade Exceeding warm weather from ye wind S.W. – Three Carts with wood returned from Finzean by 8 A.M. – One Plough Summer fallow Haugh, and P.M. Flited the Cattle Flakes Thornylee. Observed it as warm as I ever Observed in N Brittain.

Thursdy 2d Thick Misty weather & P.M. Broke out very warm – at Garvock fair more people & Cattle than usual, & the Engaging & Buying prices Brisker than usual. Bought 2 Cattle for Turnips about 3/ per stone –

Fridy 3d Very warm dry weather wind N.E. moderate – One Plough at fallow south End Cottar Bank, and Carts driving 16 Oats to Mill of Kair – Sundries –

Saturdy 4th Very warm weather from the S.E. – 2 Carts to Dava for Stones to the Well, and a plough finished first fur Fallow South End Cottar Bank – driving Earth to the Dung Court & forming road North side Well.

Mondy 6th Yesterday very warm weather from the Southw[d], Th[r] 62° this day same, 2 Carts for stones to Dava for Well, One Cart to repair to the Buss Coast side, and one driving Earth to Dung Courts, at fliting Flakes – 4 Men Cutting Hay.

Tuesdy 7th Th[r] 62° shade noon Dark Cloudy weather from S.W. – One Plough Fallow in Haugh, & One Plough taking away first furr from first sown Turnips Bridge Shade, Men Turned over the Hay Cut Yesterday & P.M. begun first dressing of above Turnips – Pasture very good &c.

Wedy 8th Th[r] 64° noon Very warm & pleasant dry weather from Southw[d] – One plough at Summer fallow West side Dickies Haugh, One plough three yoking at Turnips, Men & Wewmen at dressing Turnips, Wheat promiseth Tolerably –

Thursdy 9th Heavy fogg in the night, and day fair but dark weather from S.W. – One Plough summer fallow Haugh, and one amongst the Bridge shade Turnips, Seven Hands at first dressing Turnips &c Sundries.

Fridy 10th dry weather and little drought, a small shower at one P.M. – Coling Hay till above time – at Thining Turnips & Summer fallow – Sundries.

Saturdy 11th Th[r] 62° Closs foggy weather till one P.M. Broke out sunshine and drought from the Westward – after dinner all Hands at Hand Coling of Hay, a Cart to Johnshaven for Coals Scotch, and at sundries on the Farm –

Mondy 13th Th[r] at Noon shade 68° Yesterday fair & very warm weather Th[r] at noon 68, this day same, 14 people Hired & own at the Bridge Shade Turnips till Brakefast after at Gathering & Coling Hay Robs Hillock, but stopped by very Heavy rain a little after one o'Clock with Thunder &c, it Continued more or less all the rest of the day – P.M. at Turnips but stopped before 5 – at sundries &c –

Tuesdy 14th Bleak weather from the Eastward but fair and little drought, all people & hired ones finished out the first dressing of Turnips Bridge Shade, and after at first sown Haugh Turnips Horse & Hand Houghing. Observe some of the Oats, Barley & Lint Lying – also that the Pasture is begun to fail which it Commonly does about this Period, Sundries.

Wedy 15th Bleak dark weather from the Eastward, own & Hired people at dressing of Haugh Turnips, and P.M. turned up the Hay on Robs Hillock, which by the rain & warmth would soon have been Injured.

Thursdy 16th Bleak weather & rainy particularly afternoon very Heavy at times, one Plough whole day at the Turnips Haugh taking away first fur, & Men thinning D^{o} – Spread some lime on the Turnips not ready before, at the Mill of Kair with a Chalder of Oats for Family Meal. Some of the Rankest Oats & Barley lodged, the Earliest of the latter filling & turning, as is the pease, but the Oats Scarcely full shott – Hay in General light Crop & at present in ill Condition selling 6^{ds} & 1/ per stone.

Fridy 17th Almost Constant rain & very bad weather from the N.E. – One Plough at Haugh Turnips and people at D^{o} thinning – Brought Back from 16B Oats Kair Mill 15B 9P Meal –

Saturdy 18th Fair weather from the North but very little drought, Flited the Cattle Flakes morning – and at Horse and Hand Houghing of Turnips nearly done for the First dressing – after Brakefast all people at Turning the Hay Robs Hillock & after Dinner at Turnips again – Indifferent weather.

Mondy 20th Yesterday Coldish dry weather from the Northward White frost in the night, but day fair and some drought from N.W. – finished first dressing of Turnips Haugh, and after 16 Hands at Coling of Hay all day of which did about 6 Acres in tolerable order, Some of the Oats & Barley lying.

Tuesdy 21st a shower Early in the morning – day fair and not unpleasant except little drought – two Carts spread some lime on the two ridges west side of Thornylee at 32B per Acre, Second dressing Turnips on Bridge Shade – Sheep Markett L. fair very cheap –

Wedy 22d Warm weather and after Brakefast turned out drought – about thirteen people at the Hay Turning & after Coling – which finished by 6 P.M.in good order – Appearance of rain.

Thursdy 23d fair weather most of the day with some showers – at Laurence fair a Great many Cattle but not much Sale Except Amongst the best Cattle – Observe more lying Corns to the Markett & from that to Bogton Lord Inveruries Shooting Quarters –

Fridy 24th Bleak dry weather from the N.W. till 4 P.M.- rain moderate, 4 Carts to the Hill for Peats, and others at 2^{d} dressing of Turnips Bridge Shade.

Saturdy 25th last night and all this day almost Constant rain but not very Heavy from S.E. – sent two Carts to Corsbitt Hill for Peats, and one Cart to Laurieston for 30 feet rigging stone for new Kitchen, the Oats much layed and the rankest of Barley, very Bad weather.

Mondy 27th Yesterday Moderate weather – this morning Came on a Heavy rain from the Northw^{d} which lasted till past 10 A.M. – people flited the Flakes before dinner and P.M. at Turnips and at sundries – the High wind raised ye Corns.

Tuesdy 28th Frosty dew in the night & morning, Pulled the Lint tolerable Cropt, and after at sundries – at Lord Inveruries Shooting Quarters a kind of Fete Champitre – plenty of Drink – The Crop in that part not behind ours

Wedy 29th Fine dry weather from the Northward, yoked three Carts after

Brakefast to lead in ye Hay in pretty good order, 16 people, left the Bottoms of the Coles very little dammage – frosty at night –

Thursdy 30th Coldish Bleak but dry weather from the Eastw^{d} – 3 Carts Early at leading in Hay, about noon only two, got in all by 5 P.M.- The Sow is 60 feet long & 14 Breadth – rather low about 3 Thousand stones on 13½ Acres – about 18 people Thatched the Sow Immediately with Sprots &c. appearance of rain.

Fridy 31st In the morning Early Came on Heavy rains from the Eastward, more or less all day – people at making ropes, Barns, and P.M. Flited the Cattle Flakes – Oats much layed & filling. There is no Harvest yet or to be soon here –

Diary for the Month of September 1792

Saturdy 1st a Shower Early morning, day fair & warm, all Hands & Hired at finishing thatching Hay Sowe – Others at 2^{d} dressing of Turnips – & Sundries –

Mondy 3d Yesterday mostly Constant rain, and this day the same from the Southw^{d} – Early flited the Cattle Flakes Thornylee, after sent 4 Carts to Flemington for slates, and six others Tennants & Borrowed, at 2^{d} dressing Bridge Shade Turnips & P.M. at Barns – Painter at House

Tuesdy 4th Frosty dew morning which stinted the Pease, all spare Hands amongst the Bridge shade Turnips – the Carts returned with the slates before 6 P.M. – 8 long score of them on Ten Carts – roads Excessive Bad &c.

Wedy 5th Strong drought from the North West till 3 P.M. some rain from the Westward – 2^{d} dressing Bridge Shade Turnips nearly done, at Cutting sprots & rashes & at dighting Oats sold to A Young Mill of Conveth at 13/6 per Boll. Observe the Pease fading fast &c.

Thursdy 6th fair & dry weather from the Westw^{d} – at dighting Oats sold, sent 20 Bolls of them to Mill of Conveth at 13/6- begun laying to a fur West side Bridge shade Turnips – Others 2^{d} dressing E^{t} side of D^{o} &c.

Fridy 7th Very strong drought from the Westw^{d} – all people at Fliting the Cattle Flakes, and at Hand and Horse Houghing of Turnips, well through, dighting Oats sold, and at variety of sundries – &c Jobs –

Saturdy 8th Strong drought & Blowing weather from the S.W. – 2 Ploughs nearly finished laying to a fur to the Bridge shade Turnips, Sent 17½ Bolls Oats to Mill of Conveth before Brakefast, after took out of water 11 nights all the lint & spread it P.M. &c.

Mondy 10th Yesterday Bleak weather & rain P.M. – this day variable with some showers from N.W. – about 11 A.M. begun Harvest by Cutting up Pease Blacklands, Spoilt as in General by frost, but not Materialy – South of Bervie Observed some Bear Cut down, Bad Turnips –

Tuesdy 11th Cold stormy dry weather from the N.W. – finished Cutting all Pease by Dinner time 2 o'Clock – and after at some of the Ripest Barley South end of Bank – most people begun hereabout.

Wedy 12th Early Came on rain & stormy weather from the South East which Continued till one P.M. – before Brakefast flited the Cattle Flakes & Cut some rushes, after at dighting & delivered at Conveth Mill 18B Oats –

Thursdy 13th Cold stormy weather from the Northw^{d} – People at 2^{d} dressing of Turnips Haugh the new Ground very promising – most of our Hands at Idle Marketing at Bervie – Sent 3 Bolls Oats to Andrew Young – Observed towards Johnston a good deal Bear Cut.

Fridy 14th White frost morning day stormy but dry from N.W. – all people at Second dressing of Haugh Turnips Some of them very Growthy – at Sundries &c –

Saturday 15th last night & this morning a strong white frost which however hath not spoilt the Potatoes shaws – People at 2^{d} dressing of Haugh Turnips, and others for Thinning of Fir Wood Pittengardner about 18 year planted at 16^{ds} per two Horse Cart load, about 10 feet long & thickness of Mans arm.

Mondy 17th Yesterday morning and this Both Black & White frost which hath Destroyed the Potatoes stems – last night much rain, this day fair but not much drought – People at Turnips & flitting of Flakes till noon, when Turned the Pease, and afternoon Cutting up Barley upper End of the Kavels, tolerable Crop – at sundries – Bad weather.

Tuesdy 18th This morning Came on severe rain & High wind from the South. People in Barns and after at Turnips Haugh – Monboddo for Coals &c

Wedy 19th Variable weather from the Southw^{d} – People at Turnips till Brakefast, after at shearing Barley Kavels till 1 P.M. when a shower, when Cut up Rushes &c till dinner time dispersed by Heavy rain from the Southward – Home people at Barns &c Jobbs.

Thursdy 20th Frosty dew in the night; morning at 2^{d} dressing of Turnips, after at Cutting of Rushes &c for Thatch till 11 A.M. then at Barley Cutting till 3 P.M. stopped by rain & Hail, after at the Rushes – 3 Ploughs at laying to fur to Hg Turnips –

Fridy 21st very Heavy White frost which prevented the shearing till Noon, Home People at Turnips dressing & three Ploughs at D^{o} Haugh, the Grass sown in Spring very fine amongst the Corns –

Saturday 22d White frost in the night & morning – day Drought from the N.W. & Cold dry weather, at Turnips Hand & Horse Houghing till Brakefast, after the Shearers at finishing Kavels Barley – & after took up part Lint, & then S. side Bank Barley – A Cart to Montrose for Lath to New Kitchen.

Mondy 24th Yesterday afternoon Came on rain but the preceeding night when late a strong Storm of rain & Gale from the N.E. – this day rainy dark Bad weather particularly in the Evening – the People Flited the Flakes morning, after did out all the Turnips Haugh, and at sundries &c.

Tuesdy 25th last night much rain & stormy weather from the Northward – Shearers at Bank Barley and at Cutting up rushes &c in Intervals – Turned the Pease P.M. – and at sundries,

Wedy 26th White frost morning, day fair & dry weather – all people at Cutting rushes till Brakefast, after at Bank Barley, & the Strip North side of Bleaching Green 7th^{s} $\frac{1}{2}$ stook – the Crop after Pease pretty Heavy of Barley

Thursdy 27th a small shower Early – own people Flited the Cattle Flakes before Brakefast, after Bank Barley till 11 A.M. stopped by light rain from Southw^{d}, after at Cutting Rushes &c till Dinner, dispersed the Shearers, Own people leading rushes & Sundries –

Fridy 28th Fine weather in the morning, but about 9 A.M. Came on rain & dirty weather from the Southward all the day – people Cutting Rushes &c till Dinner time when Dispersed Shearers till the Beginning of the Week, at sundries &c –

Saturdy 29th last night dark dirty weather and this day the same. People Employed at sundry Jobs – but no Material work done; at Smithy &c and at various mixed Jobs –

Sundy 30th This day dark & Disagreeable weather but fair till 4 P.M. when rain light, but towards late much rain & stormy weather from the N.E. which Brought down rivers &c – From all the Accounts from the different parts of the Country South & North, the Crop is in a very Indifferent Condition, much dammaged by the Constant rains, lodging, frosts, and every other Mischief – it is said that Meal risen 3^{ds} per peck in the North in ten days time. Bear now 15/ to 16/ – Cattle good keep up well in the price – Pasture quite done & Cattle wane the Ash & Elm trees Shading ye leaves –

Diary for the Month of October 1792

Mondy 1st Early this morning very Heavy rain & stormy weather from the N.E. which brought down the Waters much, whether worst for the Corns or the Cattle cannot be said – flited the Cattle Flakes A.M. – after Making Fold at Robs Hillock.

Tuesdy 2d This day Bleak but dry weather from the N.E. – Shearing Oats Upper Burns till 11 A.M. then at Barley Bank all day – Sundries &c.

Wedy 3d fair weather from the Eastward, Sheared out the Barley Out Bank Threaves 320 – after the Patch benorth the House D^{o} Th^{r} 12 and after at Loup Yards Oats – 10 Th^{r} – then at Wheat & Men Turning of Pease for third time, Midling Order.

Thursdy 4th Cold dry weather from the S.E. all day – By daylight yoked 4 Carts to lead Pease from Black lands, lead Home 6 Stacks by 7 P.M. – not Hollow – & others Shearing Wheat very good order Grains –

Fridy 5th Strong drought from the Eastward – Early 4 Carts lead Home the remainder of Pease 7 stacks, and then lead all Kavels Barley and South side Bank into 4 stacks moderate size in good order, loosed at 7 P.M.- 8 Sheaves finished Wheat Kavels Threaves 55 –

Saturdy 6th a smart shower Early which stopped shearing till Brakefast time, Thatched all Pease & two Barley stacks, and at 11 A.M. Begun & lead in the whole of the Remaining Barley in good order 8 Stacks – Wewmen Cutting up Oats long Haugh very ripe –

Mondy 8th Yesterday Bleak weather from the Northward & skiffs rain P.M. – this day same – People at Barns & thatched all the Barley stacks & Eddering them, 9 Shearers at Oats in Gayte & at Sundries –

Tuesdy 9th a Considerable rain last night & this day at Intervals, people shearing & Gayting of Oats Cottar Bank & Craig shade, much in need as in the way of spoiling by lodgeing – and at Sundries.

Wedy 10th Constant rain from the Eastward – after Brakefast most part Shearers Cutting up of the worst lying Oats Cottar land, some of them spoilt, but stopped by Heavy rain afternoon; brought over all the Cattle Flakes & placed them on a dry part of Goulds Hillock – at Barns & Jobbs

Thursdy 11th last night an Immense Quantity of rain from the Eastw[d] which wet the Grounds more than hitherto & stooks also – at Drumlithy fair great Many Cattle double usual but no Sale: people at Cutting up Oats Cottar Bank in Gayte &c Jobs.

Fridy 12th rain a little last night, day fair & Cold – after Brakefast People Cutting up Oats Floors, first in Gayte & after dry stooking – at Sundries – in the Evening $\frac{1}{2}$ past 5 P.M. Came on severe rain & wind S.E.

Saturdy 13th The last nights storm Continued till 10 P.M. when frost, It hath Damaged the uncut ley Oats by Heading & Braking of them Considerably – Set up the Wheat, and Dismissed the Shearers in the morning – after at Making Cattle fold &c Jobs –

Mondy 15th Yesterday afternoon a Constant rain which lasted most of the night, this day mostly fair but some showers from S.W. – driving dung to Thornylee morning, after Begun Shearing Oats & putting up all the down Gayte, finished Floors Oats, and after at Cottar Bank Oats, flited the Cattle Flakes Goulds Hillock.

Tuesdy 16th the day fair & Moderate weather from the S.W. – all Hands at shearing of Corn Cottar Bank & in Gayte and others at sundries – Observe every person Cutting up wett & Gayting &c &c –

Wedy 17th very fine droughty day from the S.W. & Coldish, all Hands at Shearing out the Over and Nether Burns by evening – and at Sundries.

Thursdy 18th Frosty in the night – about $\frac{1}{2}$ past 11 A.M. Came on a Shower & Squall from the North West of short duration – when shearing again red Hall Park – a good deal layed – took in some Oats to Thresh for Horse Suppers Grass being finished &c Jobs –

Fridy 19th pretty strong frost in the Night, high wind about 10 P.M. – Early all Hands shearing red Hall park Oats which finished whole Crop by 5 P.M. – then set up Gayte &c Jobs – give the Horses straw nights.

Saturdy 20th about 6 this morning Came on Heavy rain & high wind from the N.E. which deluged the fields & Brought down the Waters very high by 5 P.M. faired here about 2 P.M. – Threshing Oats for Fodder to Horses which Kept in all day – Dismissed all Shearers.

Mondy 22d Yesterday Closs foggy weather and about 6 P.M. Came on rain heavy from the N.E. which Continued most of the night, this day fair mostly till Evening Heavy rain Constantly – drove some dung to Thornylee morning, Sent to Quarry for stones and at Making up a fold for Cattle below Robs Hillock.

Tuesdy 23d Cold stormy weather from the North and this day rainy – 3 Carts to Mathers for 10 Bolls of lime shells for land & Slating New Kitchen and a Cart to Dava for Stones – Cattle much pinched –

Wedy 24th Cold weather & droughty from the Northward – after Brakefast own people & some Others Set up & turned the Gayte Oats, and divided the Wheat stooks, & P.M. lead in all the Wheat in tolerable Condition into One large stack – stone stathill &c –

Thursdy 25th fair & dry weather from the Northw[d] but not much drought Early all Hands at picking & Binding of best won Oats Floors & Cottar Bank – nearly finished 4 pretty large stacks by 7 P.M. when stopped by Darkness – and at Sundries &c –

Fridy 26th Bleak dry weather from the Northw[d] – Early 4 Carts at leading Oats Cottar Bank & red Hall park into 3 Stacks – all others Binding Gayte &c Jobs, Thatched the Wheat, several Neighbours done Cutting.

Saturdy 27th Bleak weather but chiefly dry from the Northw[d], Early yoked the Carts and leading Oats Chiefly Gayte & Wyled Shocks – finished & Enlarged some of the Former Stacks, and Built two new ones in I believe as good Condition as possible but was not able to Clear any one field – Evening rain.

Mondy 29th Yesterday about 1 P.M. Came on rain from the Eastward, which after two Increased severely till $\frac{1}{2}$ past 4 – deluged the fields & Brought down the rivulets much – this day fair but no drought from the South by E. – people at Barns Threshing for Thatch, rope Making, Eddering – Fold Making.

Tuesdy 30th Blowing stormy weather with slight rain from the S.E. till 3 P.M. when Increased to a storm almost & Heavy rain, One plough Begun for next Crop Turnips Blacklands – ribbing, and three Carts twice a day driving dung to Thornylee ley and at Idle Sundries – Bad weather Indeed   Slating N. Kitchen.

Wedy 31st last night a severe storm of wind & rain from the S.E. which went further into the Corns than hitherto, this shearers put up all Gayte, and shifted the Wet side of the stooks to the drought a Considerable deal of Growing & Damage in all made; – There is no sale of Grain, Cattle &c Manufactures doing well, Harvest very late in Many places in Garvock &c &c. –

Diary for the Month of November 1792

Thursdy 1st a strong Gale from the S.W. Drought, this morning again set up & turned the sheaves to the weather and after Brakefast begun the leading of the Burns into Hollow stacks, Wyling & Setting up all Oats on the farm Out, but stopped by Heavy rain by 4 P.M. – Brockley work Indeed – a Cart for Coals to Stonehaven.

Fridy 2d Gave ye feast about $\frac{1}{2}$ past 7 P.M. the wind shifted to the N.W. with rain & High wind – this day fair & some drought from S.W. – people at Barns, setting up Corns, and finished out a stack of Oats – Thatched two of D° –

Saturdy 3d Strong Gale in the Night & morning – day fair after Brakefast yoked two Carts & lead in whole of the Crop the last in very Bad order – the stance for the Corn yard is filled & One stack Over, three Carts to Stonehaven for English Coals – 18 Barrel at 20^{ds} per B

Mondy 5th Yesterday morning a strong Breeze S.W^{y} this day same; Sent three Carts for Coals to Stonehaven and One plough at Blacklands ribbing up & down for Turnips – all other own people, and others Digging up Potatoes – took in lint very light Crop, but good Quality – two Men threshing Oats by hire.

Tuesdy 6th last night & this morning very fine drought & pleasant weather from S.W. – 4 ploughs at the ribbing of Blacklands – people at the taking up Potatoes – 2 Men Threshing &c Jobs –

Wedy 7th This morning pretty Brisk wind S,W. – this day throughout good weather – 4 ploughs at Black lands and others digging of Potatoes – at Barns &c – Observed much of the Corns Out about Fordun –

Thursdy 8th Fine dry weather from the S.W. – 2 ploughs finished ribbing Blacklands – One four horse at red Hall park D° Two of the Oat Stacks appearing to Heat, turned them over this day not very much Dammage in them as yet; – little or no sale at L kirk Market –

Fridy 9th Strong Blowing weather particularly morning Early – 3 Carts driving Dung to Thornylee and one Plough at the Burns ribbing &c – at Barns &c. Thatching the stacks turned – and at sundries –

Saturdy 10th very strong Gale from S.W. early morning abating about sunrise – fair all day – 3 Carts to Montrose for Scotch Coals & one plough at Burns ribbing at Barns – Spreading Dung Thornylee on Ley very thick – the Crop not in yet in General.

Mondy 12th Yesterday fair weather but Cloudy, this day fair and drought from S.S.E. – 3 Carts driving of Dung to Thornylee, and one Plough at N.Burns ribbing – Dighting Oats for the Mill & Sundries – Bled all the Calves, & some others to prevent that fatal Disease Commonly Called the Quarter Ill.

Tuesdy 13th Dark Gloomy weather from S.E. and about 1 P.M. rain for some Hours, 3 Carts driving Dung to Thornylee & One Plough at long Burns, afternoon sent two Carts to Mill of Pitrany with 11B:3 Oats for Meal to Family at Sundries &c –

Wedy 14th this morning Came on rain pretty Heavy more or less all day and Grounds Excessive wet – One Plough finished N Burns & after at Red Hall park – 3 Carts driving Dung to Thornylee & others spreading of it – Evening Housed for Stall Feeding 19 Cattle & Cows most of them in pretty good order – Also most of Young Beasts–

Thursdy 15th **Frosty in the night & Early morning a Guller of wind from the**

Westw^{d} – about 10 A.M. a strong flirt of Snow from N.W. of not long Continuance, after fair & drought & Cold from S.W. – three Carts at driving Dung to ley, and one plough at red Hall park Ribbing – Fields & Roads very deep – no settled prices of Grain & other Meal under 1/ per peck – the Harvest not near finished on ye Coast – Housed all other Cattle.

Fridy 16th last night a severe Gale of wind from the Westward which unroofed part of our House &c damages this day Cold & stormy from N.W. – three Carts at Dung driving & one plough as Yesterday – Barns.

Saturdy 17th very strong Gale last night from N.W. & frost – 3 Carts driving dung all day and one plough at Red Hall park ribbing – at taking in Hay for the feeding Cows – P.M. sent to Mill of Pitranie for Oat Meal – 7B $6\frac{1}{2}$ pecks – Under 11 pecks of Meal but it was Gatherings of Yard. last of Season said Mill

Mondy 19th Yesterday dark weather fresh, and this morning a strong Gale from N.W. – day showers of rain, 3 Carts driving dung to Thornylee, and One Plough, red Hall park – at Sundries P.M. –

Tuesdy 20th Hard frost morning & this day from S.W. – driving dung to Thornylee & spreading Lime on Haugh – a plough red Hall Park – at Houseing Wood &c –

Wedy 21st Fresh weather – and fair till past noon Came on Heavy & Closs rain from S.W. round by East to N.E. then strong Gale which deluged the Country – 2 Carts spread all the lime Haugh & partly ploughed it down, Others driving dung, lead Home Cattle flakes &c.

Thursdy 22d a Considerable flood in all the rivers last night which subsided this day – 3 Carts driving Dung & leading Turnips and one plough at Covering all the lime spread – at Barns & Sundries.

Fridy 23d Showers through the day Northerly – three Carts driving dung – & one two Horse plough, at Barns dighting Oats & Barley for Malt &c.

Saturday 24th Strong White & Black frost in the night, day Closs foggy weather – 3 Carts driving dung to Thornylee and one Plough ribbing red Hall Park, at Barns and took in a stack of Oats S.Barn P.M. – this is the second time it hath appeared to Heat –

Mondy 26th Yesterday dark Gloomy weather from N.W. – this day same from North & skiffs of rain – One plough ribbing & 3 Carts driving Dung to Thornylee, at Barns and Sundries – Garden &c Jobs –

Tuesdy 27th Thick dirty weather from the North, 3 Carts still at the driving Dung to Thornylee nearly done, 15 Acres and one Plough about Finished ribbing red Hall Park, at Barns – Gathering Dung &c Jobs.

Wedy 28th Slight showers of rain and Land very wett – 3 Carts at dunging of Thornylee, & one plough at finishing all ribbing here for Turnips & Barley – at Barns &c.

Thursdy 29th Dirty & Dark weather from S.W. with slight rain, 2 Ploughs Begun ploughing of East side of Thornylee Ley very rich Indeed – and one Cart leading of Earth scouring of Ditch above Bridge of Leppie for Dunghill Layer to Turnips Intended Black lands – Sundries.

Fridy 30th Cold Bleak weather but fair from S.E. – 2 Ploughs at Thornylee – and one Cart at Leading of Turnips & driving Earth Midding layer – at Auchenblae Crofts – a Good Many Cattle but no fatt ones & very little sale –

Diary for the Month of December 1792

Saturdy 1st Cold Bleak dry weather from the S.E. – One Plough at Thornylee, One Cart Leading Turnips, and one Cart for Lime, at Barns &c Jobs –

Mondy 3d Yesterday fine dry weather from S.W. & sun shine P.M. this morning rain & Stormy from Southward – Sent three Ploughs to Brake Ground on my late Purchased Estate of the Waterlairs, by ribbing of a Pease Park Intended for Bear – at Barns &c Sundries.

Tuesdy 4th last night & this morning High wind Southy – about noon Came on Heavy rain & Stormy from D^{o} which Stopped the Ploughs after 1 P.M.- Sold & delivered the Fatt Swine for Aberdeen @ £3-12 S^{tr} – at Barns & other Jobs – fair Evening –

Wedy 5th Yesterday & this morning very strong Gales from the N.W. – this Heavy showers at Intervals from said Quarter – 3 ploughs ribbing Waterlair for Bear, and One Cart leading of Turnips – Barns &c.

Thursdy 6th This day about $\frac{1}{2}$ past 10 A.M. Came on heavy rain & very stormy weather from the S.E. which stopped the Ploughs before noon – at Barns & Sundries – faired by two P.M. – Waters much out –

Fridy 7th last night & this morning a very Heavy Gale of wind from the N.W. with Heavy Gusts of Snow & Hard Frost, the Snow lay all day – People at the Barns – dighting sacking & sending to Mill 13$\frac{1}{2}$ Bolls of Oats – took in a stack of Barley S Barn – lead Turnips &c Jobs.

Saturdy 8th day pretty Hard frost & Snow, People at dighting of Oats for the Mill & sent 4$\frac{1}{2}$ B to it – leading of Turnips, and in the way to Stonehaven Observed some Ploughs going – Corn Yards very little Bulked –

Mondy 10th Yesterday Cold weather & some showers of rain N.W. – this strong wind from the Westw^{d} & after 1 P.M. Heavy Gusts of Snow which did not ly – 3 Ploughs nearly finished water furring of Waterlair Parks – leading Turnips – & sending Oats to Mill.

Tuesdy 11th Strong Gale from the N.W. and after 1 P.M. Heavy Gusts of Snow which lay an Inch or two deep only – Two Ploughs at Thornylee Ley, leading Turnips & sending two cart load Oats to Mill of Pittrenie &c Jobs.

Wedy 12th Frost & Snow but Moderate, 2 ploughs at Thornylee Ley – One Cart leading Turnips &c People at sundries – Spading of Head ridges &c

Thursdy 13th This morning pretty Constant fall of Snow all day through. Evening about 5 Inches depth – people Holing & leading of Turnips – & P.M. two Carts driving Earth to Black lands for Turnips –

Fridy 14th Early this morning a very strong Fresh from the Westw[d] & most of the snow gone by Brakefast time 3 Ploughs at Thornylee Ley & people at the Mill of Pitrany 27 Bolls of Oats – 18B-9 pecks of Meal somewhat above three firlots per Boll, and the best yet Milled in this Corner of y[e] Country –

Saturdy 15th Sleet & some light snow during the night day chiefly fair & dark Gloomy weather from the N.W. – 2 Carts for Meal to Pittranie after leading Turnips – & One four Horse Plough at Thornylee very ill to do – at Spreading Dung on Thornylee – Flited the Sheep Flakes & Jobs &c.

Mondy 17th Yesterday fresh weather &c from the S.W. – this day High Wind in the Morning from same Quarter. Two Ploughs at Thornylee and two Carts at leading of Turnips and some Turf or Feal for Covering the Stone Dyke West Side of Garden Planted Beet – and at Barns Chaffing of Barley Sold to Montrose at 18/ per Boll – Planting Quick around Bog Atry for Hard wood.

Tuesdy 18th Blowing but dry weather from the S.W. & fresh, Two ploughs at Thornylee and One Cart to Elsineur for a pair of New Wheels, at Barns dighting Barley – &c Work as Yesterday –

Wedy 19th Fresh & very pleasant day from the S.W. – 2 Ploughs at Thornylee, and One Cart leading of Turnips all day at Barns dighting Barley.

Thursdy 20th Considerable rain in the morning Early & Grounds very wett – 2 Ploughs at Thornylee Ley – and One Cart leading of Turnips, at Barns dighting of Barley & sacking of it for Montrose & Sundries.

Fridy 21st very Blowing weather from the N.W. & some snow on the Hills, but Grounds dryer by far this day – 2 Carts leading of Turnips, and 2 ploughs at Thornylee ley – at Barns & Sundries &c

Saturdy 22d Snow in the Braes & Hills, but rain here, though in little degree, 4 Carts to Montrose with 16B of Barley – Other people at Sundries.

Mondy 24th Yesterday a very Severe Gale of wind & storm from the N.W. & North which dammaged the Tiled & other Houses very Considerably – this day Moderate but hard frost from S.W. – 4 Carts driving dung from Cottar Town to the Blacklands Intended for Turnips, at Barns &c Jobs Snow &c Evening –

Tuesdy 25th Some fall of snow in the night, day hard frost Carts driving of Dung to Blacklands from the Cottar Town, Digging & leading of Turnips Haugh, at Barns and Sundries – Frost pretty deep but the River not frozen over owing to High Winds –

Wedy 26th Some Sleet here & Snow in the Hills with rather fresh – Carts driving of Dung to Blacklands for Turnips – & leading & digging of Turnips – at Barns – Ditching Bogg Atry

Thursdy 27th Frost in the night – on the Hills & Braes very deep Snow – people digging of Turnips & some Carts leading of D° – at Barns dighting Barley – and two Men ditching to Drain Bog Atry for Planting.

Fridy 28th Yesterday in the Evening a little Snow, and this frost but not very strong – people at Digging of Turnips & leading of D° – at Barns dighting & sacking of Barley for Montrose, at Ditching of Bog Atry for Planting &c Jobs.

Saturdy 29th This morning thick Daggy weather from the S.W. but towards noon rain & strong fresh – 4 Carts to Montrose with Barley – Others digging & Holing of Turnips – at Barns &c Jobbs – at St Cyrus Mr Grahames Burial   Observed the Corn Yards Scanty & not much laboured &c.

Mondy 31st Yesterday Frost pretty hard but Moderate – this day very Hard frost but fair & Clear from the S.W. – people at Barns Dighting Barley & Clearing of N New Barn Digging & leading of Turnips Haugh, and P.M. at taking Barley into the N New Barn – the prices of Grain at present Bear 17/ and Barley 18 to 19/ S^tr^Oat Meal about 16/ per Boll – Spinning High 15 to 16^ds^per spindle, fatt Cattle from 4/6 to 5 per stone – all kinds of Manufactures very thriving & high priced.

Diary for the Month of January 1793

Tuesdy 1st last night a Strong Gale of wind from the Southward with Hail & rain & frost this day – 2 Carts driving Earth to Midding layer Blacklands for Turnips – One Cart leading Turnips &c.

Wedy 2d Frosty weather from the N.W. Carts driving of Dung to Blacklands – In Some Turnips – people at Barns dighting Barley S Barn – at Sundries &c 3 Men draining at the Bog Atry for Planting.

Thursdy 3d Yesterday & this strong Frost but pleasant weather, 4 Carts driving Dung to Blacklands, Earth to Midding layer, and some Turnips – and at Barns dighting of Barley for Montrose &c

Fridy 4th Frost in the night & Skimering of Snow in the morning – day Moderate weather – 2 Carts at leading of Turnips and others digging Do – at Barns dighting & sacking of Barley sold to Montrose.

Saturdy 5th Early this morning Came on a furious Gale & fall of Snow from the S.E. or S.S.E. which lasted till about 1 P.M. – being Old Christmas no work without doors – Snow four or five Inches Depth & rather soft.

Mondy 7th Yesterday frost & pleasant weather – In this morning early a Severe Bad weather with Wind, Snow & Sleet from the S.E. – took in a Stack of Oats S Barns and sacking Barley for Montrose, & Sundries &c at Johnston Thaw

Tuesdy 8th last night a very Strong Fresh in the low Country and Braes but little in the Hills – the rivers much out – 4 Carts driving dung to Blacklands and in return Turnips from the Bridge Shade, Barns &c Jobs.

Wedy 9th Hard frost in the morning & day – Carts driving Dung & in Turnips in return, Horses at Smithy Roughing to go to Montrose with Barley &c Jobs

Thursdy 10th about 3 A.M. Came on rain & Stormy weather from the S.E. which lasted most of the day – after Brakefast took in another 2 Stack of Oats South Barns and at dighting Barley N Barn at sundries waters much out.

Fridy 11th Stormy weather but day fair weather this day – Carts driving dung all day & Bringing in Turnips from Bridge Shade in Return – at Sundries – Gardner & 2 Others draining of Bogg Atry.

Saturdy 12th Tolerable weather – Sent 4 Carts to Montrose with 16 Bolls Barley but no Coals in return the roads abominable by wide open – at Pitting &c for Planting Bogatry with Oaks &c –

Mondy 14th Yesterday very rainy & stormy weather from N.E. but fresh, this day mostly fair, but Bleak & Dissagreeable – two Ploughs at Thornylee and One Cart driving dung to Blacklands &c – Bringing Home Turnips in return from Bridge Shade. at Barns Sacking Barley Sold for Montrose @ 19/ per Boll.

Tuesdy 15th Slight frost in the morning, day fair – 2 Ploughs at Thornylee & one Cart driving dung to Blacklands & Bringing Home Turnips from Bridge Shade Morning in a Stack of Oats N Barn – &c Jobs –

Wedy 16th Frost in the night & morning day fair & clear weather N.W. – 3 Carts to Montrose with 12B Barley – a Cart leading Turnips & driving dung – Bad roads –

Thursdy 17th Slight frost in the morning, day fair, One Cart driving dung & leading home Turnips in return – at Barns dighting Oats for the Mill, and P.M. lead home 3 Cart full of Turnips for Fear of Snow – & at Sundries.

Fridy 18th Strong Hoar frost in the morning & day 4 ploughs at Thornylee ley and others at the Barns dighting Oats for the Mill – 2 Men Pitting for Planting Bog Atry daily.

Saturdy 19th last night very sharp frost & Cold weather – dighting and sent 3 Carts with 16 Bolls Oats to the Mill of Pittrenie & Brought home Turnips from Bridge Shade in return – took in Oats S B.

Mondy 21st Yesterday very pleasant dry warm weather and frost – this day same but Cloudy and rather Inclined to fresh from S.W. – Sent 4B 1f Oats to Mill – Other Carts driving dung & Bringing home Turnips in return – King of France Murdered.

Tuesdy 22d Fresh weather from S.W. – Sent 3 Carts to Johnshaven for Scotch Coal @ 11/6 per Boll, One Cart leading Turnips and at Barns &c Jobs.

Wedy 23d Fresh weather and drought from S.W. – Sent three Carts to Johnshaven for Coals Scotch and others at Barns dighting of Oats for the Mill – Hired Men pitting for Planting &c Jobs.

Thursdy 24th fine weather and fresh – 2 ploughs at Thornylee and One Cart leading Turnips – One Cart D^{o} P.M. by a violent Sedition without any reason given 4 Men serv^{ts} & two Wowmen left their work – but two of the Men Acknowledged their fault & promised to Behave properly were admitted again the rest shall not if Law & Justice is in the land.

Fridy 25th pretty strong frost in the morning & day pleasant 4 ploughs at

Thornylee Ley – 2 Men spreading dung on D° – One Man & two Wowmen at the Mill of Pitreny &c Jobs –

Saturdy 26th This day Strong Hoar frost as to stop the Ploughs – 4 Carts driving dung & Bringing Home Turnips from Bridge Shade in return – at Barns dighting Oats and at sundries.

Mondy 28th Strong fresh wind from the N.W. and Ground dryer than for some time past – 4 ploughs at Thornylee and 4 people Hired &c spreading Dung on D° – P.M. at Barns putting Turnips into Heaps &c Jobbs.

Tuesdy 29th Thick Foggy & Growing weather from S.W. – 3 ploughs at Thornylee – One Cart at leading of Turnips from Haugh – P.M. dighting & lofting Oats for Meal to family – and at sundries &c.

Wedy 30th Strong drought & frosty weather from the S,W. – all People at Taunton fair – Servants still Increasing in their wages at this term about three or four shillings or thereabouts perhaps more, the Shearers Greatly Increased about 5/ each much against farmers.

Thursdy 31st frost last night day fair & pleasant – Carts driving Dung & Home Turnips from Bridge Shade – took in a Stack of Oats for Meal North New Barn – Every Article regarding the farmer is greatly against him – Servants, Utensils, Materials Crop being very scanty though prices high –

Diary for the Month of February 1793

Fridy 1st a Considerable deal of rain in the night – and more or less all day – 4 ploughs at the Thornylee, and others at the Barns dighting –

Saturdy 2d Strong wind & fresh weather from S.W. – 2 ploughs at Thornylee, and two Carts at driving dung & Bringing home Turnips in return at Barns dighting of Oats South Barn & planting a Hedge of Thorns, roses, Sweet Briars at Brae head –

Mondy 4th Yesterday very Cold Stormy weather from S.W. – this day moderate & fresh – three ploughs at Thornylee Ley and one Cart leading Turnips & driving dung – at Barns South dighting Oats for the Mill & Sundries.

Tuesdy 5th pretty strong frost night – day Extreamly pleasant sun shine & Larks singing – 3 Ploughs at Thornylee Ley nearly finished – One driving dung & leading Turnips – took in first of Pease old Barn –

Wedy 6th This morning about 8 A.M. Came on Snow from the Westward but Moderate weather – 4 Carts driving Dung to Blacklands, and one draught of Turnips Home from Bridge Shade, Other People at dighting and sacking of Oats for the Mill &c.

Thursdy 7th pretty Strong Frost and Moderate weather Sent 26 Bolls of Oats to Mill of Pitreny, and after leading Earth to Dunghill Blacklands – at Barns dighting of Oats & Clearing of New S Barn, and at sundries.

Fridy 8th last night High Wind & Stormy weather from ye Westw^d^ this day neither much frost or Fresh – 4 Carts at driving dung & Earth to Blacklands – and Home a draught of Turnips – at Sundries filling Barns –

Saturdy 9th Black frost and about 1 P.M. Came on Sleet & rain from S by E & stormy weather – 4 Carts driving of Dung to Blacklands & Home one draught of Turnips – dighted $9\frac{1}{2}$ pecks of Barley for the Pott –

Mondy 11th Yesterday P.M. very Stormy & Snow on Hills from N.W. – this frost & fresh by sun – 4 Carts driving dung to Blacklands & home One Draught of Turnips at the Barns dighting Oats for the Mill & filling pease Barn –

Tuesdy 12th Yesternight rain & stormy weather – this day Cold Blowing weather from the N.W. & sun shine, 4 ploughs at Thornylee nearly finished & well it may – P.M. at Barns and at sundry Jobbs – little vegetation –

Wedy 13th This morning a Considerable rain from N.W. & the fields very wett – a four Horse plough at the new Limed two Old ridges West side of Thornylee not Ploughed for Half a Century – at Barns dighting & lofting of Oats for family Meal and Sundries –

Thursdy 14th Slight frost in the morning, day pleasant & drought from the S.W. – One four Horse plough finished Thornylee, & two ploughs begun ribbing South End of Haugh after Turnips – Tried the new Salmon Nett & Catched one Small Salmon – but Expect better Success by Experience afterwards.

Fridy 15th This Morning a Considerable fall of Snow which mostly vanished by noon – 2 ploughs at Haugh ribbing for Barley and one at Haugh above the Distillery for Oats, and at Barns & sundries planting with Larix & Horse Chesnut & Ash the East dyke of the Bank – planted same way North fence of Back Yards, & this day Cottar Yards – and two Men draining of Bog Atry for Planting all kinds of Forest Trees & loup Yards with Oaks – Corns & fodder from 16/8 to 21/ Hay 8[ds]& Meal Oat from 16 to 17/ per Boll, all kinds of Manufactures very Brisk & high & Servants Extravagant The French War begun may Alter them. Early spring flowers rather Backward – No Crocus or Snowdrops yet.

Saturdy 16th a Good deal of Snow from the N.W. which went off mostly by noon, 2 ploughs Clean fur on Haugh Turnips & fallow, and two Carts at driving dung & leading turnips Bridge Shade, at Barns.

Mondy 18th Yesterday a very Cold day more than ordinary so to be fresh – this day 3 Ploughs at Haugh at Turnip Steering, and Haugh above the Distillery for Oats – at Barns & Sundries – Lapwings.

Tuesdy 19th Frost in the night & morning, day Cold but dry, 2 ploughs at Haugh, and two Carts to the Mill of Pitrenie for 16 Bolls Meal rather under three firlots per Boll – at Barns dighting Gardner planting in Sundry places & others draining – Lapwings on Sunday.

Wedy 20th Fresh weather from the S.W. & dry – 3 ploughs going & One Cart driving dung & leading Home Turnips – this day lost the first "Horse" ever in my possession by a violent Colick – He had every remedy administered that I could devise more than ordinary, but being at Brechin Yesterday for Sale, well Enough, & 22 Years old, to no effect –

Thursdy 21st Cold stormy weather from S.W. but fresh & fair weather, Three Ploughs going & one Cart driving dung and turnips – at Barns dighting Oats & Barley for Malt for Family use, planting Bog Atry.

Fridy 22d a Considerable rain in the night & Waters Out also many fields very Wett – 2 ploughs at Haugh Steering late Turnip Ground, & one Plough at the Cottar Shade begun, next publick road, at the Barns dighting Barley for family Malt & seed Oats.

Saturdy 23d last night a strong wind from S.W. but fresh, this day drought & sun shine, 3 ploughs at Haugh steering & Clean ploughing Cottar Shade, at Barns dighting of Oats for seed, and one Cart driving Turnips & Dung – planting &c Jobs.

Mondy 25th Yesterday pleasant forenoon, after Coldish, and rain in the night, this day same – 2 ploughs at Cottar land, and one Steering for Barley Haugh – at Barns dighting Oats for seed, & sacking those in long loft for the Miln & Sundries –

Tuesdy 26th This morning a pretty strong frost day pleasant till near noon when Cold & Bleak & rain P.M. sent 31 Bolls Oats to Pitreany for family Meal & Brought Home turnips in return – at Barns dighting of Pease, Oats & Sundry other Jobbs –

Wedy 27th This last night & morning a strong Gale from the S.W. & drought – 4 ploughs two at the Haugh steering the Turnip Ground, and two at the Old Cottar land for Oats – at Barns and Sundry Jobbs – Ground dryer.

Thursdy 28th This morning a small frost & strong drought from the Southward, Two ploughs nearly finishing of Cross ploughing of the Haugh turnip Ground, and two D° at the Cottar Bank Award for Oats – in fine order, at Barns dighting of Barley – and the Gardner & two other Men at Planting of Bogg Atry with all kinds of Forest Trees – No Farms yet known to be sold in this Neighbourhood – Oat Meal from 16 to 17/ Barley & Bear from 16 to 19/ Seed Oats no fixed price – fatt Cattle not Saleable though Beef dear, Owing to the Turnips being Generally done. It is thought the French War will somewhat depress trade – Very little vegetation as yet but the Ground much Dried.

Diary for the Month of March 1793

Fridy 1st Early this morning Came on a severe Gale of Wind for some Hours attended with Excessive Heavy rain from the Southward – which deluged most Houses & fields – at Sundries.

Saturdy 2d Frost in the night day fair & very Cold from the S.W. till 5 P.M. Came on Snow & Milder – 2 ploughs water furring of Cottar Bank for Oats and 2 carts to Stonehaven for English Coal at 21^{ds} per Barrell – at Barns dighting seed pease &c –

Mondy 4th Yesterday Cold weather betwixt fresh & frost, but the Snow somewhat gone by the Sun – this day all Carts driving of Dung to Blacklands & Garden, at Barns &c Jobbs.

Tuesdy 5th pretty strong frost in the morning – P.M. Cold & Drought from the Northward – 4 Carts driving dung to Blacklands park & Home 4 Cartfull of Turnips –

Gardner & two others at Making of Dunghills Garden this two days, & Sowing of first Garden Pease, P.M. planting Trees &c.

Wedy 6th Pretty Strong Frost in the morning, day fair & not unpleasant – 4 Carts driving dung to Blacklands, and Gardner & others planting as before, took in a stack Oats N N Barn for Seed, and at Smithy &c –

Thursdy 7th Weather as Yesterday 3 ploughs at Cottar Shade but middling order, and One Cart driving dung to Blacklands, planting, ditching, Barns &c Jobs –

Fridy 8th Moderate weather & warm from Southw^{d} – 3 ploughs at Cottar Bank Shade and a Cart to Mundynes with Lint, other People at Smithy a New Plough, and others Planting & various Sundries.

Saturday 9th Some Hail here but snow on the High Grounds – 3 ploughs at Cottar Bank for Oats, and one Cart leading Gravel & turnips, at Barns – planting Bog Atry best part done, dunged most of the Pitts from it lying thick on the Sward, Barns – Evening stormy Hail.

Mondy 11th Yesterday morning a Considerable fall of Snow & very stormy from the S.E. – this morning the same with a Good deal of Snow & Dissagreable fresh weather from the S.E. – Carts driving of Dung to Blacklands & some Turnips in return.

Tuesdy 12th Some snow in the night & day very Cold & dirty weather – Could neither Plough or Cart this day to advantage, at Barns dighting of Oats for seed, at Planting & Pruning Trees Water Lair – Scouring & repairing Dykes West side House.

Wedy 13th last night & this day Snow & Stormy weather from the S.E. – at Barns dighting Seed Oats, and P.M. – leading of Feal for Covering the stone dyke North side Nursery for Thornylee, and lead Home One Cartful of Turnips – in Hay – Dykes &c.

Thursdy 14th Fresh pleasant & dry weather from the Southw^{d} – 4 ploughs at South side of Cottar Bank by no means in good order for Tillage – people at Mill of Pitranie other people at Barns dighting seed pease &c Jobs.

Fridy 15th This morning rain & pretty strong wind from the Southward – 3 Carts driving Home the Meal, and leading some Turnips near done – took in part of the 5th stack of Pease, Sundries –

Saturday 16th Frost last night and this day fair till near One P.M. when Came on a very severe storm of rain & wind from the S.E. which deluged the fields & Brought down the rivers – faired 5 P.M. – 3 ploughs Going & One Cart lead Home last of Turnips & obliged to Water the Cattle

Mondy 18th Yesterday pleasant dry weather from S.W. and notwithstanding of the severe rain yesterday the fields dryer & roads than before but about 4 A.M. Came on a deluge of rain from the East which overflowed the whole lower fields – Waters as much out as for some years, at Barns & Jobs.

Tuesdy 19th last night fair & this day very pleasant weather & warm. The Thermometer at Noon in the sun 61° People taking in Oats & pease, & leading P.M.

stones to a Cottar House, Pruning Trees N side House Tried the Nett Salmon Fishing without success –

Wedy 20th pleasant and dry weather from South E[t] – day fair till afternoon some drops of rain – 4 Ploughs Cottar Shade very Indifferent Order – P.M. Men at sundries – Planting &c.

Thursdy 21st last night & this morning much rain & stormy from the S.E. & rivers out – at Barns dighting Oats for seed – at dyke North side Nursery – scathing with lime, P.M. 3 Carts driving of Earth from West side of House to dress it as Approach – and leading stones to Cottar House.

Fridy 22d This morning about a little after 9 A.M. Came on severe storm of rain & high wind from the S.E. which stopped all work without doors – at Barns and sundry other Jobs

Saturdy 23d last night dry weather and this day same from the N.W. – One Cart to Montrose for coals & Sundries & three Carts driving Earth from the Green West side House to tail Cottar Bank &c

Mondy 25th Yesterday fair & warm first part – P.M. Cold from North this day Cold and dry from the N.E. – 2 ploughs at Cottar Bank, and two Carts at driving Earth from West side of House for a declining Green &c Jobs –

Tuesdy 26th This day fair & drought from the N.E. – Some Neighbours begun to Sow Pease & Oats – two Ploughs at Cottar Bank & two Carts at driving Earth & leveling West side of House for a Green.

Wedy 27th Fair & pleasant weather from the N.E. – 4 Ploughs at S.W. park Waterlair for Pease a good season and at levelling the Approach West side House, at Barns dighting Oats – delivered five Bolls of them.

Thursdy 28th Bleak dry weather from the N.E. till about noon Came on Snow pretty Heavy from S.E. and towards evening pretty deep – 4 ploughs at Waterlair for pease a good season – P.M. at Barns & Garden and Others at Levelling & dressing West side House.

Fridy 29th Slight frost in the night, day fair & the Snow all gone by a little after noon – at Barns &c – P.M. all Carts at driving Gravel to the road West side of the House, and at sundries – In Hay &c &c.

Saturdy 30th mostly fair & Cold weather from the N.E. – 4 Ploughs at Waterlair for Pease, and at Barns & sundry Jobs about the place.

Sundy 31st This day & last night Showry rain & Sleet in low Country & Snow on the Hills from N.E. – an Exceeding Backward Season, and every kind of vegetation uncommonly Backward Fodder as well Grain will be Extreamly scarce ere long & more so than is Generally Apprehended – Fatt Cattle will be in demand.

Diary for the Month of April 1793

Mondy 1st a Considerable deal of rain & Sleet last night and most part of this day from N.E. so no work without doors, at Barns dighting Oats & Gardner & Others at Sundries about the House &c

Tuesdy 2d a Good deal of rain night & dirty Cold weather from the N.E. all day, people at Barns till after dinner yoked three Carts to drive off Dung & Gutturs around the Houses, at Barns and at sundries – Bad weather –

Wedy 3d last night and this morning a Great Quantity of rain from the N.E. – rivers out, more or less rain all the day – no Horse work – People at Barns & Sundries an Excessive Bad Season Indeed & fields very Wett –

Thursdy 4th Bleak weather but fair from the Eastward – leading of Gravel to West side of House – P.M. 4 ploughs at Waterlair very wett – at Barns and Sundries –

Fridy 5th Bleak but dry weather from the N.E. – 4 ploughs at Waterlair for Pease nearly finished – at Barns and Gardner digging for Potatoes Nursery – Sundries –

Saturdy 6th Cold Bleak dry weather from East – 4 ploughs at the Waterlair – Early finished the Park for Pease, and the rest of day at N West Park for Oats on D° in tolerable good Condition, at Barns dighting & lofting Oats for Meal, Planting Potatoes Nursery Grounds, Cabbages

Mondy 8th Yesterday fair & droughty weather from the Eastward – and this day strong drought & sun shine – begun the sowing by 5B 2f 2 lippies Pease on Waterlair S.W. park – $7\frac{1}{2}$ Acres – partly Harrowed them – ploughs on said Waterlair for Oats – at Garden &c Jobbs.

Tuesdy 9th Strong drought from the S.E. and fine seed season finished Harrowing of Pease park Waterlair, and two Ploughs at the lower park for Oats – dighting rye Grass and at Garden Nursery transplanting &c.

Wedy 10th Fine pleasant & dry weather from various – One plough at Over Burns for Pease, and seven Harrows at the East side of Thornylee for Oats sowed 6 Bolls & partly Harrowed them fine season, at Nursery & Garden sowed a Number of small seeds, White frost night –

Thursdy 11th Cold stormy dry weather from the Northward, 7 Harrows at Thornylee, Have Sowed $12\frac{1}{2}$ Bolls Oats Thornylee and partly Harrowed the same, at sundries Garden sowing many of the small seeds – No vegetation.

Fridy 12th Some Appearance of Snow at Intervals but dry weather on whole, Sowed Out Thornylee with $19\frac{1}{2}$ Bolls of Oats or above 19 pecks per Acre 15 Acres – 7 Harrows at D° – One Plough for Pease Upper Burns – Sundries –

Saturday 13th Cold stormy weather but in General good seed season – 7 Harrows at Thornylee, and One Plough A.M. at Upper Burns – Sowed $1\frac{1}{2}$ Boll Pease & partly Harrowed the same, at Barns dighting Oats, and transplanting Ashes into Nursery – Pease straw done – Snow P.M.

Mondy 15th Yesterday very Cold and Bleak dry weather from the Northw[d], this day the same with snow after 4 P.M. Considerable, very strong frost yesterday morning and this – Harrowed out Thornylee, and Sowed & partly Harrowed part Cottar Bank Shade Six Bolls Oats – In Hay and at sundries on the farm –

Tuesdy 16th This morning about 7 Came on a very Heavy Snow & high wind from the N.W. which Covered the Country quite white & lay till about noon when fair & drought, after dinner yoked 5 Harrows & one Plough – filling Barns &c – There is a Great Scarcity of Fodder Meal &c and a very unpromising season for Man & Beast – Seed Season very Backward Indeed –

Wedy 17th very strong frost night, day fair & drought from the Southw[d], 4 Ploughs for Oats & Pease and P.M. sowed 4 Bolls Oats & One Pease being all of them for season – partly Harrowed same & Jobs The Milch Cows on good pasture new Grass Bank – Sowed Onions Carrots &c.

Thursdy 18th Cold Bleak dry weather from the N.E. – being FAST day no work without doors – but observed several Neighbours Harrows going –

Fridy 19th very rough Cold stormy weather & frost – all Harrows at Cottar Bank shade, sowed all Ploughed, at Barns and sundries – Garden &c.

Saturdy 20th very strong Black frost and ground Hard morning, 4 Ploughs finished all the land for Oats Waterlair and Brought Home the Ploughs – Inn Hay – and at Garden &c Fodder & Meal is now thought to be Extreamly Scanty & will need support from Other places, this was easily foreseen by some long ago.

Mondy 22d Yesterday Course weather afternoon, this day fair and drought from S.E. – at Oat seed Sowing at Waterlair & Harrowing of same – Observed yesterday towards Aberdeen on Assise Business more Fodder there –

Tuesdy 23d Strong drought from S.E. but no frost. People at Oat seed for Cottar Bank and at Barns and sundries –

Wedy 24th Fair & pleasant weather – Milder A.M. – all strength on Ploughing whole of land for Oats Cottar Bank – and at sundries – little or no vegetation anywhere.

Thursdy 25th Dark Bleak weather from the Eastward and little growth – at Laurencekirk fair a good deal of Cattle but no sale, finished the Oat seed having sown Boll on Acres – Victual rising.

Fridy 26th Some skiffs rain from N.E. – day more, 8 Harrows nearly finished all Harrowing – and planting of Potatoes late Nursery ground some Increase of vegetation, but little.

Saturdy 27th This morning very slight daggy rain from the Eastw[d] – One Cart to Stonehaven for Coals – Ploughs at steering of Bridge Shade for Barley and others dighting of rye Grass, Oats for Meal &c the Grass rising tolerably – planting Potatoes

Mondy 29th Yesterday dry & moderate weather from the S.E. – this day tolerably pleasant though white frost in the morning – 4 ploughs at steering of Bridge of Leppie shade after Turnips for Barley – and Harrowing of D[o] P.M. – Garden &c.

Tuesdy 30th Yesternight a Considerable deal of rain from the S.E. – day fair till past noon when several showers of rain from that Quarter – 4 ploughs at steering of Bridge Shade for Barley, and at sundries – Observed towards Stonehaven several people Dunging & ploughing for Bear seed and but little or no Vegetation on the whole – Hear of numbers of Young Cattle dying Whether Owing to Bad guiding or Unfavourable Fodder or Season Cannot be said a Dozen or thereabouts at Mains of Hallgreen, at Barras &c places – at the Annual Meeting of the County it was Agreed to Carry on the Business of a Bill for Turnpikes without delay – &c prices of Grain &c Increasing –

Diary for the Month of May 1793

May 1st Wedy a Considerable deal of rain in the night & most part of this day – P.M. only yoked 4 ploughs to steer the Bridge Shade for Barley – at Barns & Sundries.

Thursdy 2d Cold but dry weather from the Northw[d] – 4 ploughs at the steering of the Bridge Shade for Barley – P.M. finished it & five Harrows Harrowing it – at sundries – Observe severals yet at the Oat seed – Grass rising.

Fridy 3d Fair and tolerably moderate weather from N.W. – 2 Ploughs at Bridge Shade & at a furr for Potatoes South End Cottar Bank Shade new Ground and all Harrows at Bridge Shade for Barley – &c Jobbs –

Saturdy 4th Bleak & dark weather from S.E.- white frost morning – 4 ploughs at Bridge Shade for Barley of which Sowed & partly Harrowed same with 3 Harrows. Observe severals begun the Bear Seed &c Jobbs.

Mondy 6th Yesterday fair weather – this day pleasant & best of the season, 4 ploughs at Bridge Shade for Barley & sowed part of it P.M. – Rouped & Sold 14 fatt Cattle at from 5/6 to 6/ per stone on an average, little Gain.

Tuesdy 7th Yesterday very pleasant dry & warm weather from the Southw[d], this day same, three ploughs and three Harrows at Bridge Shade Barley now good season vegetation appearing rapidly – at a Cottar House building.

Wedy 8th Very pleasant warm & dry weather from S.E. – 4 ploughs A.M. Bridge Shade, and P.M. part of D[o] & some Harrows – nearly finished with six Bolls of Barley – Got 3 sheep strayed three Months ago – to Balmakewan they were Cheaper Maintained y[e] Home –

Thursdy 9th fine pleasant weather about Brakefast time Came on rain which lasted Moderately till near Dinner time Have Sowed on Bridge Shade 6B-five pecks of Barley and Harrows Cross Harrowed said field by $4\frac{1}{2}$ P.M. for Grass seeds – 10 lb Clover, $\frac{1}{2}$ Boll Rye Grass full of Plantain to each Acre, Braking in red Hall Park.

Fridy 10th little frost morning day fair & Growing – at sowing Bridge Shade with Grass seeds very Early Harrowed & partly rolled the same, at red Hall park Braking in & 3 ploughs at seed furr on D[o] –

Saturdy 11th Drisling rain A.M. – day fair & P.M. drought – Ploughs & Harrows at red Hall park, Sowed of Barley on D[o] partly Harrowed D[o] – and P.M. Ribbing of Bridge Shade sown with Grass seeds – Sundries –

Mondy 13th Foggy but Broke up fine dry weather – this day Coldish drought from Northward – ploughs & Harrows at the Red Hall Park Sowed and partly Harrowed D° Rolled Out the Bridge Shade, Sown with Barley & Grass seeds.

Tuesdy 14th Fine dry weather from the Northw^{d} – P.M. warmish – 3 ploughs and three Harrows on Red Hall park Have sowed 4 Bolls on it – at Barns & sundries.

Wedy 15th Cold Bleak but dry weather from the N.W. – 3 ploughs & three Harrows at Red Hall Park all day – at Barns and at sundries Garden – learn there was a very dull fair of the Rood Markett at Montrose –

Thursdy 16th Bleak weather from the Northw^{d} but dry – all ploughs & Harrows at Red Hall park, sowed it out with Barley – and partly Harrowed same, at Barns and at Sundries – Observe the Bear seed very rough, Cloddy & not very promising – some people Have about Half done – Oat Meal 20/ per Boll and not to be Had – all fatt Cattle but four delivered – turned one to Grass – being doubtful of the Purchaser taking it up.

Fridy 17th Dry & not unpleasant weather from the Northward – Harrows finished nearly all red Hall park – and after braking in at the Haugh for Barley – and at Sundries – Hay at 10^{ds}per stone.

Saturday 18th This day pleasant weather from the Northw^{d} – 4 Ploughs at Barley furr Haugh very fine, and at Quarry for stones to Cottar House One cart Sowed near two Bolls Barley E^{t} side.

Mondy 20th Yesterday full sun shine, this day same, and very warm – wind Easterly – 4 ploughs at Haugh Barley and P.M. Harrows at D° Sowed 6 firlots – at sundries.

Tuesdy 21st Coldish Bleak & dry weather from the North – all Ploughs and Harrows at Haugh for Barley fine Mould – at Barns and Sundry Jobs – Building Cottar House.

Wedy 22d very strong drought from the Northw^{d} & stormy weather – all Ploughs & Harrows ploughed & Harrowed all the Haugh land about 9 Acres last Season in Turnips and Broke in what was summer fallow said Haugh.

Thursdy 23d Fine dry weather for Labouring – 4 Carts driving Dung to West side Haugh and Others at Cottar House finished and at Garden & Sundries Jobs &c.

Fridy 24th White frost last night & this morning – Blasted Ash leaves, 4 Carts driving of Dung to Haugh nearly finished, spreading D° P.M. – several People at Casting Divots, Holing Broom &c for new Cottar House.

Saturday 25th Dark Bleak weather but warm from the Westw^{d} – betwixt 10 & 11 A.M. Came on some slight rain which was not of note, but towards the Evening Increased more, Ploughs & Harrows at the Haugh Barley sowed one Boll more & Harrowed.

Mondy 27th Yesterday moderate weather but drought, this morning stormy & turned out Excessive warm day – finished the Barley seed on this farm and sundries – towards Montrose observed the Barley seed very late, and things as Backward as here – Heared of no disturbance of Farm Business. the Bridge below this Founded.

Tuesdy 28th a strong storm of North Westerly wind from that Q^{r} – dunged, drilled & planted about three rood of new Ground South side Cottar Bank with Potatoes, and P.M. all Harrows at Braking in of a Park Waterlair for Chester Bear – Sundries.

Wedy 29th very stormy weather from North & frost pretty strong in the morning flitted two Cottars from Pittengardner to our Houses, and P.M. sent 4 Cart load of Dung to Waterlair, & afterwards finished Braking in a park for Bear there – Squalls of snow & severe weather afternoon & Evening.

Thursdy 30th Very stormy weather and very Cold from the Northw^{d} and Carts driving dung from Cottar land House about 30 Cartful spread & ploughed West side of said field very hard below – driving dung from farm to Waterlair – roup of Kair sold very High –

Fridy 31st last night & this morning a very strong frost Black & White – Blasted Potatoes shaws & many of the leaves of the Forrest Trees, driving dung to the Bear Park Waterlair, and P.M. Ploughing at same, Observed vegetation much Checkqued & some Bear Yellowed – the Bear seed not near done yet in the Country – Oat Meal 20/ per boll

Diary for the Month of June 1793

Saturdy 1st This day fair weather but came on rain more or less through the day, Ploughs & Harrows at Waterlair, at Barns – and at sundries – very bad weather.
Mondy 3d Yesterday Came on rain pretty fresh in the afternoon, this day several light showers from the Southw^{d} – all ploughs & Harrows at Waterlair & drove two draught of Dung to D^{o} – Obliged to yoke four Horse Ploughs – the two being too weak – at Sundries –

Tuesdy 4th This day mild & Moderate weather from the S.S.E.- all Ploughs & Harrows at Waterlair Sent two draught of Dung to D^{o} – Sent Cows &c to East Haugh – Grass good new Hained Bank – Sundries &c –

Wedy 5th warm & pleasant dry weather from the Southw^{d} – Ploughs and harrows & dung driving as yesterday some Neighbours Harrows – all dunged & mostly Ploughed out & sown – fields very dry Indeed –

Thursdy 6th very pleasant & warm weather – ploughs & Harrows finished all Bear seed of season at Waterlair, 7 Acres of land seed $5\frac{1}{2}$ B, at turning Dunghill Smithy park for Turnips pretty well rolled – and at Sundries.

Fridy 7th Th^{r} 63° very warm weather from Southward – Braking in the Blacklands for Turnips very fine – at Sowing all the park of Waterlair with grass Seeds & Harrowed the same, P.M. Rolling D^{o} – turning the Dunghill Blacklands and at Sundries –

Saturdy 8th fine dry weather from the Southw^{d} – Some light skiffs of rain but a fine shower after 7 P.M. – all Ploughs at giving a Cross Ploughing to Blacklands for Turnips very fine being ready for Dunging when Harrowed – People turning Dunghills –

Mondy 10th Yesterday fine showers P.M., this day very strong drought from the Westward – 4 Ploughs at Blacklands, and P.M. sent 18 Bolls of Oats to Mill of Pitrany – & after at Plough – Sowed a few Turnips tail Cottar Bank for Harvest.

Tuesdy 11th Bleak dry weather till about Noon a Smart Shower of Hail & rain, then fair till about 3 P.M. a very Heavy rain for above an Hour, then moderate all the Evening – 4 ploughs all day at Turnip land, and at Sundries – Houghing Early Potatoes – fine Pasture – saw Swallows.

Wedy 12th dry weather but Coldish from Northward – all the Ploughs at the Turnip ground, and at Turning of Dunghill – and Houghing Potatoes in the Orchard Ground, at Sundries – at Camock Muir some Sale of Cattle & Cows near Calfing –

Thursdy 13th Coldish dry weather from the Northw^{d} – 4 Ploughs A.M. and Harrows P.M. at Blacklands finished for Dunging – turning Dunghill & Houghing at Potatoes Orchard – at Sundries – learn that at Trinity Muir Cattle of size, and in tolerable Order sold as well as for years by past Contrary to Expectation – washed & sheared sheep.

Fridy 14th Moderate & pleasant dry weather Various – All Harrows at the Blacklands A.M. – P.M. 4 Carts at driving dung West side Blacklands for Turnips about 24 Cart load per acre, the weather being moderate Some Growth this day – Wewmen & others picking off Caterpillars Gooseberries

Saturdy 15th a Good heavy rain this night about Half an Inch in the Water Gage Garden – begun Sowing of Turnips West side of Blacklands & Covering D^{o} – Carts dunging and others spreading Dung &c Jobbs –

Mondy 17th Yesterday Bleak dry weather and this a perfect storm from the North – plough drilling Blacklands for Turnips – Carts driving dung – Set up the Cattle Flakes Gibsons Park – spreading Dung and at sundries. Pasture very scarce & little appearance of Hay –

Tuesdy 18th This morning Blew a perfect strong Gale from the N.W. dammaging all Vegetation – Drill plough – Carts &c at Turnip Blacklands – People at Mill of Pitreny with 18 Bolls Oats –

Wedy 19th This day a perfect storm from the Northward Blacking every vegetable, People at Dunging, Drilling &c of Blacklands, and at Mill all night something above thirteen pecks of Meal per Boll – & sundries &c by the Inclemency of weather several of our people Confined.

Thursdy 20th This day Cold & Stormy from the Northward with skiffs of rain P.M. – all people at the Turnips, dunging Sowing, drilling &c – Observed towards Aberdeen everything very Blasted & Cattle starving – & like it

Fridy 21st This whole day Constant Showry weather from the South West & Mild till P.M. when Cold & Stormy from the Northw^{d} – People at Turnips &c Hay at Aberdeen One Shilling per Stone –

Saturdy 22d Cold & Stormy Showry weather but little rain from the North – all people at Turnip ground about 9 Acres dunged & sown – the sick servants recovering of a kind of Scarlet Fever &c –

Mondy 24th Moderate weather from the Eastw^{d} – Carts & ploughs at Blacklands for Turnips – flited the Cattle and Sheep Flakes and at sundries – Observe no Growth in Grass & but little Vegetation –

Tuesdy 25th Hoar frost last night and this day fair & Moderate from the S.E. – all Carts & drill plough at Smithy park Turnips and people spreading dung – dighting Wheat for Mill of Convey Sundries.

Wedy 26th Strong drought from the S.W. and Checking all vegetation, drill plough and Carts driving dung to Smithy park all day – a Cart to Mill of Convey with 4 Bolls Wheat & Brought home 6 Barrel of English Coal, at sundries spreading Dung &c Jobs.

Thursdy 27th some light skiffs of rain from the Southw^{d} and strong drought otherwise finished all Smithy Park from 11 to 12 Acres of Turnips & rolled same out Harrows Braking in N Burns for Turnips &c.

Fridy 28th last night pretty brisk showers of rain from the S.W. – this day fair & strong drought from the S.W. – 2 four Horse Ploughs at the Nether Burns for Turnips very stiff & lumpy at Collecting Dung for D^{o} – & Jobbs.

Saturday 29th This forenoon about 10 to 11 A.M. Came on pretty smart rain from the S.E. which lasted till about 5 P.M. and wett the Grounds & fields Considerably – and Men & most others in the Neighbourhood assisting in putting up the Coomb for the new Bridge of the river below this place, it appears when sett to be very High & will require much Making up at both Ends – will be a Roomy Arch.

Sundy 30th This day much rain P.M. from the S.W. – the Quantity fallen during this Month by my PLUVIOMETER is 2 Inches & four Tenths of an Inch, The Hay Crop is very light in General and pasture more than Commonly Scanty – the Corn Crops also on the whole very Blate & not looking well particularly the Bear on stiff & wett Grounds, all the above here tolerably good but late – all kinds of trade & Commerce very dull & numerous Bankruptcies –

Diary for the Month of July 1793

Mondy 1st Blowing weather from the Westward – ploughs did out 2^{d} furr Nether Burns & P.M. Harrowing of D^{o} very rough, flited the Cattle & Sheep flakes, and at Gathering Dung and sundries Garden &c Jobbs –

Tuesdy 2d fair & dry weather but little growth from all ye rain Owing to the Impetous winds – all Harrows and a Horse roller at Nether Burns for Turnips but betwixt wet & dry CLAY is not Manageable –

Wedy 3d tolerable weather from Southw^{d} – and some drops rain from D^{o}- all Harrows, roller & 4 Ploughs P.M. at Burns, at Cammock Muir very few Cattle 150 to 200 of low value sold about 2/6 per Stone & no demand. Sundries.

Thursdy 4th Some showers last night – and this P.M. a Considerable deal, 4

Ploughs at N Burns nearly done, and at Clearing Hedge North side House, Threshing Wheat, There Appears a Considerable Growth this two days –

Fridy 5th Fair weather but Blowing from S.W. – the Ploughs did out & partly Harrowed N.Burns – P.M. all the Harrows 9 at said field but not fine Enough, the Crop Mending & pasture, Our Gooseberries Bushes quite Striped by Caterpillar & Crop will be lost.

Saturdy 6th Strong drought from the S.W. & warm weather – all Harrows & two Rollers on the Nether Burns for Turnips remarkably Bad to Make – flited Sheep Flakes, Threshing Wheat N N Barn, dressing Belt planting W side Garden.

Mondy 8th Yesterday very warm weather and this day same and 4 Carts driving Dung all day to Nether Burns – at Gathering & Collecting D^{o} Dung Courts – at Sundries.

Tuesdy 9th Foggy weather & Growing from S.E. – 4 Carts driving Dung four Carts A.M. & 2 P.M. – nearly finished – after Dinner begun drilling Burns & others spreading of Dung all day – at sundries flited Cattle Flakes &c Jobbs.

Wedy 10th Fine dry weather & Extraordinary Warm Th^{r} Shade 72° at 11 A.M. – a Plough at drilling Turnips BURNS and nearly finished whole of 14 Acres – Roller &c on D^{o} – Considerable Growth of all kinds &c Jobs –

Thursdy 11th very warm weather & strong drought from the Northw^{d} – Being Fast day all People at Church, Observe the Pasture here Amongst the best I ever saw and the Hay Crop Mending Amasingly the Bees swarmed yesterday but either went off or returned to Hive again –

Fridy 12th Cold stormy weather from the Northw^{d} with dry Bleak weather finished all the Burns & at Paving at the Dung of the Dung Court & driving of stones to D^{o} – Sundries.

Saturdy 13th Warm and very dry weather from the S.W. – Carts driving stones to Pave the Dung Court, and at Church P.M.at D^{o}Observe the Drought Setting of the Crop on thin Grounds & those not in high order – Observe the Caterpillars Stopped in their ravage.

Mondy 15th Yesterday very warm weather and this day Came on rain from the S.W. – People at Paving of Dung Court, and at Church, P.M. Driving Gravel to Cover D^{o} -a Considerable rain all the day – &c – Lightning

Tuesdy 16th last night and this day a Considerable deal of rain from the Southward – & Mending of the Crop – 4 Carts to Mathers for Lime shells to Gibsons Park Ley, One of them Broke down and Obliged to send another Cart for the Lime P.M. – Hear the Cattle Marketts are doing well – the Crop very Backward – The Hay will be late & but Scanty & thought to be very dear –

Wedy 17th Pretty frequent showers of rain from the S.E. – 3 Carts for Lime Shells, and one D^{o} at driving sand &c Jobbs –

Thursdy 18th This day fair in General, 2 Carts to Johnshaven Coals to the Smith, and One 4 Horse Plough at fallow Ground Cottar Bank – at Paldy fair the Dullest

Markett I ever saw – a very small Proportion Sold and those at Inferior prices – Crop better than here.

Fridy 19th a Considerable deal of rain in the morning day mostly fair & Cloudy Growing weather – 3 Carts to Mathers for Lime Shells to Gibsons park. Two Ban Men Threshing Wheat and Houghing some Turnips.

Saturdy 20th Variable weather & rather Cool from the Northward – 3 Carts to Mathers for Lime to Land, and others at sundries – Little sale amongst the Horses yesterday & prices much reduced – some People begun to Cutt Hay, Turnips promising –

Mondy 22d Yesterday fine dry weather & Drought S.E. – this day same & very pleasant, the Bees swarmed yesterday – 3 Carts to Mathers for Lime to Land, and at taking a trench from Corn Yard Head to Dung Court – Vegetation very luxurious at present – Lord Gardentones Death

Tuesdy 23d Foggy in the night day warm & pleasant, 3 Carts to Mathers for Lime – and one Cart to Quarry for stones P.M., at sundries &c.

Wedy 24th Thick foggy weather till 8 P.M. – 3 carts to Mathers for Lime Gibsons Park – Other Men at the Barn dighting of Wheat sold, and at sundries Th^{r} 70 noon –

Thursdy 25th foggy weather and through the day very warm – 3 Carts to Mathers for Lime to Gibsons land, One Cart lead Home Whins &c A.M. and a Draught of stones from Dava P.M. – dighting of Wheat. Observe the Crop setting by the Great drought.

Fridy 26th Fine pleasant dry weather from the S.W. – 3 Carts to Mathers for Lime – and others at Thistling and P.M. at Brewing preparations & Sundries.

Saturdy 27th rather frosty Dew this morning day warm & Drought from the Eastw^{d} – 3 Carts to Mathers for Lime – Others at Sundries Brewing for Hay – Cottar Houses and Sundries &c Jobbs –

Mondy 29th Yesterday dry & warm weather from the Eastw^{d} – this day same – 2 Carts driving Earth to Dunghill Layer Cottar Town, from road scourings – One Plough taking away first fur from Turnips West side of Blacklands – and others fliting Cattle and Sheep Flakes – drought stunting vegetation much.

Tuesdy 30th Fine dry weather from the N.E. till 4 P.M. rain but very Moderate – Horse & Hand Houghing of Turnips Smithy Park – a Cart to Mill of Cove with 4 Bolls of Wheat but no Coals in return – at Sundries – Cutting Hay 2 Men

Wedy 31st dry weather & strong drought from the Westward – a Plough at taking away a fur from Turnips & 7 people at first Thinning of D^{o} – Quantity of rain fallen this month is Only ONE INCH – and Grounds Excessive dry – no sale of Cattle of any kind – the Turnips look well – the Oats & Barley only offering to shoot field pease Bloomed & some of them good –

Diary of the Month of August 1793

Thursdy 1st Very pleasant summer weather from the S.W. – all People at Garvock fair – Servants wages are fallen & no wonder – no kind of Sale amongst the Cattle of any kind – and other things proportionly.

Fridy 2d Very pleasant warm & dry weather from the S.E. – 2 Men Cutting Hay, and own people & hired turning out of Bout what was Cut before, and at Horse and Hand Houghing of Turnips promises well –

Saturdy 3d Very pleasant weather from the S.E. – before Brakefast at Turnips, after all Hands at Gathering & tramp Coling of Hay Kavels large Ones, P.M. thick fogg and drissling rain, Observe the Crop Good at Fordun.

Mondy 5th Yesterday afternoon several showers of rain & very Closs foggy weather – Early stronger showers, this day very warm, Th[r] noon shade 70° – at Turnips, Horse & Hand Hoeing and turned the Hay Cut last week –

Tuesdy 6th This day very Pleasant weather from S.W. – all People alternately at Turnips and Coling Hay – in fine order, at sundries &c Jobbs –

Wedy 7th Very pleasant and dry warm weather Th[r] 65° – at Turnips and after Brakefast 12 people at Gathering & Coling of Hay Tramp Coals -The Crop promising fine in Many places

Thursdy 8th dark weather from the Eastw[d] – people at Turnips till Brakefast – after Begun Hay Gathering but Stopped by Heavy rain which begun about 11 A.M. and Continued till about 4 P.M. – at Sundries.

Fridy 9th a Considerable rain first part of last night day fair & mild, some of the Grain lodged all people at Turnips, then Turning Hay – & after at Turnips promising better at present than for years by past, a Cart to Quarry for stones to some triffling Houses.

Saturdy 10th very warm weather from y[e] Eastw[d] people at the Turnips till Brakefast, after at gathering & Coleing of Hay Kavels in good order a Cart to Montrose with Wheat, & for Coals – Sundries, Excessive Heavy rain at Fordun little here –

Mondy 12th Yesterday afternoon Heavy showers of rain, and this afternoon some D[o] – all People & Hired at Turnips first 12 Acres gone through & at Turning Hay – flited Cattle & Sheep flakes P.M. –

Tuesdy 13th about 5 A.M. Came on rain & stormy weather from the S.E. which lasted till past One P.M. and brought the Rivers much out, Highest about 4 P.M. – In the afternoon begun to spread lime Gibsons park but not Slacked Enough, driving dung Feal to Cottar Town being scouring of Ditches for Benefit of publick road.

Wedy 14th Fair weather till about noon when Came on very Heavy rain from SSW – which lasted till 3 P.M.- people at Turning of Hay Some second & third time – and at sundries, Observed the Crop to Woodstock pretty Good.

Thursdy 15th Fair weather and very warm from S.W. till 2 P.M. a shower – after fair till 4 P.M. rain so as to Stop Coleing of Hay Kavels – after at Turnips but Observed Heavy rain Hills after Miday &c Jobs.

Fridy 16th this day moderate weather from the Northward, people at Turnips before Brakefast – after our own & others to Ammount of 12 at Hay Coleing did most of the Kavels in good order – a fine Crop – at sundries – Observed a Good Crop of Bear & Pease on the farm Waterlair –

Saturdy 17th dry & fine weather from the Northw^{d} good for Hay , at Turnips Morning, after Coled all Hay Kavels 22 most of them large Coles full 150 stones in an average, and will run above 250 stones per Acre – P.M. Gathering & Coleing of South Side Bank Hay.

Mondy 19th Yesterday fair & dry weather from the Eastward – this day also fair variable – people at Turnips Hired ones till Brakefast after at Gathering & Coleing of Hay Bank 14 of them – Sundries.

Tuesdy 20th a Slight Shower in the morning day fair & afternoon drought from S.W. – at Turnips till dinner time & fliting the Cattle & Sheep Flakes and after at Gathering & Coling Hay Bank &c.

Wedy 21st This morning about 8 Came on rain & stormy weather from the S.W. which stopped work after Brakefast till dinner time, fishing with the Net for Salmon & Caught two small ones, P.M. at Ploughing the fallow Cottar Bank & at Turnips. Sold Kavels Hay.

Thursdy 22d rain morning, day fair Except some showers about dinner time – people at Turnips & Hay Cutters at Waterlair, at Laurence fair Some Sale of Cattle those fit for Stall feeding at 3/6 per stone – but on the whole but a dull but Indifferent Markett –

Fridy 23d Fair weather and but little drought from the Westw^{d} morning at Turnips, and after Brakefast 11 people at Threshing Hay for Seed Waterlair all day the rain hath Discoloured it Considerably &c Jobs

Saturdy 24th This morning some rain, day fair till about noon Came on Heavy rain for an Hour & stopped Threshing of Hay – after at Turnips Smithy park.

Mondy 26th Yesterday fair & pleasant weather – this morning a shower of rain through the day fair & Mild, at Turnips and after pulled all the Lint very fine Crop, 37 Threaves & one Stook – Sundries.

Tuesdy 27th Shower in the morning day fair & Rather Stormy from the Westward, flited the Flakes, at turning Hay – Horse & Hand Houghing of Smithy park Turnips a very promising Crop – at Sundries.

Wedy 28th last night about 9 P.M. Came on rain which lasted most of the night & morning – day fair & drought from the S.W. – 3 ploughs at Turnips, and own & Hired People dressed out 12 Acres Blacklands before 11 A.M. after a Gathering & Coling of Hay Bank all day –

Thursdy 29th Frosty white Rhime in the night & morning day fair & drought from S.W. – three Ploughs amongst the Turnips & people Houghing D^{o} till Brakefast when went & Coled out all Bank Hay –

Fridy 30th White frost every night – fair till past 10 A.M. and a little after noon

Came on very Heavy rain which lasted so for several Hours – & stopped partly work – at a Roup of Corns & Fodder at Kair they all sold from perhaps twelve to 18/ per Boll – about 3£ 10 per Acre 38 D° – Bear Early – Oats late –

Saturdy 31st White frost in the night – day rain at Intervals – Sent three Carts to Corsbitt for Peats – others Hand & Horse Houghing of Turnips long Burns – some others Holing of Broom Thornylee & Sundries – Quantity of rain four Inches fallen this month – Observe the Crop Extreamly Green & Backward over the whole Country – the ship Meal However fallen from 18/ to 16/6 – and the Cattle Marketts rather on the rise for Good Bestials – Small fruits nearly done.

Diary for the Month of September 1793

Sundy 1st Bleak lowring weather from the S.E. – and towards 7 P.M. Came on rain & stormy weather which Increased.

Mondy 2d This day fair about 2 P.M. Came on showers of rain which now & then was the Case for the rest of the day – flited the Cattle Flakes morning – after Horse & Hand Houghing of Turnips – Others Cutting of Sprotts & rushes for Hay Thatch, Cutting Broom Thornylee.

Tuesdy 3d This day fair but stormy from the N.W. – morning Cutting Rushes for Hay Thatch – after Brakefast 13 people Begun Gathering Waterlair Hay not being worth the Threshing after so much rain – 7 Coles – finished Hay Coling.

Wedy 4th Fair & dry weather from the N.W. – Own & Hired people at Cutting rushes for Thatching Kavels Hay till dinner time after at Burns Turnips & Sundries –

Thursdy 5th Early own & Hired people to the Ammount of 21 begun leading in Bank & Waterlair Hay into a Sowe or rick 14 feet by 60 long – the weather good till ½ past 5 P.M. rain – Got the rick however pretty well up & in good Condition – Eclipse of Sun 10 deg

Fridy 6th Morning Strong Gale from the N.W. which prevented finishing the Hay Sowe till after dinner, people own, Hired &c Making ropes, Cutting Rushes &c A.M.- needed about 5 Cartfull of Kavels to finish it –

Saturdy 7th pretty strong White frost in the night, day Cold & some showers P.M. – finished out & Thatched the Hay Sowe in very fine order took all the Lint & spread it on Long Burns Ley &c 2 Carts to Johnshaven for Scotch Coal @ 7/6 per B –

Mondy 9th Yesterday very pleasant dry warm weather from S.W., this day fair & Moderate till 10 A.M. slight rain which stopped own, Hired & Neighbours from weighing & Building a field of Hay of 13 Acres Sold to my Lord Monboddo – weighed & Sowed 420 stones 60 feet by 14½

Tuesdy 10th This day fair weather & middle of very warm, own & others to ammount of 26 people weighing & Coling of Hay Kavels mostly done by night – Observe severals begun Cutting Bear, Pease &c –

Wedy 11th Fair weather but louring & foggy all day – finished out the Hay sowe for Monboddo total 2,050 stones & Thatched the same in fine order, at Sundries – Crop promising well but very late –

Thursdy 12th Frosty dew in the night, day very pleasant weather and fine filling weather from the S.W. – finished out the Eddering of Hay Sowes, dighting Rye Grass saved from Putting up them, at Burns Turnips & Sundries –

Fridy 13th This morning frosty weather day fair & very pleasant, 3 Carts spreading lime on Gibsons Park and Middle of day at Burns Turnips – After dinner spreading lime again in D^{o} park – at sundries.

Saturdy 14th Closs foggy weather & very warm – Carts at driving Dung to Gibsons Park & spreading of same on D^{o} & lean part of Old Town Park – at Turnips Burns – 2 slaters Small Houses, little Court –

Sundy 15th This morning about 7 Came on rain pretty strong the whole day till 7 P.M. – which brought down the rivers Considerably more so than this summer.

Mondy 16th last night & this morning very Heavy rain from the N.E. day fair & Moderate till about 6 P.M. a light shower – Carts driving dung, and spreading of Lime on Gibsons Park, all spread – at fliting Flakes Cattle & Sheep – at Sundries – Observe several of Neighbours begun Harvest this day.

Tuesdy 17th Fair & pleasant weather from the N.E. – Carts driving dung to Auld Toun ley over Lime now spread, & Cottars spreading of same, at Sundries, Grounds very wett –

Wedy 18th pleasant fair weather & drought from ye Northw^{d}, At Spreading dung on Gibsons Park – 4 Carts at driving dung to D^{o} from Brakefast to Dinner and at sundries – took many fish.

Thursdy 19th Fair & dry weather from the Westw^{d} North^{y}, spreading & driving dung 4 Carts to Gibsons Park all day – and at sundries – Observe most of the people Neighbourhood shearing.

Fridy 20th pleasant weather from various – Carts driving Dung to Gibsons park – and at sundries – people spreading dung and at sundrys &c Jobs –

Saturdy 21st This day moderate & Mild weather – Carts driving Earth from top of Corn Yard into Dung Court and at Holing of Broom Thornylee for Firing and at taking Small Salmon, Grilse (21 taken) &c with Netts.

Mondy 23d Yesterday very strong Drought from the N.E. and very Cold – Early this morning much rain – day showers – Begun Harvest by Burns Pease, little fodder but well filled, only the people on the farm Employed –

Tuesdy 24th Early this morning very Heavy rain from Eastw^{d} which deluged the fields very much, people at Digging up Broom &c from Thornylee &c – this will Dammage the Crop Considerably –

Wedy 25th Fair weather but little or no drought – sent three Carts to Mathers for lime to Gibsons park – One leading of Broom Thornylee – Cottars digging – at sundries –

Thursdy 26th Very pleasant & warm dry weather from the S.W. people flitted the Cattle & Sheep Flakes, and after took up & Gayted the Lint – P.M. digging & leading Feal to protect the Hedge Back of House, Caught 8 Young Salmon this day & 42 last week.

Fridy 27th very fine weather and strong drought from the S.W. Carts leading of Feal for the protection of the Hedge Back of the House – & people digging of D^{o}. Observe severals leading of Bear – &c Jobs –

Saturdy 28th Frosty dew but no Hurt to Vegetation of any kind – people at driving Feal &c 2 Carts to Montrose for Coals and sundries, lead in the Lint 2 Cartfull and Built into a stack & Thatched it Immediately.

Sundy 29th dark weather but fair till 10 P.M. some very light rain of no duration – Crop Coming very slow.

Mondy 30th This day pleasant & dry weather from the Southw^{d}, all Shearers 20 after Brakefast at Cutting up of Pease Waterlair a very fine Crop – if right got in – Quantity of rain during this month is Two Inches & Eight tenths- Manufactures of some kinds as Osnaburgh &c pretty much in demand & Angus & Mearns.

Meal falling – prices of Bear about 14/ per Boll – though some Bought for the new Great Distillery at Bridge of Don Aberdeen hath Bought as high as near 20/ per Boll – The Harvest is not yet General nor will be so on this or any other farm in this Neighbourhood for several days yet – no sale of Cattle fat or lean nor of sheep – the Crop will not be bulky.

Diary for the Month of October 1793

Tuesdy 1st This morning betwixt three & four Came on a strong Gale from the Southward with Heavy rain which lasted till betwixt 6 & 7 o'clock A.M. – no shearing till after Brakefast when did cut Waterlair Pease & after 1 P.M. at Bridge Shade Barley stopped by rain at $\frac{1}{2}$ past 5 P.M. – a Bad Beginning of Harvest.

Wedy 2d fine droughty day from the Westw^{d} – early all Hands at Cutting up Bridge Shade Barley a Moderate Crop 100 Threaves on 8 Acres but richer in Grain than Fodder – finished $\frac{1}{2}$ past 5 P.M. – then Turned the Burns Pease – very light Crop Generally.

Thursdy 3d Dark Cloudy weather from the S.W. – all people at Cutting of Rushes Sprotts &c for Thatch and after Brakefast lead Home all Burns Pease about 3 Acres into one small stack & thatched the same, Dismissed them for a week for want of ripe Grain – Hard at this Season

Fridy 4th Very strong drought from S.W. – Carts driving at Feal for the Dyke along Brae Head Back of the Mansion House – nearly Compleated, Sent 5 Bolls of Wheat Over Head for family use &c Jobs –

Saturdy 5th This day strong drought from S.W. & pleasant weather – sent three Carts to Mathers for Lime to the Haugh – One Cart sundries finished Dyke Back of the House & scouring road by it, Sundries. the Bridge over this Water below this finished this day.

Mondy 7th Yesterday strong drought Westerly and this day same with a Strong Gale from Westw[d] which Hath shaken the Oats Considerably though not full ripe – a Second Growth amongst the Barley in short most articles are against y[e] Farmers at present – 4 Carts for Turfs.

Tuesdy 8th Closs Foggy moderate weather – and some appearance of rain – Two ploughs N Burns Turnips, and 3 Carts to Brownies Ley for fine Turfs and at sundries. Erecting a temporary Bridge over Leppie &c Jobs.

Wedy 9th Closs weather and little drought from Any Quarter after Brakefast begun leading Waterlair Pease but by the wett Fog & Heavy Crop Obliged to stop by the first fair Cart load – Built them in Hutts after all Shearers at Red Hall Park Barley pretty good Crop.

Thursdy 10th Fine dry weather from the S.W. – all Shearers at Red Hall Park Barley and nearly finished – at Drumlithie Markett a Great Many Cattle but no Buyers – and on the whole very Bad sale – Harvest General –

Fridy 11th pleasant dry weather & no Dew – after Brakefast Came on a light shower till noon – Obliged to Gayte the Oats Cottar Bank shade – Greens amongst them – In General Harvest very far Back – fair P.M. –

Saturdy 12th Fine pleasant filling weather but little drought. people shearing Oats Cottar Shade till Brakefast and after at E[t] side Thornylee D[o] – dry at sundries.

Mondy 14th Yesterday good weather – at night very Heavy frosty dew but did no Dammage to Potatoe stems yet quite Green – this day shearing & Gayting of Thornylee Oats till 11 A.M. when yoked 4 Carts & drove in all Bridge Barley into nearly 3 stacks by 7 P.M. – Others Turning a second time the Waterlair Pease in good order.

Tuesdy 15th Yesternight fine drought & dry shearing morning Early yoked 4 Carts to Lead Home the Pease from Waterlair in fine order – & after Brakefast 2 Carts from a Neighbour very Busy got Home mostly whole by 7 P.M. there will be 7 large stacks on as many Acres – Shearing Bear Waterlair.

Wedy 16th Fine dry weather & strong drought from the Westw[d] – lead Home the remainder of Pease Waterlair and after Cutting up Bear D[o] a good Crop but a good deal shaken on the Old Crowns of ridges riper than the rest – sent for Flower Meal 7B-6 pecks –

Thursdy 17th This day & every morning at Thatching pease stacks, Cutting rushes, and after Brakefast at Cutting up all Bear Waterlair – finished – and after at Thornylee Oats, and at a roup at Castleton &c –

Fridy 18th very fine Harvest weather, but a Heavy White Frost in the morning which rendered Shearing in Gayte till 11 A.M. – Men then Binding Gayte and three Carts lead Home one Stack Oats Cottar Bank on stathill, wewmen shearing – also 2 Cartfull Thornylee Oats S Barn.

Saturdy 19th Fine dry weather from S.W. – People at shearing of Oats Cottar Bank – Thatched out Pease Stacks and at sundries – Many leading &c not good order

Mondy 21st Yesterday fine dry weather – and this day same wind Northerly – 2 carts leading the West side of the stack red Hall Barley Setting up the East side, and all others shearing at Cottar Bank Oats &c.

Tursdy 22d about 8 P.M. Came on Thick fogg & Small drisling rain – lead some Oats Cottar Bank – and Sheared out D^{o} field till 1 P.M. then at Barns Threshing Barley for Malt & Oats for Thatch &c Jobs.

Wedy 23d The Dew Continued till past Miday – all people however shearing Thornylee Oats in Gayte When dry stooked – none fit for leading this day – Sundries.

Thursdy 24th Fine dry weather from the S.W. till betwixt 6 & 7 A.M. a light shower – people finished shearing all Thornylee Oats by 10 A.M. – and after most of those of Waterlair very ripe – a few in the Neighbourhood done shearing, but most others very late –

Fridy 25th a strong white Frost and some light skiffs of rain at Intervals – Barometer falling below rain in the Evening – finished shearing of all Oat Crop – after wewmen Shearing Haugh Barley not being dry & some second Growth Gayted it though not Common – Others lead Home the remainder of Red Hall p. Barley into 2 stacks &c

Saturdy 26th Drisling rain & Appearance by sinking of the Barometer to much rain of other wind or much rain it accordingly turned out the first luckily. Early begun to lead Home Waterlair Bear very fine Grain & in Good Condition into 2 stacks stathil after finished 2 stacks Oats – Women at Haugh Barley. Showers 4 P.M.

Mondy 28th Yesterday fine dry weather from S.W. – this day same Early yoked four Carts 2 leading from Thornylee and two from Cottar Bank into three stacks – and others shearing of Haugh Barley quite ripe in Gayte Afternoon a pretty strong Gale of wind but dry.

Tuesdy 29th Strong Gale in the night from S.W. with little rain, day fair & fine weather – all People at Shearing Haugh Barley till Brakefast – after Men leading Home Oats & mostly in sundries –

Wedy 30th Frosty Rhime in the morning, day fair & very pleasant weather – wind West – Morning lead Home all the Oats Out – and Shearers Cutt up & finished the whole before Dinner – after at setting up Barley Gayte Blowed down and shearing of Rushes &c for Thatch &c Jobs.

Thursdy 31st Very Strong Frost last night & this morning, Ice near Half Inch Thick – quite Blacked the Potatoe stems, Indie Cresses &c Flowers – all People at Barns Threshing for Thatch, and others at Digging of Potatoes 10 of them, all Cottar Bank tolerable Crop. The Quantity of rain this month one Quarter Inch. a Considerable number of Towns in this Neighbourhood Have got all in but more Hath not done so – Price of Grain not yet Settled – Bear about 14/6 and Oat Meal 15/ to 16/ per Boll – plenty of Beasts Fogg

Diary for the Month of November 1793

Fridy 1st Strong frost & drought from the Northw[d] – at Barns morning, after three Carts lead in two stacks first shorn Haugh Barley into 2 Hollow stacks in tolerable order – Wowmen Gathering – Thatched 3 stacks.

Saturdy 2d very strong both Black & heavy Hoar frost which lay till betwixt 11 & 12 o'Clock – people at Barns Threshing, Thatching &c – and others taking up Potatoes & Housing them, but not quite dug up in the nursery – finished 2[d] Barley Haugh stack. rain at 5 P.M.

Mondy 4th Feast Yesterday fair & dry weather from S.W. – this day the same – all Shearers at Thatching, Threshing & all Thatched in the Corn Yard in fine order – Hear many too rash farmers are suffering by it.

Tuesdy 5th Strong frost – day fair but little drought, yoked 2 Ploughs to Rib Waterlair pease Park angle wise – and a four Horse Plough at ground fallow below the Cottar Bank, flited Cattle & Sheep Flakes –

Wedy 6th Some little Dragle in the night day fair and a little drought from the Northw[d] – Ploughs at Cottar Bank – and at Waterlair ribbing – 2 Men Threshing of Barley for Malt & sent 4 Bolls to J.A.

Thursdy 7th Dark day with some little drought from the Northw[d] – after Brakefast all own people & some of the shearers Binding Gayte Barley in tolerable order & lead Home a stack of it. At L kirk fair a good Many Cattle but little sale and prices low – demand for victuals –

Fridy 8th about 2 morning Came on Heavy Showers of rain from the Eastward which lasted more or less all day – 4 ploughs ribbing & Water Furring of Waterlair park and at sundries about Barns; flited Sheep Flakes.

Saturdy 9th last night Came on very Heavy rain from the Eastward which lasted till 2 P.M. and Brought down the Waters much-flited the Cattle Flakes P.M.-and at Sundries &c

Mondy 11th Yesterday about 1 P.M. Came on a furious gale of wind & rain from the N.E. which lasted all night & Brought down rivers much, this day fair – Sent five Carts to Montrose for Coals & Sundries, Others driving Turnips to 17 Cattle put up to Feed –

Tuesdy 12th Frost in the morning and day warm & very warm & pleasant – 4 ploughs ribbed out all of Waterlair Pease land & Brought Home all the Ploughs – taking in Coals & at Sundries &c.

Wedy 13th Cold & pretty hard frost till about 4 A.M. Came on much rain from the N.E. which lasted till 9 A.M. – yoked 3 Ploughs at Cottar Bank ribbing for Turnips & at Haugh above Distillery – dryer – Sent a Cart to Stonehaven for Coals to Mrs Kn – Sundries –

Thursdy 14th Moderate weather and dry very warm sunshine – 3 Ploughs at Cottar Bank shade ribbing – and one Cart Brought 8B 20f – nearly full Meal – prices of grain not fixed – Meal very Cheap –

Fridy 15th Frost in the night day fair till P.M. some rain. Ploughs at ribbing for Turnips Cottar Bank, Observed a Good deal of the Crop Out to Shear betwixt Muchalls & the Causeway Moss –

Saturdy 16th a very stormy day & Heavy rain from the S.E. which Brought down the rivers much at Ploughing as above and Sundries – Prices of Grain at Aberdeen 16/ per Boll –

Sundy 17th a Severe Storm of Wind & rain from the S.E. and very High Sea from D° Quarter – this morning a Ship of Considerable Value of the Port Aberdeen of Ten Thousand p^{ds} S^{tr} lost in the Cove, but People Saved by the Falling of ye Masts –

Mondy 18th This day a Severe Storm from S.E. with much rain – at sundries and various work.

Tuesdy 19th very Dissagreeable weather & stormy S.E. with rain all day – Ploughs Ribbing Cottar Bank and leading Turnips & Barns &c Jobbs –

Wedy 20th Closs Foggy weather with drissling rain from the S.E. – all Ploughs at Cottar Bank nearly Ribbed out – but very Wett – at Sundries.

Thursdy 21st Thick drissling rain & Everything very wett – 3 ploughs at ribbing out Cottar Bank land for Turnips, and after Begun ley Gibsons park the Thornylee being to wet to Rib for Bear – Got a Horse Severely wounded on nearer fore foot by Carelessness of servant going with a plough.

Fridy 22d This day very thick Daggy weather & Calm – 3 Ploughs at Gibsons Park Ley dunged & well Limed – a Cart to Aberdeen with my Oldest sons Apparel for the East Indies – Sundries.

Saturdy 23d Closs Foggy weather – 3 Ploughs at ley – and one Cart leading Turnips, and driving dung from Cottar Town to Gibsons park Ley – Sundries.

Mondy 25th Yesterday and this Closs Foggy weather & Mild – very Bad for the stacks & victual out all growing – lead Home Turnips, and sent 18B Oats to Mill of Garvock – took in a stack of Bridge Shade Barley P.M.

Tuesdy 26th Closs Foggy weather & Stacks Heating everywhere – three Ploughs at ley – & 0ne Cart Leading of Turnips, at Barns and Sundries.

Wedy 27th Fair & Clear weather morning slight frost – day fresh and very variable – 3 Ploughs at ley – & One Cart leading Turnips & spreading lime on D° – at Barns & Sundries – prices of Cattle &c on Increase.

Thursdy 28th dark Gloomy weather & no appearance of Alteration – 3 ploughs at Ley as yesterday and at leading Turnips, spreading lime on Ley – Barns dighting Oats N Barn &c

Fridy 29th Closs Foggy weather but drought P.M. from the Northw^{d} – 3 Ploughs at Auld Town Park ley – and one Cart leading of Turnips – at Barns & Sundries – Many stacks in the Country Heating &c –

Saturdy 30th last night & this day pleasant weather with sunshine for this long time by past – 2 ploughs at ley & two Carts leading of Turnips – Observed one of our largest Oat stacks Heated on the N E side and obliged to draw out the worst of it which Hopes will stop its progress further. Many in the Country are said to be Rotting. The Quantity of rain fallen this month is $4\frac{1}{2}$ Inches – Have Pitted the Potatoes in Corn Yard & Garden – also some of the Carrotts a very Great Crop large Sowes.The prices of Cattle Considerably up of late both Fatt & lean – Oat meal about 1/ per peck – and Bear 15/ Barley 16/6 but prices seems not fixed Have not sold any as yet & few others as far as I know.

Diary for the Month of December 1793

Mondy 2d Yesterday very pleasant dry weather from W.S.W. this day dark Gloomy D° with drissling rain from N.N.E. 2 Ploughs at Ley and two Carts leading of Turnips at Barns dighting of Oats lofted some 22 for Meal P.M. dighting Barley South Barn &c Jobbs –

Tuesdy 3d pretty strong frost wind little but Northerly – 3 ploughs at ley – and one Cart lead Home most of the Flakes & Housed them – took in a stack of Oats N Barn – dighting Barley & Sundries.

Wedy 4th very fine weather & pretty strong drought from N.W. One Cart to Montrose with Barley & for Coals One plough at Ley & two carts driving dung and Home Turnips – at Sundries.

Thursdy 5th very pleasant dry weather from N.W.- Mornings Breeze One Plough Ley – and three Carts driving of Dung & Home Turnips – at Mill of Garvock – 18 Bolls Oats 16B 5p Meal – and at sundries –

Fridy 6th Fine droughty weather from the Westward – 2 ploughs at Ley – sent two Carts for the Meal and after at driving Dung to Ley – and leading Home some Turnips – P.M. at Binding Haugh Barley.

Saturdy 7th Dry fair weather but no drought – all People at Binding the Out Barley & lead it Home to Two Compleat Hollow stacks & thatched the same Only fitt for the Horses use – Sundries –

Mondy 9th Yesterday pretty strong Frost all day – this morning Early Heavy rain from the Westward till 10 A.M. fair – Lead Home some Turnips Sent Corn to the Mill Garvock at Barns &c Jobs.

Tuesdy 10th Early morning pretty Brisk wind at S.W., day dirty spitting weather – 2 Ploughs at North Ley Parks, and 2 Carts driving Dung to D° – Spreading of it – Sent the rest of Oats to Mill of Garvock P.M.

Wedy 11th a Considerable deal of rain in the morning & some through the day – One Plough at Ley and Water furring of red Hall Park for Oats – and two Carts driving dung to Ley & spreading of D° – One Cart to Montrose for Coals & Sundries – Wind N.E. pretty strong P.M. – and Heavy rain night.

Thursdy 12th This whole day Heavy rain & stormy from N.E. no work without doors – dighting & lofting of Barley from S Barn – Waters much Out and the fields perfectly Deluged – Sundries.

Fridy 13th faired in the morning but began Again about 9 A.M. and lasted Extreamly Severe till past 5 P.M. brought down the rivers as high for five years by past – lead Home 2 Cartful turnips and at the Barns Chaffing Oats N B – Evening wind shifted from East to S.W. & fair Blowing Weather –

Saturdy 14th Fair weather till about 5 P.M. Came on rain pretty Heavy – 2 ploughs at Ley – and two Carts at leading of Turnips – two last days $1\frac{1}{2}$ Inch of Water in the Gauge – the River last night full as High as for several years by past – took in a stack of Oats S Barn – Sundries –

Mondy 16th Yesterday Cold Bleak weather & showers Evening a Strong Gale of Wind S.E. – this day a still stronger Gale from S.W. and drought. Three Ploughs finished Out North Park Ley 12 Acres and Begun ley of Goulds Hillock five years Old & well Enriched with Flaking with Cattle & Sheep the later for years by past – The prices of Grain Meal Oat 1/ – Bear $7^{ds}\frac{1}{2}$ Bear 14 to 14/6 – at Barns dighting & Chaffing of Oats – leading Turnips &c Jobs –

Tuesdy 17th Fine dry weather from S.W. – three Ploughs at the Ley as Yesterday and One Cart leading of Turnips large sized – at Barns dighting of Oats for the Mills and at sundries –

Wedy 18th In the morning some very heavy showers of rain & wind high S.W. – three ploughs at Ley – and one leading of Turnips-at Barns dighting Oats for the Mill & Sundries-

Thursdy 19th In the morning some showers day fair and drought – 4 ploughs at the Ley – and at Sundries on the Farm –

Fridy 20th Moderate Frost in the night, day very Pleasant sunshine – 2 ploughs at Ley and One Cart leading of Turnips – People at the Mill of Garvock 22B Oats 20 Bolls Meal –

Saturdy 21st pretty strong Frost till 5 A.M. a shower of rain – day fair but bitter Cold from S.E. & about 3 P.M. Came on rain & stormy weather – One Plough at Ley – two Carts leading Home Meal and after Turnips – Flited the Sheep Flakes – S side Garden and a Cart to Montrose for Coals & sundries –

Mondy 23d Yesterday rainy weather from the Eastward & P.M. Waters much out – this day mostly fair but Bleak from N.E. – 2 Ploughs at Ley – and One Cart Leading of Turnips – at Barns sacking Oats &c –

Tuesdy 24th last night & this day much rain from S.E. – One Plough at Ley – One Cart filling up sand Hole & Two Carts leading Oats to the Mill and after at D° Turnips – Ploughed down above Hole –

Wedy 25th a Considerable deal of rain but mostly fair from S.W. but dark & Gloomy moist weather – One Plough at Ley, and three Carts driving Oats to Mill of Garvock & Dung & Turnips – Sundries.

Thursdy 26th This day fair & sun shine till past 2 P.M. when dark & Cloudy – 3 Ploughs at Ley and One Cart leading Home Turnips – Sundries.

Fridy 27th Showers at Intervals, but in General Fair & open weather from the

Eastw^{d} – 3 Ploughs at Ley and one Cart leading of Turnips – Barns Chaffing Oats for the Mill S B – and Sundries.

Saturdy 28th last night a Good deal of rain & morning Frost – this day very Pleasant but frosty – 3 Ploughs at Long Burns ley 8 years old & one last leading of Turnips – Barns &c Jobs

Mondy 30th Yesterday very pleasant Clear & Mild weather Early A.M. a strong Gale from Southward bitter Cold – about 3 P.M. Came Heavy Sleet & rain from D^{o} which lasted till 12 P.M. – 3 ploughs at Burns & Back Yards Ley – and one Cart leading of Turnips – at Barns &c Jobs.

Tuesdy 31st This morning pretty strong Frost and all the Hills & Braes Covered deep with snow Here some Ploughs going & two Carts leading first part of the Hay sowe to Monboddo – Hear that the Barley is On the rise and much demand – prices of Beef &c much lower & owing to Want of Salt for Curing it – the Manufactures & Trade tolerably Brisk – Salt 32^{ds}per peck – Quantity of rain 4.25 four inches and one Quarter –

Diary for the Month of January 1794

Wedy 1st rain & sleet last night morning & day Frost and Clear pleasant weather – lead Home Turnips before Brakefast & being a Thanksgiving day at Church for the Good Harvest & Collection for ye Soldiers abroad –

Thursdy 2d pretty strong Frost all day – Horses at Smithy and at Barns dighting Oats – took in a stack of Barley red Hall park – Garden &c –

Fridy 3d Stormy weather & rather fresh from the Westward – 2 Carts leading Turnips & other Horses at the Smithy – at Sundries sacked 4B Barley for Montrose.

Saturday 4th This morning a strong fresh from S.W. & frost out 2 Carts to with 4B Barley – and Brought Home Coals &c One Cart Leading Turnips & one Plough at the Back Yards Ley – and at Sundries –

Mondy 6th Yesterday fresh weather & fair from S.W. – this day same & Ground dryer – Ploughs at Ley & leading of Turnips – at Johnston & to it Bad roads –

Tuesdy 7th Fine Pleasant and dry sun shine weather wind S.W. 3 ploughs at Back Yards Ley – and One Cart at leading Turnips – at Mill of Garvock 33B Oats Meal 30B $\frac{1}{2}$ lofted $17\frac{1}{2}$ B – Oats &c.

Wedy 8th very Blowing Cold weather from the S.W. – the Carts for the Meal, and two Ploughs at the Back Yards Ley – Chaffing Barley N Barn –

Thursdy 9th Fine Pleasant dry weather – 3 ploughs at Back Yards – One Cart leading stones to Causeway below the new Bridge East side – Sacking of Oats for the Mill 24B – and at sundries

Fridy 10th Very pleasant dry weather & drought from S.W. – Sent Two carts to Mill of Garvock with Oats and after leading Turnips – 2 ploughs at tilling out Back Yards Ley & afterwards at the Stuble Ground for Oats & Grass seeds Dickies Haugh –

Saturdy 11th This day very Pleasant & fair weather – 2 Ploughs at Haugh – One Cart leading of Turnips – and one D° to Montrose for Coals & Sundries – and at different Jobbs on the farm –

Mondy 13th Yesterday fine fresh weather & wind Southerly – this morning some frost and day dark & Bleak with fresh wind S.W. – 3 ploughs at the Haugh, and one Cart leading Turnips – at Barns dighting Bear – took in part Oat Stack.

Tuesdy 14th Strong wind S.W. and some spitting wet – 3 Ploughs at Haugh – and one Cart leading of Turnips – took in a stack Oats S B – Sundries –

Wedy 15th Fine dry weather from S.W. – 4 ploughs at the Haugh for Oats very fine order – at Barns and Sundries – got Home Malt Bridge of Leppie &c Jobs –

Thursdy 16th Slight frost in the night day fresh wind from S.W. 2 ploughs at Haugh, and One Cart spreading of Lime on stuble Ground West side of last years Fallowed Haugh in very fine order – One leading of Turnips. and after some stones to Fence the Haugh Bank below the Bridge East side – Sundries –

Fridy 17th last night some frost but light – day fair and strong drought from S.W. – 3 Ploughs at Haugh and One Cart Leading of Turnips & spreading of Lime on D° – at Dighting Barley N Barns & Sundries – Ground now pretty dry – Frost in England –

Saturdy 18th Fine droughty weather from S.W. – 2 Ploughs at Haugh nearly finished – One Cart leading of Turnips – and one spread out all the Lime on Haugh & leading of stones to Bank the Water below Bridge, and others Dighting & lofting of Barley &c –

Mondy 20th Yesterday fine dry soft weather from S.W. – this day same. 3 Ploughs finished Out Haugh tillage by noon, after at the Upper Burns after Pease & for Oats & Grass seeds – at Barns dighting Barley & lofting it – Flited the Sheep Flakes S side Garden Shade – Sundries.

Tuesdy 21st light frost morning day fair & fine Drought from S.W. – 3 Ploughs at Upper Burns – good season for Oats – One Cart Leading of Turnips – took in first Pease, at Sundries.

Wedy 22d Fine Fresh and dry weather from S.W. – 4 Ploughs did out Upper Burns for Oats by noon – after at North side of Red Hall Park for D° -at Barns and Sundries- My Mill

Thursdy 23d very strong Gale & High Wind from the Westward about or before 9 A.M. Came on some snow and stormy weather – 2 Ploughs at red Hall Park – leading Turnips – One Cart to Montrose for Coals and at Barns dighting & lofting Barley – Sundries.

Fridy 24th last night pretty strong Frost & skimmering of snow from S.W. – Two Ploughs at Red Hall Park – & two Carts with Hay to Monboddo -lead Home some Turnips-Sundries.

Saturdy 25th last night about Nine Came on Snow & Hail Moderate at first but Increased into a very strong Gale from the North with Snow & drift and which

Continued all day but Increased to a great degree from 3 P.M. – all the rest of the day & nigh no work without – at Barns Chaffing Oats, Pease &c.

Mondy 27th Yesterday severe Frost & Strong Wind N.W. which filled up the roads & Hollows to many feet and rendered traveling of Carriages and also Horses Impracticable almost – this day more Moderate – People Digging & leading of Turnips Nether BURNS – at Barns & Sundries.

Tuesdy 28th Moderate weather and very strong Frost – Carts driving of Dung to Layer West side Cottar Bank and at Barns and Sundries – All roads stopped –

Wedy 29th Strong Frost & fine weather if the roads was Passible all People at Taunton Fair – Engaged most what we wanted by the Grieve & paid the former the wages very high.

Thursdy 30th Strong Frost & a little Snow last night and the day same – 3 Carts driving Dung to Cottar Bank For Turnips – and P.M. Lead Home some Turnips – and at Sundries –

Fridy 31st Early this morning Came on a Strong Thaw with showers of rain from the Southw^{d} which Carried most of the Snow Except in Blown Places – & roads – Carts Driving Dung, and at Barns dighting Oats and other Jobs – The Ice not Broke on the river Evening Quantity of Rain in the Pluviometer four Tenths of an Inch. The Prices of linen Yarn & Sacking Fallen above 20 per C^{t} – and Butchers Meat more – Servants Wages not in the least so – Grain rather rising

Diary for the Month of February 1794

Saturdy 1st Pretty strong Thaw all last night and this day with Drisling rain – most of the Snow gone. The River Had Broke Early morning and Had been High – Carts Driving Dung & Turnips –

Mondy 3d Yesterday soft weather & thaw – This day same with rain from the N.E. at Intervals – 1 Cart to Montrose for Coals, and two Ploughs at Red Hall Park & One Cart leading of Turnips from Burns &c Jobs –

Tuesdy 4th This day Chiefly rain from the N.E. – all People at the Barns Dighting Barley – Gardner Pitting at Waterlair Planting for Blanks Forrest Trees & Sundries –

Wedy 5th a Great deal of rain from the Eastw^{d} – day fair no Ploughing – People at Barns dighting of Barley for Montrose, and at Sundries Jobs –

Thursdy 6th Cold Shirpy weather from S.E. & Grounds very wett No Ploughing of the Stubble Grounds and People at the Barns dighting Pease &c – Some Showers A.M.

Fridy 7th This morning about daylight Came on much rain stormy from SSE – Yoked the Carts but loosed at 11 A.M. – at Barns and sundries about the Farm Jobs – P.M. Came strong drought from S.W. – Grounds Wett –

Saturdy 8th Fine Blowing Soft weather from S.W. – 3 Carts to Montrose with 12B Barley & for Coals – leading Turnips One Cart and at Sundries – roads deep –

Mondy 10th Strong Blowing weather from the Westw[d] all day – sent 2 Cart Load Hay to Monboddo – their Carts prevented by wind – 2 Ploughs at Red Hall Park – Home some Turnips and at sundries – Gardner Pitting Waterlair.

Tuesdy 11th Strong Drought from S.W. drying Ground much – 4 Ploughs at Red Hall Park – took in stack Barley N Barn – at Delivering Hay – Barns Chaffing Barley &c –

Wedy 12th Strong drought & Blowing weather from N.W. all day – 4 Ploughs at red Hall Park – and others at the Barns Chaffing of Barley S B – and sowed the first GARDEN PEASE – at sundries &c –

Thursdy 13th Thick foggy weather & drisling rain at Intervals – 3 Ploughs at Red Hall Park, and One Cart leading of Turnips Burns & Black lands picked Turnips of different Colors to lay Nursery Ground for Seed – Sundries –

Fridy 14th This morning Strong Gusts of Wind Northerly – about 11 A.M. Came on rain pretty Closs but not Heavy from N.E. which lasted all day – three Ploughs nearly finished all Tillage for the OAT SEED of season red Hall Park – and One Cart leading of Turnips for Cattle, and also some of the best kinds Picked to Transplant to the Nursery Ground for Seed – Gardner at it all day – at Barns dighting Pease & Lofting &c Jobbs –

Saturdy 15th rain & stormy weather Continued from N.E. till near 2 A.M. when fair & frosty Drought from the same Quarter – 3 ploughs finished all the Oat seed furr but the Nether Burns Turnips – leading of Turnips from Diff[r] places for Seed & for Cattle – at Barns dighting Barley and at sundries – Snow deep on the Hills & Braes. The roads & fields not wetter than before y[e] rain –

Mondy 17th Yesterday pretty Strong Frost and this day same till about noon Came on drissling Small rain Constant – 3 Carts to Montrose with Barley and Out Coals – One Cart driving Dung for Turnips and at sundries Barns &c Jobbs – Dunging Garden &c.

Tuesdy 18th Cold Bleak but dry weather from the S.E. and three Ploughs at ribbing Thornylee for Bear – and one Cart leading of Turnips Burns – Planting Larches Waterlair Park – either neglected or smothered by Broom – at Barns and at sundry other Jobbs –

Wedy 19th Strong Gale from the S.E. and about $\frac{1}{2}$ past noon Came on very Heavy rain & severe Stormy weather from S.S.E. which Stopped the Ploughs Thornylee and all work without doors – Deluged all the Fields & Brought down the rivers much, at Barns & sundries – Faired up before midnight.

Thursdy 20th raw wet weather & some showers from S.W. – lead some Turnips Burns, and all other People at the Barns lofting Barley, Sacking D[o] – & P.M. Took in a Stack of Oats for Seed South Barn in fine Order – Gardner planting of Larchs &c Waterlair Planting &c.

Fridy 21st Fine dry weather from the Westw[d] & roads some better 3 Ploughs at ribbing of Thornylee for Bear, and one Cart leading of Turnips from the Burns – at Barns Chaffing Barley further draining of Bogg Atry-Gardner only –

Saturdy 22d This morning several showers of rain from the Southw[d] day fair till Evening rain again from S.S.E. – at Barns before Dinner lofting Barley – and after it lead Home a Draught of Turnips – Grounds very Wett –

Mondy 24th Yesterday much rain & very stormy weather S.W. This day same till past noon Mild wind South Westerly – No Horse work for Wetness – the People at Turning part Dunghill for Turnips Cottar Bank – at Barns and other Jobbs – P.M.lead some Turnips Burns &c Bogg draining 2 Men.

Tuesdy 25th This day pleasant & dry Clear weather from S.W. – One Cart to Montrose with 3B Barley to Mr R.Napier & Out $\frac{1}{2}$ B Coals – Carts driving of Dung and Turnips – and at Sundries-

Wedy 26th Very fine dry weather from N.W. – three Ploughs at Ribbing of Thornylee for Bear – and One Cart Leading of Turnips Black lands & Gardner &c Draining further Bog Atry for Planting, at Barns – Turning Dunghill for Turnips &c

Thursdy 27th Strong drought and rather stormy from S.W. Being a General Fast of the Church of Scotland no Work Done – Observed it to be well Kept everywhere –

Fridy 28th Some light rain from S.W. till morning all day variable and three Ploughs at Ribbing of Thornylee for Bear – and One Cart leading of Turnips Blacklands – P.M. Dighting & Sacking of Pease to loft – Gardner & others at Sundries – Observe a Considerable vegetation on the Fruit & other Trees well Budded & some of the Early Flowers Blown – Quantity of RAIN this Month Three Inches – Hear of no Farms yet Sold or much Grain or Demand in the Markett – Fatt Cattle no demand whatever and in General everything dull in Trade and Manufactures – Whether owing to War or Otherwise.

Diary for the Month of March 1794

March 1st Saturday Morning White Frost after Cold from Southward – 2 Carts with Hay to Monboddo – two Ploughs at Thornylee ribbing Gardner & others Planting Willows Bog Atry Planting – before three P.M. Came on a severe storm of wind & rain S.S.W. which lasted all night.

Mondy 3d Yesterday very pleasant weather from S.W.- this morning about day light Came on Snow & rain & stormy weather from S.E. which Stopped work without Doors – at Barns Chaffing Seed Oats, &c Sundries –

Tuesdy 4th Fair weather from S.W. till 3 P.M. Heavy rain & Bad weather from S.E. – 3 Ploughs at Thornylee and One Cart Leading of Turnips from Black lands – and at Barns dighting seed Oats – Cutting Setts, 4 to 500 of them, of the Red Willow Tree Kair to Plant Bog Atry &c –

Wedy 5th All this day fair, sunshine & strong drought from S.W.- 3 Ploughs finished Ribbing and Water furring of Thornylee, One Cart leading Turnips, Planted all the Willow Setts – and after at Cutting of the Whins & Broom E[t] side Dike Waterlair planting for Fewell and also to preserve the Fence –

Thursdy 6th Bleak but dry weather Mild from S.W. – 4 Ploughs at Seed fur on the part of Waterlair in my own Hand for Pease tolerable good season – People at Cutting off Whins & Broom for Firing D^{o} – at Sundries –

Fridy 7th Fine pleasant dry weather from South – 3 ploughs at Waterlair, and One Cart leading of Turnips from Blacklands – Other Men Cutting up Whins &c at Waterlair Evening a slight shower from SSE – Vegetation –

Saturdy 8th This day warm but little drought – about noon slight rain & foggy – 3 ploughs at Waterlair and One Cart driving Dung & Leading of Turnips – at Barns & at Sundries.

Mondy 10th Yesterday dark weather but dry from S.E. – this day the same till noon when some pretty Hard showers from SSE – 4 Ploughs at Waterlair for Pease nearly finished – and Digging up & Carting off some large stones from it – at Barns and at Sundries –

Tuesdy 11th last night and this morning a Good deal of rain and very High wind from the Southward – day fair & stormy SSW – 3 ploughs did out Waterlair for Pease – One Cart leading Turnips & out Dung &c –

Wedy 12th Strong drought from S.W. drying the Ground – 4 Carts driving Dung to Cottar Bank for Turnips, the Gardner Planted some Trees N side Corn Yard Larches and Red willow Trees Thornylee – laying Curleys &c for seed –

Thursdy 13th Strong White Frost till Brakefast time – day pleasant, pleasant sunshine – 4 Carts driving Dung and other People at Barns & sundries –

Fridy 14th Bleak dry weather from S by W – 4 Carts driving Dung as before and some Turnips – Gardner putting young Elms 2 years Transplanted into the Nursery – and at sundry other Jobs Garden &c –

Saturdy 15th fine dry weather & warm from SSW – 3 Carts driving of Dung for Turnips & Home some Turnips – One Cart to Montrose for Coals & Sundries –

Sundy 16th Cold droughty weather from SSW – Heard some few were sowing Oats near Montrose & Pease –

Mondy 17th a Considerable fall of Snow from S.W. and Cold but strong drought – the Snow Gone by Brakefast time, 4 Carts driving dung & Turnips all ye yoking – Planted one Plot Garden with Red Cabbages – Observe first sown Pease Garden above Ground & some of the Gooseberries Foliating – not much of vegetation as yet, Turnips offering to shoot from 2 to 3 acres of them –

Tuesdy 18th Blowing weather from Southw^{d} P.M. from S.E. and Morning sowed Waterlair N.W. park with 4$\frac{1}{2}$ Bolls of Pease & Harrowed same before loosing, and two Carts driving dung & leading of Turnips – at Barns and Sundries – Gardner &c Nursery.

Wedy 19th Some showers last night – this day a smart shower from 10 to 11 A.M. – 2 Carts to Monboddo with Hay, and 2 ploughs at N Burns for Oats after Turnips rather Wett – P.M. at Barns Chaffing Oats – and at Sundries in the Garden & Nursery – Potatoes.

Thursdy 20th Slight Frost in the morning – day pleasant and fine drought from the Westw^{d} – 4 ploughs at the Burns – all the day – and others at Barns Chaffing Oats for seed – Sowed some small seeds garden.

Fridy 21st Very fine droughty weather from S.W. – 4 Ploughs at finishing out all Oatseed of Season – and leading of Turnips – at Barns and Sundries.

Saturdy 22d Uncommonly warm & Pleasant day – Sowed on Upper & Nether Burns $7\frac{1}{4}$ Acres with 8B 1f Cottar Bank Oats & mostly Harrowed same pretty good season – Gardner Transplanting 5,000 Larchs into the Nursery – at Barns dighting Barley – Observe several Farmers Just Beginning but not General –

Mondy 24th Yesterday & this day very warm & pleasant weather but Foggy every night – at 7 A.M. yoked 7 Harrows to Cover the Burns Sown with 63 lb of White & red Clovers two thirds of the first, and 7 Half Bolls rye Grass mixture of Plantain – did out & Water Furred same – Sowed 4B Oats Back Y^{ds}- partly Harrowed D^{o} –

Tuesdy 25th Very warm weather from S.W. – Sowed all Back Yards 8B 1 firlot and partly Harrowed at leading Turnips One Cart P.M. at Barns dighting Pease & Sundries – Thermometer Noon Sun 90 degrees Compleat Shade 53° – fine weather.

Wedy 26th Closs Foggy weather till noon when Broke out warm & very Pleasant – 9 Harrows all day at Oatseed finished Back Yards & did part of Golds Hillock Ley D^{o} – at Barns – and Sundries – Strong Vegetation –

Thursdy 27th Foggy weather from S.E. & very warm middle of day and uncommon Clear sunshine – all Harrows at the Ley Golds Hillock Sown with 10 Bolls Oats – and P.M. lead some Turnips – at Sundries &c.

Fridy 28th Fine Pleasant dry weather from SSW – all Harrows did out Golds Hillock & part of Auld Town park, Sowed nearly whole with 13B Oats, and partly Harrowed the same – at Barns – Sowed Onions Carrotts &c principal Seeds – and at Barns &c Jobs.

Saturdy 29th about 6 A.M. Came on Some Showers of rain about 8 Heavy & stopped the Harrows – after 10 lead one draught Turnips yellow 5 Cartfull to Cattle P.M. – 6 Harrows Auld Town Ley – One Cart leading Turnips & One to Stonehaven for Coals but obliged to Mr Low there for lend there being none in it for sale – very strong drought & High wind afternoon.

Sundy 30th This day fine dry weather from S.W. and the Ground no worse order by the rain.

Mondy 31st last night a White Frost this morning a Heavy Shower about 8 A.M. of short Continuance – all the Harrows finished out the Ley Field & partly Harrowed Six Bolls of Oats Sown on the Red Hall Park – Sundries – The Quantity of rain this Month 1 Inch 3 tenths. The Fodder will not be scarce as was Expected but not much Vegetation as yet. The Gooseberries partly Foliated – Fatt Cattle some more in Demand & Grain on the Increase though the Markett is near over – Several Farms – Kinaber Northesk – Glenbervie Sold lately –

Diary for the Month of April 1794

Tuesdy 1st Early this morning Heavy Showers of rain from S.E. – fair till 11 A.M. much rain from S.W. – 6 Harrows at Red Hall Park sown with total 10B 1f Oats & Harrowed D° P.M. Two Ploughs at Ribbing of Blacklands after Turnips for Barley a Cart to Stonehaven for Coals, at Barns Garden &c –

Wedy 2d Early this morning very stormy & Heavy rain from S.E. – day fair – Sowed the dry Haugh with 12B 1f Oats and Harrowed the same – this finishes all Oats seeds total Being first done in the Country & Well – Barns &c.

Thursdy 3d Early Sowed all the Haugh with Grass Seeds 3 Men & Harrowed same A.M. – One Cart leading of Turnips P.M.- 4 Ploughs Ribbing Smithy Park – and at Sundries Barns – Wind at S.E. to SSW – Coldish.

Fridy 4th Cold Stormy weather from Northward & P.M. Heavy rain to stop work without doors – 4 Ploughs at ribbing Smithy Park – at Barns dighting Pease –

Saturdy 5th Very Strong Gale from the North with showers of rain – P.M. fair – more Moderate – 4 Ploughs at ribbing nearly out Black lands – and driving of Gravel and to road above House – & Home the Turnips many of them Flowered – Sundries –

Mondy 7th Yesterday very Bleak but dry weather till about 4 P.M. Came on rain from the Eastward which Increased with strong stormy wind all day – & Brought down the rivers very High Covering most low Grounds – no work without – all People at Barns dighting &c –

Tuesdy 8th The Storm of Wind & rain lasted without Remission till 11 A.M. when it Faired & the Rivers very High began to Subside – no work without doors at Barns dighting of Oats & Sundries; three Inches of RAIN Have fallen in 42 Hours-Stormy weather & Showers Continued all day Northy

Wedy 9th This day Cold Stormy weather from North with flirts of snow – 4 Carts driving Gravel all day to the road above the House, at Barns &c. Observe the Haugh Field last sown Oats Greatly Damaged the Soil being Carried off with the Seed in Many places & the Banks likewise fallen in so that more Seed will be needed –

Thursdy 10th Cold Bleak dry weather from Northw^{d} – Carts driving of Gravel to road above the House, leading in of Whin & Broom from Thornylee, at Barns & Sundries –

Fridy 11th Cold Bleak weather from Southw^{d} with Small Showers of rain at Intervals – Sent 4 Cartload of Hay to Monboddo- and One Cartload large of FLAX to Mill of Mundynes – leading Broom, Graveling.

Saturdy 12th Fair & Pleasant dry weather from Westw^{d} – Carts leading of Turnips and driving Gravel A.M. – Afternoon at Sowing again the Damaged part of the Haugh by the late Flood – Compute betwixt two & three Acres Hurt & Resowed 2B 1f Oats & near a Boll of Rye Grass & Harrowed the most of the field again –

Mondy 14th Yesterday and this pleasant dry & fair weather from Southw^{d} – 8 Harrows at giving a streak to the red Hall Park A.M. – The Wooden Roller Covering the Haugh Oats all day – P.M. 8 Harrows Braking in the Waterlair Park to Steer it for Bear – at Barns & Sundries –

Tuesdy 15th a light shower in the morning, day fair and good labouring weather – the Ploughs at steering of Waterlair Park for Bear, and Rolled out the Haugh Oats – at Sundries – Hear that most of the fatt Cattle in the Neighbourhood are sold at Good prices.

Wedy 16th Very fine dry weather from S.W. – three Ploughs at Waterlair – and One Cart leading of Turnips – at Sundries – little vegetation – fatt Cattle selling very high & Demand to Glasgow &c

Thursdy 17th very fine dry weather from SSW – 4 ploughs at Waterlair for Bear as before – and at Barns and sundries Covering y[e] House WELL –

Fridy 18th Moderate dew night, day strong drought from Southw[d] – Steered out Waterlair park for Bear and P.M. Harrowing of it – a Cart to Montrose with 4B Pease for Mr C Gillies & out Coals & Sundries.

Saturdy 19th Cold Harling weather from S.E. – 2 Carts Helping a Neighbour to lead Bought Barley from Cairnbegg and other Horses at Harrowing of Waterlair Park – at Barns and sundries –

Mondy 21st Yesterday and this very fine Seed Season a SSW to SSE – Broke in Waterlair Park for Bear & water furred it – & P.M. 4 Carts driving dung to said field 3 draught at a yoking although a Scotch Mile by the Temporary Bridge over Leppie –

Tuesdy 22d Very strong drought from S.E.all day – 4 Carts driving of Dung from Town to Waterlair – and at Barns & Sundries – Lean Cattle in demand –

Wedy 23d Some Slight rain through the day, but at times pretty strong as at 5 to 6 P.M. & later – 4 Carts driving dung to Waterlair all day – and at Barns dighting & sacking Barley sold for Montrose.

Thursdy 24th Very Blowing weather from S.W. – 4 Carts all day at driving of Dung to Waterlair – Barns dighting of Barley for Montrose – at Laurencekirk Tryst a Great Many Cattle but not Great Sale.

Fridy 25th Stormy weather from Southw[d] & Some flying showers – 3 Carts to Montrose with 10B Barley and five of Pease – to Sundries – One Plough at steering remainder of Smithy Park Out Coals.

Saturdy 26th Very fine pleasant weather from the South – One Plough at steering last of Smithy Park and P.M. 2 Carts driving dung to Waterlair and 3 Harrows Braking in Smithy Park for Seed Fur.

Mondy 28th Yesterday fine dry weather from Southward – this day rather stormy & showers at Intervals – Two Carts driving Dung to Waterlair all day and one Plough and three Harrows Braking in Smithy Park for Bear – at Barns and Sundries &c

Tuesdy 29th Stormy kind of weather in the night day moderate P.M. several showers of rain from Southw[d], 2 Carts driving all day dung to Waterlair now judged sufficient – 5 Harrows did out the Smithy Park & after at Braking in of Thornylee for Bear – at Barns & Sundries &c.

Wedy 30th Frosty in the night day fair but Cold & stormy from the Westward – all Harrows at Braking in of Thornylee for Bear – and at sundries on the farm – towards Stonehaven Observed the way to Stonehaven Bear seed begun and pretty well advanced they to the East are farther advanced than in the Hollow of the Mearns – at a County Meeting an Appearance of Turnpikes being brought to Effect though I much suspect it will not be in a Hurry Owing to the Money Matters – Cattle fit for driving Have been in demand for some time and fatt Cattle very scarce. The vegetation much advanced, but few fruit trees Blossomed – Grass a tolerable Bite – &c – Quantity of rain Pluviometer four Inches & One Half.

Diary for the Month of May 1794

Thursdy 1st Frost last night day fair till near 6 P.M. some showers of rain – Harrowed out Thornylee P.M. 3 ploughs begun seed fur for Bear W^{t} side of Blacklands – One Plough at long Haugh for Potatais – at Mill of Garvock with 16B Oats Meal –

Fridy 2d very strong drought from the Westw^{d} – two Ploughs at Blacklands for Barley – & one D^{o} for Potatais – One Cart to Montrose with Barley & Pease – &c –

Saturdy 3d very strong drought from the S.W. – all Horses at Ploughing & Harrowing first part of Barley seed 3B 3f West side of Field – at Ploughing for Potatoes as before – and at sundries Jobs –

Mondy 5th Yesterday Bleak dry weather – this day better but drought – 3 ploughs for Bear, and One Plough at long Haugh for Potatoes – a Sale of Fatt Cattle here today, little demand & few Purchasers 5 : Sold 10 Cattle for Montrose about 4/9 per stone – Kept three –

Tuesdy 6th Moderate weather with some light showers from N.W. – Sowed 3B Chester Bear & three ploughs at seed Furr – One Plough & Harrows for Potatoes &c Jobs Sowed the Annual Flowers Garden – Sundries –

Wedy 7th Strong drought from N.W.- & S.W.- three Ploughs at Blacklands & One Cart for Meal &c Mill of Garvock – at Sundries – Ground very dry –

Thursdy 8th a Fine Shower of rain at daylight – day strong drought from S.W. – Ploughs nearly did out all Blacklands and Drilling & planting of Potatoes on the long Haugh – at sundries –

Fridy 9th This day very strong drought from Southw^{d} and slight frost in the night – 2 ploughs did out Smithy Park 2B 3f Barley & 6B 3p & 2 Lipp of Chester & Harrowed same for Grass Seeds – did out the Potatoes near 7 firlots about an Acre of ground – Sent a Cart to Whitston for 5 Bolls Rye Grass –

Saturdy 10th Very early morning Came on Heavy rain from S.E. which lasted till near 8 A.M. & went deep into the Ground much in need of it – all Ploughs at seed furr E^{t} side Thornylee now in fine Condition – at Barns dighting Seed Barley &c –

Mondy 12th Yesterday windy & some showers – this day moderate with last morning very White frost and Injuring the Grass & tender vegetables – Early Sowed all Smithy park with $5\frac{1}{2}$ Bolls Rye Grass – and 59 lb Ams red & 29 D° White Clovers and Instantly Harrowed the same – P.M. Rolling of it – three Ploughs at Thornylee as before, Barns &c.

Tuesdy 13th Frosty clear every night day very strong drought from S.W. – all Ploughs and Harrows at Thornylee pretty ill to do but in good order, at Barns dighting Seed Bear – delivered 6 fatt Cattle this day –

Wedy 14th This day strong drought from South Westw[d] – Ploughs and Harrows at Thornylee – and at Barns & sundries, Bleak P.M.-

Thursdy 15th Pleasant dry weather and warmer than for some time past – all at Work as yesterday on Thornylee for Bear easier to do on the West side, and a Cattle Markett at BERVIE revived but no Great Sale though a number of Dealers –

Fridy 16th Warm pleasant dry weather from S.E. – all people at Work as before at Thornylee Bear seed – at Barns dighting Bear last year Waterlair sown the 6th June Weighed this day 18 Stone per Boll Montrose Measure & no dightings – at sundries – &c –

Saturdy 17th Pleasant dry weather & an appearance of a shower towards Evening – finished sowing of Thornylee 11B-2f & nearly Harrowed it out – a Horse Roller all day at Waterlair all day-Turning a Dunghill beside Cottar Bank

Mondy 19th Yesterday morning a fine shower of rain, day growing weather – this day Strong drought from S.W. – did out Thornylee – & P.M. 4 Carts & Seven Men begun Dunging W[t] side of the Waterlair Park for Bear – at Barns &c &c.

Tuesdy 20th a very strong storm from N.W. which Cheequed vegetation very much – 4 Carts driving out all Dung on Waterlair & some from the Farmhouse – P.M. One Plough at seed furr & 3 Carts – Others spreading Dung – Sundries.

Wedy 21st Cold & Stormy weather – day more Moderate & Evening a little rain – Dunged out WaterLair and Ploughs at D° – Sowed about 6 firlots on West side Park & Harrowed partly D° – Sundries.

Thursdy 22d about Midnight Came on Strong Stormy Wind from North with some rain, day Squally &c – 3 Ploughs & 3 Harrows all day at WaterLair Sowed & partly Harrowed 2B more – at Sundries

Fridy 23d About 4 A.M. Came on pretty strong rain from S.E. which lasted till past noon & wet the Grounds – Ploughs & harrows did out the whole of the BEAR SEED of season in fine order – Sundries – 5B-11p Being 2 pecks on Yards & sowed them with Grass Seeds.

Saturdy 24th Cold & Stormy weather from Northw[d], but very Early Two Men by 3 in morning sowed the Park of WaterLair with Grass Seeds, Harrowed, Water furred, & Rolled the Whole – also Yards – Eight Harrows Braking in for Turnips &c.

Mondy 26th Yesterday and this Cold Bleak but dry weather till P.M. some showers from North – all Harrows at Braking in Cottar Bank for TURNIPS and

P.M. 4 Horse Plough at rough Ground new benorth the Bog Planting – Put Horses on Grass &c

Tuesdy 27th Bleak Coldish dry weather from the N.E. – One four Horse Plough at Cottar Bank – Harrows – roller stone on the Burns Oats – One Cart for Lime shells to Mathers for Pointing the House &c Jobbs –

Wedy 28th This morning about 3 A.M. Came on very heavy rain (1 Inch rain), and High Wind from the N.E. which lasted till 8 A.M.- then showers at Intervals – 3 Carts Fliting a Cottar from Boghall to this – and sundries &c

Thursdy 29th a very fine pleasant day & warm middle of it – two four Horse Ploughs at West side of the Cottar Bank for Turnips – in Barn last Bear, and at sundries – no great vegetation as yet –

Fridy 30th Moderate weather & Growthy from N.E. – 2 Carts to Flit a Cottar here & two Ploughs at C Bank for Turnips – At Sundries a little more vegetation –

Saturdy 31st This day very pleasant & warm weather wind N.E.- 2 Ploughs at Cottar Bank – and at Sundries. The Pasture tolerably Good & Hay not amiss here, though very light as yet Generally: The Oats & Bear Breer looking well here. The Quantity of rain this month $1\frac{1}{2}$ Inches. The Fruit Trees Cannot be Said as yet Sett and the Gooseberries though rich yet Generally ravaged by the Caterpillar – and an Earth Worm Called Tom Vulgo Hath Spoilt much Oats & Lint in some places – the prices of Fatt Cattle under 5/- Victual High & demand.

Diary for the Month of June 1794

Mondy 2d Yesterday very warm Therm^{r} at Noon 67° Shade. This day same all Ploughs at Cottar Bank – Warm Th^{r}62°- Wind NE- very great drought –

Tuesdy 3d Strong drought from the Eastw^{d} – Ploughs at the Cottar Bank &c – at Sundries on the Farm – The Bear Seed is now done in general –

Wedy 4th Very dry weather & much need of rain from North – 2 Ploughs at Cottar Bank and others at sundries – the Caterpillars ravaging small fruit –

Thursdy 5th very strong drought from North – Sun warm – Ploughs at Cottar Bank – P.M. preparing & Dunging a little bit of Haugh for Harvest Turnips –

Fridy 6th Showery weather & rather Cold from the North – 2 carts to Montrose with 7 Bolls Bear & Out two Bolls of Scotch Coal – 2 Ploughs at Cottar Bank – Sowed y^{e}Turnips –

Saturdy 7th Strong drought from the Northw^{d}, ploughs and Harrows at Cottar Bank, and others Cottars at Gathering & turning dung for Turnips &c –

Mondy 9th Yesterday & this day weather Inclined to Frost in the Night, Harrows &c at Cottar Bank – the drought will ruin the Hay in most places & Hurt pasture –

Tuesdy 10th a Slight Shower with wind from the S.W. – day mostly fair – 3 Ploughs & three Harrows at the 3^{d} furr Cottar Bank – rebuilding a Cottar Byre – Observe the Ground for Turnips far Back in General.

Wedy 11th Blowing & very droughty weather from S.W.- with some showers of rain scarcely felt – all Ploughs and Harrows at Cottar Bank – Heard that Cammock Muir went well &c.

Thursdy 12th very warm dry weather – Ploughs and Harrows at Turnip land, and stone roller on D^{o} – At Trinity Muir a More than usual number of cattle and a Considerable Sale at Higher Prices than usual Except Fatt Cattle not Saleable at all – Had two there so – Crop to Markett tolerable.

Fridy 13th very great & warm drought Continues from the Southw^{d} – at the Turnip ground as before – Hear there was very little Sale of Horses –

Saturday 14th Very Early this morning a tolerable shower of rain but of short Duration and warm & severe drought from N.E.- Harrows on the Turnip Ground and two Carts at Johnshaven for Smiths Coals –

Mondy 16th Yesterday very warm & dry weather Thermometer in the shade at Noon 59° degrees – this day very warm wind Easterly and Harrows at Cottar Bank A.M. and four Carts driving dung to East side of Cottar Bank afternoon – Holing Broom and Sundries – The Crop Suffering much –

Tuesdy 17th Uncommon strong drought from the Eastw^{d} and very warm Therm^{r} about 60 degrees – 4 Carts at driving out Dung to Turnips & Cottars Spreading it middle of the day on East side of the Field – The Season for Sowing these will be very late in the General, little dung being yet out and scarcely any yet sown – The Hay Crop will be very light Generally my own amongst the Best in the Country as well as Pasture & my Cattle uncommonly thriven at this Season –

Wedy 18th Frosty dew in the night – morning very Hott – 4 Carts driving dung A.M. and, P.M. begun Drilling E^{t} side Cottar Bank & sowed it – & three Carts driving Dung on D^{o} – Spreading dung Middle of Day.

Thursdy 19th very warm weather & Excessive drought – parching every vegetable – drilling Turnips Cottar Bank and three Carts dunging at D^{o} very rich – Observe the Corns Spotting in many places but little of it here on this farm – Sundries –

Fridy 20th This day dark dry weather from the Eastward – all People at Turnips & Horses, & towards Aberdeen Observed the Crop looking worse than with us – and about Dee side the Grass quite Burnt up – much rain required.

Saturday 21st Weather as yesterday – drilling & sowing of Turnips & finished all the first Half of Cottar Bank about 9 Acres in fine order. Observed many of the Turnips Sown towards Aberdeen much earlier than formerly and almost Generally in drills –

Mondy 23d Early this morning a tolerable shower of rain which went about one Inch only into the Ground – Ploughs & Harrows at West side of Cottar Bank for Turnips and others Holing Broom &c Jobs – will refresh the Crop –

Tuesdy 24th Fair in General, some showers with Thunder about noon from S.W. – 3 ploughs at Turnip Ground & three Harrows Sett up the Cattle Flakes Robs Hillock – Mending Dykes &c Jobbs

Wedy 25th very warm weather – 3 ploughs & Harrows at Cottar Bank- and about

4 P.M. Came on very Closs & Heavy rain from the Eastw[d] which lasted for Several Hours till near 9 P.M.

Thursdy 26th dry weather & Moderate from the Northward – all ploughs at Cottar Bank finished out the Ploughing before the Dunging & also Most of the Harrowing of D° – Sundries –

Fridy 27th Fine pleasant dry weather a little frost in the night – 3 Carts driving dung to Cottar Bank all day Harrows & roller on roughest of D° – Others Spreading Dung – at Sundries &c

Saturdy 28th very strong drought from S.W. – 3 Carts driving Dung from the Town all day, and one Plough at Drilling two Yoking – Sowed & rolled the same P.M. in fine Order &c

Sundy 29th Excessive Hot weather – Breeze S.E.- Therm[r] 10 A.M. and at Noon 70 degrees which is as high as I ever Observed it in North Britain –

Mondy 30th Very warm weather Th[r] 67° shade noon Wind a Breeze from S.E. – 3 Carts driving Dung all day for Turnips Cottar Bank – Others Spreading it – drilling Sowing and rolled what was Sown – the Major part Breerd and looking fresh & well – The Quantity of rain this month is 0.75 Inches – from first of June 1793 to the last of May 1794 was 29 Inches & three fourths – There is great demand for lean Cattle & Cows as also Small Milch D° to the Southw[d] – prices of all Grain high – Oatmeal 18/ – Barley from 18 to 20/ per Boll – all kinds of Manufactures very Brisk & prices high – Caterpillars on Gooseberries Stopped.

Diary for the Month of July 1794

Tuesdy 1st This day very warm weather from various Th[r]68° – Ploughs and Roller &c finished sowing &c of the Turnips Extraordinary Dunged from 13 to 14 Acres – P.M. Horse & Hand Houghing Haugh Potatoes &c Jobbs – Pease & Strawberries a few.

Wedy 2d last Evening a small shower – this day very warm Therm[r] Noon Shade 71°- real Hott in the House – Ploughs & Harrows at piece fallow South End of Cottar Bank for Liming – Houghing Potatoes –

Fridy 4th Very strong drought from S.W. – Th[r] noon 72 degrees – 6 Men at White Water Moss, from 4 to 5 Miles South of this, belonging in part – four or five Acres of it on East Side – to this Estate, Casting & landing very Good Sods, real vegetable Moss – seemingly mostly Alder trees Mossified –

Saturdy 5th very severe drought from Southward – Th[r]70° noon Three Carts to Mathers for Lime to Cottar Bank for 11 Bolls – Others Cleaning & Weeding of the Hedge around Bogg planting thriving – Sundries –

Mondy 7th Yesterday very Hot & Sultry weather – this day same Th[r] at noon 72° – 3 Carts to Mathers for Lime and others scouring ditches & Cleaning Hedges –

Tuesdy 8th Th[r] 73° very warm & Excessive dry weather – 3 Carts driving Gravel to the Publick road statute Labour all day – Day labourers at assisting the Gardner in weeding & aligning of Rose planting West side Garden &c Jobs.

Wedy 9th very Hott & Sultry weather – 3 Carts to Mathers for Lime as before – Scouring Ditches, and flited the Cattle Flakes Robs Hillock – Sundries –

Thursdy 10th Blowing weather & very strong drought with some light skiffs of rain S.W. – being Fast day no work without doors – Th^{r} 60° – much Cooler –

Fridy 11th Still Continues very strong drought from the Southw^{d} – 3 Carts to Mathers for Lime to Cottar Bank and Scouring of Ditches – towards Montrose Observed the Crop very stinted & but poorly – the same alongst the Coast from it –

Saturdy 12th Strong drought Continues – leading Home Broom &c for Fire morning – Midday Church, and P.M. driving Gravel to Cottar Town road – Th^{r} 66 degrees – never was more Occasion for RAIN.

Mondy 14th Yesterday warm & dry weather – Strong drought S.W. Th^{r} 67° – this day same but the Air moist both yesterday and afternoon & Evening – Carts driving Gravel from river side to the road leading from the Publick one to the Mansion House – People at Church and P.M. dighting & sacking Barley & Pease Sold for Johnshaven and Montrose – need of rain –

Tuesdy 15th Bleak and dry weather from the Southward Th^{r} 65° three Carts driving Gravel to road as yesterday and Others dighting & sacking Barley – Scouring of Ditches and at Sundries on the farm –

Wedy 16th Th^{r} 68° Uncommonly Hot weather and drought Scarcely Observed so Severe before – Weathing almost every vegetable – the Barley in some places Whitening before being Shott & Pease same also Garden Plants – the Hay &c most People are only begun to Cut here as yet People driving Earth to Dung Court and others Casting Sivers Cottar Bank & filling them with stones – Gave second dressing to the Whole Potatoes Haugh – Turnips looking well –

Thursdy 17th Very strong drought as before – Sent three Carts to Johnshaven with 12B Barley and Brought Home rigging Stone from Laurestoon – at Paldy Fair not so many Cattle as usual & very little Sale at reduced prices Owing to ye drought.

Fridy 18th Drought severe as before and no appearance of Alteration – One Cart to Montrose with 3 Bolls Pease Sold & for Sundries – Two Carts driving of Stones to Sivers foot of Cottar Bank chiefly off sowed Field and river side – Others Carting D^{o} –

Saturdy 19th Uncommonly warm weather from South – Therm^{r} in shade noon 76 degrees – much higher than I ever saw in this Country before – 2 Carts to Johnshaven with Barley 8B 3f and brought Home Coals in return – Others filling & Covering Sivers – finished D^{o} – P.M. flited the Cattle & Sheep Flakes R Hillock – rain at 6 P.M.

Mondy 21st last night about 10 Came on pretty smart rain which lasted most of the night day fair but Thunder in the Northw^{d} – layed up fur to Potatoes Haugh – Harrowed the Drained S Side Cottar Bank begun Cutting Hay.

Tuesdy 22d very strong drought various – Sent to White Water Moss for 3 Cartful sods – Others at Hay and after Dinner Begun Coling of West Side Bridge Shade and at Sundries-

Wedy 23d Drought & very warm Th^{r} 68° noon – people at the Hay Cutting & Coling, rain Came on about 7 P.M. from the Northw^{d} – Sundries.

Thursdy 24th Very heavy rain & Stormy from Northward most of the night & this morning which brought down the river a little & layed some of the rankest Oats – pretty deep in the ground – Brought Home some Cart load sods – Others Cutting & Coling of Hay – the Turnips advancing very slowly – Th^{r} noon 59°

Fridy 25th Cooler weather from the Southw^{d} – Cutting Hay till Brakefast own people & one Hired one all day at Waterlair North park – All others Gathering & tramp Coling D° Bridge Shade & Kavels – The K Mare Foaled – Vegetation not Increased –

Saturdy 26th Some light skiffs rain but very light from the S.W. – all People & some others either Cutting or Coling of Hay till 7 P.M. – most part in very good order; weather now Cooler – at Sundries.

Mondy 28th Yesterday Cooler & very dry weather from the S.W.- this day same, all people and Several Hired Ones at the Hay Gathering & tramp Coling of D° – did out Bridge Shade 22 Coles and what of the Kavels was Cut – 7 Men Cutting.

Tuesdy 29th Very dry and ARID weather – all people & Hired at Cutting & Coling of Hay at Home and at Waterlair after Brakefast – at Garvock few Sheep and those not Cheap Bought a Dozen Yield Ewes at 3/9 each –

Wedy 30th Dark Gloomy weather from the E.N.E. and Moderate but much drought – people finished Cutting of all Hay for Season 35 Acres, and all tramp Coled but from 6 to 7 Acres – and at Sundries.

Thursdy 31st Bleak and dry weather from S.E. – all people at Garvock fair Engaged all my Servants the present remaining at the former Wages although Heared they were giving high wages to Angus – Quantity of rain 1.60 Inches – On the whole as the most of the rain fell on the 21st and the drought great everything was burned up & the Crop stinted – the Pease near ready to Cut, Oats well shot, & Bear turning-Hay a light Crop more so than last year.

Diary for the Month of August 1794

Friday 1st About 5 A.M. Came on pretty Smart rain from the East which lasted most of the day & stopped work without doors after Dinner – Horse & Hand Houghing of Turnips begun – and at Sundries –

Saturday 2d The rain was Heavy till near midnight & this morning Strong Gale from the N.W. & very Cold about Miday moderate & drought – People Hand Houghing Turnips Cottar Bank – Flited the Cattle and Sheep Flakes – & P.M. three Ploughs at the Fallow on West side Cottar Bank and at sundries Garden &c. –

Mondy 4th Yesterday very Heavy rain & river much out – this day Cold Stormy from N.W. – Sent three Carts to Garrols Wood for Fir Bought – and a Plough at Taking Away fur from Turnips – P.M. Men dressing of D^{o} – Showers Evening – at Sundries – Sowed Spring Plants.

Tuesdy 5th very strong drought last night – Early begun with our own & Hired Hands to number of 20 Threshing Hay Waterlair & tramp Coling D^{o} – Got from 10 to 12 Bolls very fine rye Grass Seed – Horse Houghing Turnips.

Wedy 6th very fine pleasant dry weather from S.E. – about 20 people at Threshing remainder Water Lair Hay 6 Acres which did out about 8 Bolls, & put into very large Tramp Coles 7 – a Plough at taking away a fur from Turnips –

Thursdy 7th dry weather till 4 P.M. showers from the Northward finished out all Hay of season – P.M. Flited the Cattle and Sheep Flakes Robs Hillock – Houghing Turnips. Thatched the Cottar Houses – and at Sundries.

Fridy 8th Cold Bleak dry weather from the North – Sent 3 Carts of my own and about One Dozen Borrowed from Neighbours for Slates to Flemington for long Loft – a Plough Horse Houghing of Turnips & 4 Hired Men D^{o} –

Saturdy 9th Pleasant dry warm weather from S.E. – The Carts returned with the Slates from 6 morning to before 2 P.M. – One Plough at Turnips own Hired people at Turnips &c Jobs – Observe the Turnips in Broad Cast rather better than Drilled.

Mondy 11th Yesterday strong drought & this day same from N – People & Cottars Cutting Sprotts & rushes for Thatching Hay Sowe 70 feet by 15 – & lead Home them – a Plough at Fallow Cottar Bank &c Jobbs.

Tuesdy 12th very strong drought from various N a SE P.M. and very warm – 4 Carts Early begun to lead Home the Water Lair Hay into a Sowe 70 feet by 15 – did not get Home quite the 10 Coles-though busy – Sundries.

Wedy 13th very Hott weather & drought from S.E. – 4 Carts all day at leading all Bridge Shade Hay & part Kavels – 20 people Employed busily & loosed near 8 P.M. – Horse Tramped both days & at sundries &c Observe some People Cutting Pease a withering.

Thursdy 14th very warm and dry weather from the S.E. – 4 Carts & 20 people leading Home the remainder of Hay in fine Condition & finished the Hay Sowe – & a stack over – flited the Cattle Flakes P.M. – and at sundries –

Fridy 15th very pleasant dry weather from the Southward Thatched the Hay stack, dighted up the Rye Grass Seed Saved in putting up the Hay about 6 Bolls of very fine seed full of Plantain and at Sundries, Bedded the Pidgeon, Hen & Horses.

Saturdy 16th This morning some showers of rain from S.W. day pleasant – thatched out the Hay Sowe, Flited the Cattle Flakes – the Carts returned with the slates at 3 P.M. and sundries – The Turnip looking well in some parts of mine but in General the Crop quite failed – got all the Slates 2000 at 6 score to the Hundred – a second Growth amongst the Oats – Turnip seed Sowed 39 lb English.

Mondy 18th Yesterday pleasant weather from the Southward and this morning a small shower – Ploughs at summer fallow – and after at Banking of the River above the Mill Dam with whins stones, Gravel, Old Banks &c, at sundries –

Tuesdy 19th some light showers at Intervals – three Carts driving stones to Bank the river below the Bridge morning & P.M. – a Plough at taking away fur from last sown Turnips & Sundries –

Wedy 20th very pleasant & warm weather various – One Plough at taking away fur from the whole Turnip now promising pretty well & own and Hired Cottars at dressing the most Luxuriant &c.

Thursdy 21st very pleasant drought from the Northward – people Hand & Horse Houghing of Turnips – at Laurencekirk fair Sold five Cattle at tolerable good prices better than heretofore, and bought two small ones reasonable about 3/ per stone – not above One Third formerly in the Markett.

Fridy 22d very pleasant weather – all people at Turnips and P.M. flited the Sheep & Cattle Flakes Robs Hillock – and at sundries – rather showery towards Evening – a second Growth becoming pretty much amongst best Oats –

Saturdy 23d about 7 A.M. Came on a Smart Shower of rain from S.E. – people at Turnips – and at Nursery & Belt of Planting West side of the Garden and Planting Coleworts &c – more or less rain all day.

Mondy 25th Yesterday generally fair with skiffs – this day same – Sent three Carts to Stonehaven for Scotch Coal – Others at Houghing Turnips Holing Whins & dressing Belt of Planting W S Garden.

Tuesdy 26th Some light showers P.M.-3 Carts to Stonehaven for Coals and Others at Hand Houghing of Turnips and at Planting Coleworts Belt – digging & Holing Whins &c Haugh –

Wedy 27th Some Showers Heavy at Intervals all day – three Carts to Stonehaven for English Coals and others at Turnips Houghing – and planting Coleworts & Swedish Turnips on Blank Ground of the Turnip Drills – and at Sundries –

Thursdy 28th Coldish drought from N.W.- People amongst the Turnips 2^{d} dressing – and One Cart to Stonehaven for English Coal – P.M. 5 Harrows at the summer Fallow – and Flited the Cattle and Sheep Flakes – many People begun Harvest.

Fridy 29th Dry weather from the Northw^{d} – 2 ploughs all day at laying to furr to Turnips – and 8 others at 2^{d} dressing of D^{o} Previously – and at sundries.

Saturdy 30th dry Bleak weather from the Northw^{d} – Sent three Carts to Montrose for Wood to roofing the Long Loft – and One Plough at Laying to fur to Turnips – Others Hand Houghing D^{o}

Sundy 31st Dark thick weather but dry – not very much yet Cut down and the Crop very light – Greens in most fields either of Bear or Oats – The Quantity of rain in this Month is Two Inches & One Half – Observe the Potatoes Scabbed or Spoilt on the surface owing to the rains – The Turnips in General not above One Third or fourth of a Crop – No Second Growth of Clover or other Grass – Yet a demand for Cattle – Manufactures & trade Brisk –

Diary for the Month of September 1794

Mondy 1st This day dark weather from S.W. – after 2 P.M.a slight shower – about 11 A.M. begun the HARVEST with Water Lair Pease – at the Turnips morning & Brought 4B Lime for Slating –

Tuesdy 2d Bleak weather about 6 A.M. Came on rain at Intervals very Heavy from the North – after 10 A.M. fairer till after Miday when Heavy rain most of the rest of the day & brought down the River Considerably towards night – at petit sundries. Flited y[e] Flakes.

Wedy 3d Strong drought from the Northw[d] & very Hott in the Middle of the day – all People after Brakefast at Cutting up Smithy park Bear rather ripe & but light although of fine Quality – Pulled the LINT –

Thursdy 4th White frosty dew morning, day fair, but not much drought from the S.E. – after Brakefast all Shearers at Bear as yesterday – morning at putting lint water, leading Sundries &c – Jobbs –

Fridy 5th Variable weather from the Eastw[d] – after Brakefast at Barley Cutting Black Lands, but stopped by rain before 11 A.M. – after at Rushes &c Cutting & after Dinner at Oats Cutting on Knowe Goulds Hillock 2[d] Growth &c Jobs-

Saturdy 6th Fair weather from the N.E. but dampish – all Shearers at Cutting up Oats Back Yards & Haugh, by Second Growth two thirds of them Green so as to spoil the Crop – 2 Carts to Montrose for Wood –

Mondy 8th Coldish dry and very fine weather for Harvest – flited the Cattle & Sheep Flakes before Brakefast and after at Shearing Haugh Oats – Sundries –

Tuesdy 9th Bleak dry weather from N.E. – finished the Haugh Oats & after at Back Yards Oats – Observe severals Leading Pease & Bear – Sundries –

Wedy 10th Cold Bleak weather little drought from Northward – after Brakefast at Shearing Oats – Cow folds – and Turned the Pease Water Lair – Sundries –

Thursdy 11th Strong drought from Northw[d], Early begun Leading of Black Lands Bear into 2 stacks and after Lead Home Water Lair Pease into 3 Stacks in good order – Others Shearing Barley, Oats &c.

Fridy 12th very fine Harvest weather from Northw[d] – all People Shearing Oats Goulds Hillock which finished – The Oats by no means Equally ripened & ill to Make Clean work of – Closs Harvest General –

Saturdy 13th a Slight Skiff rain at 8 A.M. – day fair & pleasant all People at Oats Back Yards & N. Side red Hall Park an Excellent Crop – Thatched Pease & Bear Stacks P.M. – Flited Cattle & Sheep Flakes robs Hillock Ley – Ploughing down Lime new Ground Cottar Bank.

Mondy 15th Yesterday fair weather – but White frost morning which Touched the Potatoes – This White frost morning & a shower of rain at 4 A.M.- day fair & drought from S.E.- at Horse Houghing out of Turnips & spreading out Lime on Fallow N.End of Fallow Cottar Bank – and most of the day at Water Lair Bear fine Crop, but not good Shear – and at Sundries &c –

Tuesdy 16th Early this morning a very strong Gale of wind from S.S.W. and a good deal of rain fair till past 2 P.M. when Came on Heavy rain attended with stormy weather from the S.E. which lasted all Evening & Stopped all Work without Doors – at Barns &c – Cutting rushes &c morning & Ploughs ribbing Summer Fallow & D^{o} down Lime on D^{o} – about half Cutt down –

Wedy 17th Severe stormy weather from the Southw^{d} and much rain till 8 A.M. when fair – Cutting rushes &c till near noon – when Cutting up Oats red Hall Park finished – flited the Flakes &c Jobbs –

Thursdy 18th morning fair – begun Cutting Oats Old Toun park and Gibsons D^{o} – about 8 rain but fair & shearing till 2 P.M. when Came on Heavy rain till 4 P.M. and showers afterwards – at Barns & sundry Jobbs –

Fridy 19th Moderate weather & drought from Southw^{d} – Shearing Sprots &c for Thatch Bog Planting till Brakefast after at Old Toun & Gibsons Parks till night, Oats a rich Crop but not weighty – at Fallow morning Laying up in the rib, as well as the new Limed Ground benorth the Bog planting -added new to the Cottar bank – which Makes ye ridges about Five Hundred Yards long from the Publick road –

Saturdy 20th about 2 A.M. Came on very Heavy rain from the Eastward which Brought down the rivers and Deluged the fields – before noon begun Shearing of Oats Gibsons park which finished before Dinner. Threaves on it Old park 240 Th^{s} – about 3 P.M. Came on rain when dismissed all shearers, flited Flakes.

Mondy 22d Yesterday fair dry weather till afternoon some showers. This day about 8 A.M. Came on rain from the S.E. Excessive Heavy & stormy till noon – no Shearing & Dismissed the people till better weather. Souring Lime for Slating and variety of sundries.

Tuesdy 23d Early this morning Came on Heavy rain & stormy weather from S.E. which stopped all work without doors & Brought the river down very High – Obliged to Shift some of the Corns on the Haugh Floating on Higher Grounds – Water begun to subside before 2 P.M. – faired before noon –

Wedy 24th Fair weather and Drought from S.W. a N.W.- all People at Cutting up Water Lair Bear which finished by 4 P.M.- Threaves 143 – after at West side of Thornylee D^{o} – both too ripe – and much loss by Heading &c – Setting up Oats &c

Thursdy 25th a very strong Gale from Northward which Hath shaken the Oats I should suppose to Equal the Seed – and all day at Thornylee Bear very thin Crop & much lost in Shearing – the stooks overturned &c –

Fridy 26th very Strong Gale from the North with flying showers at Intervals – all People Shearing whole day & finished the Thornylee & all Bear season – Th^{rs}193. Observed several Neighbours Leading Bear & Oats without Picking which was Surely not good work.

Saturdy 27th very strong drought & cold weather from the Northward Turned over Haugh Oats Early & lead them after Brakefast 2 Stacks & Black Lands Barley 56 Threaves with some Thornylee Bear Top & a stack of D^{o} Bear – Others shearing of Back Yards Oats very much shaken several Bolls per Acre in some places One Half-Sundries

Mondy 29th Yesterday very strong Frost in the morning which totaly destroyed the Potatoe stems, Indian Cress this day very warm weather – Early yoked 4 Carts & Lead Home all Water Lair Bear into 2 large Stacks by 1 P.M. – after leading Ley Oats Golds Hillock & Back Yards into mostly 4 large Stacks – Shearing not dry morning –

Tuesdy 30th Foggy weather & frosty rhime morning – day fair and drought from S.W.- very warm – 4 carts leading Oats whole day – and Others Shearing of the Oats Burns whole day Mostly Gayte – Quantity of RAIN this month is Inches 3 Inches – Some few Have done with their whole Harvest, and others not much more than Half – the Oats in some Places Half the Crop Shaken – Others less &c There is an uncommon demand for Cattle for England & High prices – fatt Cattle Scarce.

Diary for the Month of October 1794

Wedy 1st Strong dew Early morning – All Women &c Shore Out all the Crop by 4 P.M. – and the Carts lead Home all the Bear Crop before dinner – One of the stacks of Haugh Oats lead last Saturday begun to Heat & Obliged to Turn it – y[e] other will follow –

Thursdy 2d Fine droughty weather from S.W. – people Cutting up of Rushes &c for Thatch to Staks till Brakefast, then 16 people with two Iron rakes at least Teethed so – at Thornylee Bear all day Gathered 2 large Cart load of it about 4B – Turned the other Oat stack & Oats on Timbers & thatched 2 Bear stacks.

Fridy 3d Fine dry weather – Shearers at Cutting rushes till Brakefast after it 14 people & two rakes at Waterlair Gathering Bear till 3 P.M. – 1 Cart Load rain S.E. Came on & Employed Barns – Others Thatched all Bear Stacks & some of the Oats before the rain & Build up a Stack Oats –

Saturdy 4th rain Continued from S.E. all night & morning – the wind to the N.W. about 11 A.M., but the rain Increased & stormy – the river much out by 2 P.M. no work without doors – People a few Generally Dispersed at making ropes – Barns &c.

Mondy 6th Gave the Feast Yesterday very raw weather & some rain P.M. – this morning Early Excessive Heavy rain from Northw[d] which Brought down the rivers very much – flited the Cattle & Sheep Flakes – at Barns dighting up Brocks of Oats & threshing D[o] of Bear & sundries.

Tuesdy 7th Fair dry weather & tolerable drought from S.W.- Cutting rushes Thornylee Brae & Mill Lead & Brought them Home, and after Dinner Bound & lead Home Nether Burns Gayte in tolerable order & Thatching of Stacks –

Wedy 8th very strong Frost in the Morning Half an Inch into the ground & quite White till 8 A.M. – Thatching Stacks all done but one, & Brought in whole Crop – several showers after 3 P.M. – and at sundries –

Thursdy 9th very fine pleasant droughty weather from N.W.- the Ploughs at Fallow Cottar Bank nearly Laid up in rib – at Drumleathie Markett a Good Many Cattle & Many Buyers – Bought two Small Cattle very reasonable about 2/6 to 3/ per stone when fatt – much Corns out.

Fridy 10th Frost in the morning, about 9 A.M. Came on rain with stormy weather from East but not Cold – Ploughs finished Laying up Fallow Cottar Bank in the rib by Brakefast time – after at Barns threshing Gatherings of Bear, and at sundries within Doors –

Saturdy 11th much rain & very stormy weather from S.E.- rivers out day various – Heavy rain after 2 P.M. – flited the Cattle & Sheep Flakes Robs Hillock – at Barns – Cutting 2^{d} Growth Bridge Shade for Horse Suppers – Pittarrow, Polburn & several farms done shearing this day –

Mondy 13th Yesterday pleasant weather from S.W. – this day same but Frosty in the morning – 3 Carts driving Dung to Upper part of Robs Hillock Ley very thick & spreading of it – Several Men, Women & Children at digging up the Potatoes Haugh & Pitting of them Corn Yard – and Many Leading this day –

Tuesdy 14th Fine dry weather but P.M. Stormy from the East with appearance of rain – 3 Carts moving dung all day from past 7 A.M. to 1 P.M. to Ley – Others at Digging & Pitting Potatoes about 10 Bolls finished – Horses on Hay Foggage very fine Grass – Sundries –

Wedy 15th more wind than rain & Mild weather from S.E. – three Carts driving Dung to Robs Hillock ley and Cottars & others Spreading of it – after Sent over the Cattle Flakes to Water Lair & all the Yield & feeding Cattle to Eat up fine Foggage there – Clipping Garden Hedges – and Threshing some Oat straw for Horse Suppers.

Thursdy 16th Early this morning Heavy rain from the Southward, Middle day fair, & after 2 P.M. again very Heavy rain – 3 Carts driving Dung to Ley as Yesterday – Others Spreading D^{o}- Observe the Crop is not yet Settled at Dava in Garvock & Sundry other places, no prices yet of Grain &c.

Fridy 17th Variable weather from the S.E. & Grounds very Wett – 3 Carts driving dung to Ley as before & Cottars spreading, Cutt last of Grass for Horses WaterLair & others digging & Casting of Earth Dung Court &c.

Saturdy 18th From 7 to 11 last night severe storm of wind & rain from S.E.- this day variable – 3 Carts driving dung to Robs Hillock & Cleared the Dung Court – Spreading D^{o} – at Barns, and at sundries – Grounds very wett –

Mondy 20th Yesterday fair, this day Same – 3 ploughs at Ribbing Angle wise the Pease park of Water Lair, and One Cart driving Dung Robs Hillock and at Barns & sundries – Crops not all inside.

Tuesdy 21st Frosty morning day drought from the Westw^{d} & pleasant weather – 3 ploughs ribbed out and Water furred the N.W. or Pease park Water Lair – Cleared the Cottar Town of Dung on Ley – Spreading D^{o} & Sundries –

Wedy 22d Dark weather & wind Westerly drought – 2 Ploughs Ribbing of Thornylee for Pease & Turnips – and two Carts driving & Gathering stones on said Field to Bank the river below the new Bridge – all the Bestials Here & Water Lair on as full Grass as Season.

Thursdy 23d very fresh Breeze from the North & showery at Intervals- 2 Ploughs at Thornylee ribbing for Turnips and two Carts leading off stones from D° to Bank the Water below the Bridge, at Barns & Sundries Jobs –

Fridy 24th Fair weather from the Westw^{d}, but still most of the Grounds are Wett – 2 ploughs ribbing Thornylee & two Carts Gathering & leading of stones to Bank the river – at Barns dighting Oats & flited ye Cattle Flakes on Water Lair – Gardner Clipping Hedges &c.

Saturdy 25th Fair weather from the Westward till 1 P.M. Came on rain which stopped work without doors, at Barns, threshing &c Broke up a piece of Old Sward at N.W.Corner Thornylee, Bog & Stones – Other Ploughs ribbing for Pease & Turnips D° field – In Hay for Horse Suppers &c.

Mondy 27th Yesterday pleasant dry weather from the S.W.- and this day same – 3 Ploughs Ribbing at Thornylee & at Casting drains new Broke Ground, at Barns dighting Bear – Sent $4\frac{1}{2}$ B Oats & One D° Bear to Mill of Kair –

Tuesdy 28th pretty strong Frost – day fair & Coldish dry weather from N.W. – 3 Ploughs at Thornylee ribbing nearly finished – and One Cart driving out Dung & Home Earth for Dung Court – and at Barns dighting, Sivering Ground Thornylee.

Wedy 29th Frost this morning day fair & drought from N.W.- One Plough finished ribbing Thornylee, and three D° at Ribbing Red Hall Park For Turnips – Formed a road by Plough North side of the House Well – Sundries –

Thursdy 30th In the night a Heavy Shower of rain from the N.W. day fair & rather Stormy Cold from D° – 4 Ploughs at Ribbing of red Hall Park – and P.M. Chaffing & dighting of Bear for Malt to Family use. Slaters dressing slates for Long Loft, and the Wrights Squaring of the roof for D° – at the 2^{d} Markett of Drumlithie very few Cattle but great Number of Buyers & High Prices.

Fridy 31st Fine dry weather from the S.W.- 4 Ploughs did out all the ribbing of season, and begun Tiling for seed Fur North side Gibsons Park – in fine order – weighed some of the Blacklands Bear 18 stone & 3 lb Montrose measure. The People in the Neighbourhood at Publick roads for first time for many years – Quantity of rain this month is three Inches – The Pasture Grass as good here as during the Season almost though the reverse in General – All Trades & Manufactures Brisk. great demand for Grain.

Diary for the Month of November 1794

Saturdy 1st a Slight Shower in the Night, day fair from S.W.- 4 ploughs at Gibsons park – and flited the Sheep Flakes – at Sundries – Sent to Stonehaven for Lath nails.

Mondy 3d Yesterday Cold dry frosty weather – this morning a strong Frost Ice near Half an Inch thick – 3 ploughs at Gibsons Park – One Cart leading stones & sand to Causeway the swine House – Sent 4B Barley for Malt to Bridge of Leppie – Slating long loft &c.-

Tuesdy 4th very Hard Frost in so much that it was Stiff for the Ploughs till towards noon – 3 ploughs & One Cart to Mathers for Lime to Slaters – flited the Sheep & Cattle Flakes, and taking in Hay, and at other Jobbs –

Wedy 5th Strong Gale from the N.E. & Showers at Intervals after 9 A.M. – 4 Ploughs at Gibsons Park, and after at N. side Old Town park for Oats – flited the Cattle Flakes Water Lair & Sheep D^{o} Robs Hillock &c.

Thursdy 6th Constant rain all night a little fair morning – betwixt 10 & 11 A.M. Came on a severe storm of wind & rainy sleet from the N.E. which lasted all the Day – Obliged to House the whole Cattle on the Farm – Waters very much out Evening – no work without doors.

Fridy 7th It Faired last night about 7 P.M.- Waters very high – This day fair & some drought – 3 ploughs at Old Town park, and One Cart Leading of Turnips from long Haugh – took in stack Bear – at Auchenblae Markett few Cattle – Bought a Calver cheap.

Saturdy 8th This day fair and pleasant weather from S.W.- 4 Ploughs at Old Town Park seed furr – Sent a Cart to Mill of Kair for Meal Oats whole Meal & 1 Boll Broken Old Barley for Swine &c – Sundries –

Mondy 10th Yesterday fair & dry weather from S.W.- this morning Came on rain whole day no work or little without doors -lead some Turnips and at Sundries –

Tuesdy 11th Strong drought from the N.W. – 3 ploughs at Gibsons park Award & one D^{o} leading of Turnips – and at Sundries – at Slating long loft &c. The Cattle 12 on Turnips Thriving – In some Bear –

Wedy 12th very strong Gale from the N.W. & drought – 3 ploughs at finishing out Gibsons & about 12 Acres for Oats & Begun D^{o} on the South End of Long Burns – a Cart at leading down of Tiles to Malt Barn, and Home Turnips a Heavy Crop &c.

Thursdy 13th very strong Gale from the N.W. abated Early in the morning – day showery – 4 Ploughs at Burns for Oats & one ribbing Potatoes Ground – at a Tryst Laurkirk few Cattle & little Sale, Sold to Montrose 30 to 40 Bolls Chester @ 19/6 per Boll – very high price –

Fridy 14th very fine dry weather – 3 ploughs & one Cart leading of Turnips Sent Lint to the Mill Mondynes – took in 2^{d} stack Black lands Bear and at sundries – Slating – Grounds pretty dry –

Saturdy 15th This morning rain and till near the Middle of the day – three Ploughs at Golds Hillock Award for Oats, and One Cart leading of Turnips, the Slaters at Long loft – and the Gardner dressing Nursery Hedge; Sundries –

Sundy 16th & 17th Monday This day fair weather and Coldish from Southw^{d}. Monday Cold Bleak dry weather from S.S.W. – three Ploughs at Stubble Ground seed furr for Oats fine season – One Cart leading Turnips from Haugh above Distillery – Chaffing & dighting Bear N New Barn for Montrose. Cattle on fine Pasture & Eat little or no Straw – Several of the Horses Ailing – &c.

Tuesdy 18th Dark weather & Cold from the S.W.- about 3 P.M. Came on rain which lasted all the Evening. 4 Ploughs at work all day – dighting & Sacking of Bear for Montrose – finished Slating &c.

Wedy 19th Fresh Gale & a little Hail from the S.E. – Sent three Carts to Montrose with Bear and One D^{o} Carting off rubbish from the new Slated Houses & Home Turnips &c

Thursdy 20th Strong Gales from the Eastward with Flying Showers of Hail & Cold weather – Lead Home a Good deal of Turnips & out rubbish of Slates – and Dighting of Bear for Montrose &c Jobbs –

Fridy 21st Early this morning much rain & very strong stormy weather from the East all the day – no work without doors – Except Flitting Sheep Flakes – dighting, sacking of Bear for Montrose –

Saturdy 22d Dark Gloomy weather with slight rain from the Eastward – 3 ploughs at seed furr as before, One Cart leading out rubbish from Houses and down Tiles into Malt Barn – Sundries. The Grounds no Wetter for all the rain. in Barley stacks.

Mondy 24th Yesterday dark Gloomy weather & showers from S.E. this day same but Heavier rain & more stormy. Yoked three Ploughs & one Cart to lead Turnips but oblidged to Loose them by noon – at Barns dighting Bear for Montrose &c

Tuesdy 25th last night and all this day very stormy weather from the S.E. & much rain – the Fields deluged – at Barns dighting Bear, Cleaning doors & sundries.

Wedy 26th last night or early this morning Came on a Severe Gale of wind at S.E. with Heavy rain which brought down the rivers very much – 2 Carts Gathering Earth &c for a Dunghill to red Hall Park for Turnips and one D^{o} leading of Turnips & Sundries –

Thursdy 27th Frost in the morning day fair & drought from S.W. – 3 Carts to Montrose with 12B Bear – One D^{o} at driving out Gathering & Cleanings about Houses to some thin Ground Goulds Hillock – Carts by Open road very deep & 18 Hours absent from 4 to 10 P.M.

Fridy 28th This morning about 7 A.M Came on a very severe storm of wind & rain which lasted with violence till past 5 P.M. no work without doors – dighting Barley N Barn, and at sundries – the rivers much out but faired Evening. Took in part stack of Haugh Oats S N Barn.

Saturdy 29th Fair weather and drought but very Cold at S.W. – Carts leading Turnips & Brought Home the Cattle Flakes from Water Lair, and took in a stack Oats morning – the Grounds surprisingly dry Considering the Great rain. Observed no Ploughs Going Neighbourhood.

Sundy 30th very Cold weather from S.W.- and about 6 P.M. Came on very Heavy rain for some Hours when Cleared up again with strong winds. The Quantity of rain in Novr is Inch 2:75 and the fields and roads are Considerably drenched – the Open Weather however hath Considerably Saved the Scanty Crop of Fodder and what is of more importance the Great demand for Cattle Have Thinned this Country of them Amazingly – Fatt Cattle are however not in demand and the best under 5/ per stone

Oat Meal 1/ per peck – pease 16/ – Bear 18 to 19/6 – Wheat 21 to 22/ – Manufactures high –

Diary for the Month of December 1794

Mondy 1st This morning about 9 a smart shower of rain but no great Continuance – 2 Carts to Montrose with Bear $6\frac{1}{2}$ Bolls & Got Coals – driving Earth to Dunghill red Hall Park – flited the Sheep Flakes –

Tuesdy 2d This morning a very Heavy rain for some Hours from S.W. – day fair & milder – 2 Cart at leading of Earth for red Hall Park dunghill – and at Barns, Grinding new Malt & at sundries –

Wedy 3d a Considerable deal of rain last night & Obliged to stop the Carts before 11 A.M. for it – lead Home some Turnips, and at Barns dighting Barley & Jobs.

Thursdy 4th Fair pleasant and dry weather from N.W. – Carts at driving of Earth to Red Hall Park – leading of Turnips- In Hay and P.M.a stack of Oats N N Barn and at sundries

Fridy 5th Early this morning rain & stormy weather N.N.E. and no work done without doors – at Barns Chaffing & dighting Oats for the Mill, & Sundries.

Saturdy 6th last night & this morning rain & rivers out – 3 Carts driving Earth to red Hall Park Dunghill Layer & Home a draught of Turnips in return – P.M. at Graveling a piece of road behind the House &c Jobs.

Mondy 8th Yesterday fair weather but dark till 3 P.M. Came on rain from S.E. which lasted all night and all this day afternoon with storm severe & Brought down the rivers very high – no work without doors, dighting & sacking Haugh Oats for the Mill – Sold 6 fatt Cows –

Tuesdy 9th Fair weather from Midnight till past noon this day when Came on Heavy rain from Southward – Sent 2 Carts with $14\frac{1}{2}$ B Oats to Mill of Kair, and others leading Turnips & driving of Gravel to the road behind the House – took in part stack Oats &c –

Wedy 10th Fair weather all this day for a wonder – One Plough Begun E[t] side Robs Hillock Ley Flaked – The Carts driving Earth to red Hall park Dunghill & Home Turnips in return & Sundries Jobs –

Thursdy 11th Slight Frost which leaped vulgo – day very Cold from S.W. – One Plough Ley & two Carts at driving of Earth red Hall park & Home Turnips – and about 5 P.M. Came on rain pretty Heavy rest Evening –

Fridy 12th Fair weather & Moderate – 2 Carts driving Earth to red Hall Supposed about 100 Cart load & Home Turnips, and one Plough at R.H. Ley, at Erecting of the Cattle Flakes before the Byres as Shade to Hold the Turnips from y[e] weather & at Sundries.

Saturdy 13th fair weather & Cold from the Southw[d] – 3 Ploughs at Ley as before and One Cart to Pittengardner for some Thinnings of Firr for a dead Hedge N.W.

Corner Garden, at Barns Chaffing Oats N Barn, at Garden Transplanting Gooseberries &c Jobbs –

Mondy 15th Yesterday slight frost but agreeable sunshine an uncommon thing for time by past – This day frosty & rainy A.M.- rest fair – 4 Ploughs at Ley – part after Sent one of them to Mill of Kair with Oats & One Cart lead Home some Turnips – at Barns dighting and sacking Oats for the Mill & at Sundries –

Tuesdy 16th last night and this morning Strong Frost which however did not stop the Ploughs at Ley – One Cart to the Mill with Oats and after leading of Turnips & another Cart leading of them all day – 2 Fatt Cows went to Montrose sold at 4/6 to 5/ per stone part of six sold – took in a stack of Oats last load Upper & Nether Burns – into N Barn – People Chaffing Oats SB & Cutting Broom for Covering Turnip Shades before the Byres & Cattle Flakes.

Wedy 17th very strong drought from the S.W. & Cold weather – 4 Ploughs at Robs Hillock Ley – at Barns and Broom Thatching Shades for Holding Turnips, at Garden sundries – Grounds drier, Bushes Budded.

Thursdy 18th Rough weather from S.W. with rain not Heavy A.M. – 4 Ploughs at Ley as before and flited the Sheep Flakes P.M.- at Barns Chaffing Oats and at Sundries – fair P.M. and frosty &c.

Fridy 19th Cold dry weather from the S.W. – 3 Ploughs at Ley and One Cart leading of Turnips – at Barns &c. Bought a Farry Cow about 2/6 or 8^{ds}per stone at Auchenblae a good Many Cattle & many people.

Saturdy 20th Cold dry weather from the Southward, 3 Ploughs at Ley – and One cart leading of Turnips – the Grieve in Montrose sold some Barley 20/3 – and at Barns dighting P.M. Oats for the Mill –

Mondy 22d Yesterday Excessive stormy & Cold weather but dry S.E. this morning Came on rain from same Quarter but not very Heavy, but stopped the Ploughs after 10 A.M. – after dighting & lofted Oats for the Mill and P.M. took in a stack Burns Oats & Sundries –

Tuesdy 23d a kind of stormy weather from the East with little rain 3 ploughs at low field Award dry – and one Cart leading of Turnips, at Barns dighting or rather Chaffing of Oats N Barns – Sundries –

Wedy 24th Dark Gloomy weather but dry from the East – Sent 2 Carts at 4 A.M. to Montrose with 9 Bolls of Barley sold at 20/3 per B & Out sundries – two ploughs, and after Lead a draught of Turnips – at Barns Chaffing &c Oats – Garden &c.

Thursdy 25th Early this morning Came on Snow from N.E. but mostly fair till towards Evening Stormy & Snow with frost – 2 Carts driving dung, and Others lead some Turnips – pulling & Heaping D^{o} –

Fridy 26th Moderate weather from the North, little frost the Snow about 2 Inches depth average – 4 Carts driving Dung to Midding layer for Cottar Bank & Home One draught Turnips – People at Mill of Kair P.M. dighting Oats N Barn & at Sundries –

Saturdy 27th Fresh weather and some rain from the Northward the Snow mostly gone this morning – 2 Carts Carried Oats to the Mill & Brought Back ye Meal – 2 Ploughs at the Ley R Hillock – Sundries.

Mondy 29th Yesterday Moderate weather – little frost – this day same wind Northerly – 3 ploughs at R.H Ley and One Cart leading of Turnips – morning took in a stack of Ley Oats – at Barns Chaffing & Sundries.

Tuesdy 30th Very keen frost all the day & clear weather – 3 Ploughs at Stuble Ground Cow folds nearly done, One Cart at driving Stones & Gravel to a Water Conduit across road, Head of Bogg Planting & Mending said Approach to ye House – In some Turnips.

Wedy 31st very strong frost all this day a little Snow last night – 4 Carts driving Dung as yesterday – and Home some Turnips – digging more, at Barns dighting Oats for the Mill – The Quantity of **rain** this Month is 2 Inches – The Prices of Grain Increasing – Chester Bear is said to be now worth 20/ per Boll and Oats & Oat meal 16/ – Wheat 23 or 24/ Pease 16/ per Boll – Fatt Cattle are not Better than 5/ per stone – from 3$^{ds}\frac{1}{2}$ or 4^{ds}per lb – Manufactures Brisk – Coals plenty but dear 8/ per Boll &c

Diary for the Month of January 1795

Thursdy 1st Th^{r} Noon 28° very Hard Frost – 4 Carts driving Dung and some Oats to the Mill – dighting, sacking & lofting of Oats for the Mill – lead & digged some Turnips –

Fridy 2d Th^{r} 30° Hard frost wind West – Carts driving dung some Turnips, and Corn to the Mill – begun to Turn a Dunghill for the Fallow Cottar Bank &c – The frost Considerably in river not Frozen over as yett –

Saturdy 3d Strong frost Wind S.W. – 3 Carts driving Dung all day - and Home 3 Cartful of Turnips dug with the Clats – Th^{r} noon 31° – Appearance of Change of weather either fresh or Snow.

Mondy 5th Yesterday a very strong S.W. wind & fresh weather – this day same and the Frost & Ice on river mostly Gone – being Old Christmas no work without doors &c

Tuesdy 6th Fine pleasant warm weather at N.W. – 3 ploughs at Stuble Ground, frost quite Out – leading Turnips – and P.M.took in a stack Oats S B – Sundries –

Wedy 7th Fresh & foggy weaher – 3 ploughs at Back Yards and One Cart leading of Turnips and at Sundries – Ground as dry as for three months by past

Thursdy 8th Thick Hazy weather all day – 3 ploughs at Back Y^{ds}and one Cart leading Turnips – Flited the Sheep Flakes R Hillock near Cottar Town & Sundries –

Fridy 9th Brisk Gale from the Northw^{d} with some flying Skiffs of Snow – 3 ploughs at Back Y^{ds}and one Cart Leading of Turnips – at N Barn Chaffing of Oats all day – Grounds in good dryness &c –

Saturdy 10th pretty Brisk Frost but did not stop Ploughing 4 Ploughs did out Back Yards seed furr and all of the Stuble Ground for following Crop, Only about 9 Acres of Ley to Plough for Whole Oat Crop – P.M. at Barns dighting Oats for seed at least loft &c – at the Mill –

Mondy 12th Yesterday and this Frost but not very strong – 3 Carts driving dung for Cottar Bank Fallow and one D° Conveying Oats to the Mill & Brought Back the Meal – Digged & lead some Turnips &c –

Tuesdy 13th Frost & Dark Gloomy weather from the Westw^{d} – 4 Carts driving Dung to Water Lair late Pease Park three load each – Home some Turnips digged morning took in a stack Oats N B – Sundries –

Wedy 14th Frost & thick dark weather from the Westward – 2 Carts Driving Dung to Water Lair & one Cart Carrying Oats to Mill & one leading of Turnips-Men digging D° & Sundries

Thursdy 15th Th^{r} 33° Cold Bleak weather from the N.E. with some Hail – Carts driving Dung to Water Lair and some Corn to Mill & Home Turnips – Men P.M.Digging & throwing into Heaps Turnips for dread of Snow, at Barns dighting & Chaffing Oats South Barn – and Sundries.

Fridy 16th Th^{r} 34° Frost very moderate, but very Cold & Showers of Hail & Snow from the N.E. – 2 Carts driving dung to Water Lair One D° leading of Turnips and one D° to Montrose for Sundries Beef &c Others dighting Oats for Lofting &c.

Saturdy 17th Th^{r} 34° Some Showers of Hail &c from the N.E. all day at Intervals & but very little frost dark lowring weather - 3 Carts driving Dung to Water Lair and P.M. lead Home two Cartful Turnips & In two D° of Hay – at S Barn sacking & lofted last threshed stack of Oats, no Ice on the river – the snow not above $1\frac{1}{2}$ Inch deep – the Manufactures Except Hosiery very Brisk &c.

Mondy 19th Th^{r} 32° Yesterday a Considerable Fall of Snow and Early this morning more-from 5 to 6 Inches depth on the level, but little frost – all the Carts driving Dung to Water Lair & Home some Turnips – P.M. People Turning Dunghill &c.

Tuesdy 20th Th^{r} 27° Early this morning a very Considerable fall of Snow – now above 8 Inches average – 2 Carts to Stonehaven for Coals & Oak Garrons – Several of our People Indisposed with Colds &c Jobbs –

Wedy 21st Strong Frost and a Considerable additional Snow, by the best Medium I could take the Depth is about 10 Inches at a Medium average – Although the People call it Knee deep – Lead some Turnips previously Digged-took in Oats S B and at a variety of Jobs. The Mill of Kair Could not go for want of Water although not Frozen over &c.

Thursdy 22d a very Strong Frost all this day – Although the sun Shone yet the Thermometer Farenheit at Noon Exposed on a North Wall no wind or snow was as low as 13 degrees or 19 degrees below Freezing, lower than I ever Observed it this 20 years – People digging of Turnips first spading of the Snow &c.

Fridy 23d very strong Frost All last night & this day – at 11 A.M. The Therm^r at 17 degrees above 0 or 4° higher than yesterday sensibly felt difference – People at Barns Chaffing Oats N Barn – at Sundries &c –

Saturday 24th very strong Frost Therm^r at 11 A.M. at 17 degrees – People at Barns dighting of Oats N B and P.M. Digging and lead Home some Turnips – P.M. the wind got into S.W. a Breeze –

Mondy 26th Th^r Noon 34° Yesterday Moderate though the wind S.W. drifted some of the snow into the roads – Frost in the Night and after 2 P.M. a good deal of snow from N.E. & very stormy Indeed – People Employed digging and Leading of Turnips 7 Cartful – so far Lucky – Fodder traveling Apace.

Tuesdy 27th last night and this morning a severe storm of wind & snow drift & which Hath rendered the roads almost Impassable, it Continued whole this night. People at Barns dighting & lofting Oats &c.

Wedy 28th Severe storm of Wind & Snow last night which Hath rendered the roads almost Impassable for Horses & totally so for Carriages – at Taunton Fair but few People, Servants Extravagant, Engaged none and but few I believe many did same.

Thursdy 29th Therm^r at Noon 16° very strong Frost all day-but warm sun – at Taking in a stack of Haugh Oats N Barn and after Chaffing one in part South D° the Gardner Carrying in Dung Garden – and sundries – The Publick Post is Stopped & the Fly &c –

Fridy 30th Therm^r 27° pretty Strong Gale from the Westward & frost, day Clear, but near as Cold as before. People at the Barn South Chaffing Oats, and Sundries – In the Evening the weather much Milder Wind S.W. –

Saturdy 31st Therm^r noon 19° Strong Frost all night & this day but More Moderate Evening – People at Barns dighting of Oats, P.M. Inn more Hay – The Country is Covered with a Greater load of Snow than I have Observed for many years by past – but by Blowing Cannot be Ascertained – The Frost & Snow in the Southern part of the Island more Severe than here, by many degrees – No Carriages can travel – the Quantity of **rain**, Pluviometer is Half an Inch. The prices of Grain Oat Meal & Oats 15/6 the last purchasing for the London Markett &c.

Diary for the Month of February 1795

Sundy 1st Early this morning a Considerable new fall of Snow from N.E. to about 2 Inches, no Frost – Therm^r Noon Shade $32°\frac{1}{2}$ – Evening more snow –

Mondy 2d Moderate frost & very warm sun shine the Th^r noon 31° – all people digging of Turnips and lead Home 3 Cartfull, it was very deep Snow to the Carts & Horses more than Belly deep –

Tuesdy 3d a Considerable fall of Snow from North & High wind morning – day moderate with partial showers of Snow – & frost Th^r 29° – lofting Oats from S Barn – and P.M. took in a large stack of Oats S Barn – Snow very deep –

Wedy 4th At day light Came on a very Heavy fall of Snow from the East which lasted most of the day to 4 or 5 Inches additional depth & filled up all the Pathed Tracts – Fresh Th^{r} at 33° shade noon, frost at night – People at the N Barn Chaffing the stack of Haugh Oats &c.

Thursdy 5th very stormy wind & frost from the Northw^{d} – People at dighting of Rye Grass $2\frac{1}{2}$ B and at sundries – I do think SNOW average 18 Inches and deeper than for 20 years by past – no carriages go.

Fridy 6th last night & this day very strong Frost – The Them^{r} as before at 14 degrees or 18 below freezing – People Employed in Digging 2 Cartful of Turnips & after Dinner lead them Home with Difficulty – no Carting on the Aberdeen road &c.

Saturdy 7th A Considerable fall of Snow & Sleet morning which through the day turned out Fresh with sleety rain at East – no work without doors, dighting & lofting Oats – a Servant to Bervie for Sundries –

Mondy 9th Yesterday a tolerable & Moderate Thaw from the Eastw^{d} & dark weather – this day same and Evening rain & sleet – Sent 3 Carts to Gurdon for 2 Bolls of Scotch Coals at 8/3 per B – lead Home 2 Cartful of Turnips – Sold 60 to 100 Bolls Oats to Mr Smart at Burnton delivered at Montrose at 15/9 ready Money at Leisure.

Tuesdy 10th Stormy weather with rain & fresh at NNE – all day at Barns lofting of Oats & drew in some from another large Stack on Stathill N.Barn – much of the Snow gone but yet very deep – fields Braked & dismal.

Wedy 11th very Early this morning Came on Snow, and Wind from North with frost & filled up the Tracks as before – This day throughout a Constant Storm of wind & snow & no Person Could Travel almost – at Barns Chaffing & Dighting of Oats S B – and at sundries.

Thursdy 12th last night and this morning an Excessive Severe Gale of Wind North, Snow & drift with Hard Frost which filled the roads worse than before if Possible – took in the Oat Stack N.B., after some Hay, & after Dinner Holed & lead Home two Cart full of Turnips a Cart with three Horses very deep – This will again stop the Post and all Carriages – Snow from 4 to 5 feet depth about ye doors.

Fridy 13th very Stormy Blowing weather from North, with Hard Frost Th^{r} 28° – People Digged & had Home 3 Cart load of Turnips three Horses in a Cart Two was unable do it which I never SAW before – at taking in some Hay – and at Barns – Clear & fair all day – The Usual Adage I Hope will be Falsified – Old Candling.

Saturdy 14th Th^{r} 33° Very Stormy & Blowing weather with Hard Frost at North^{d} People at the Barns dighting Oats &c. The POST Hath been Stopped from the Southw^{d} since Tuesday last and the Mails Obliged to be Carried by Men – Observe a very Great Mortality at Montrose & amongst all People in General – No Thaw, but Sun Shine &c.

Sundy 15th Fair weather and rather fresh – Wind at W.N.W.Therm^{r} noon Shade 37° little Alteration on the rivers – the Fodder very Scanty 6 stacks Oats & Bear & 3 of pease only remaining – most others much worse.

Mondy 16th Th^r 32° Frost & Clear weather from West – Digging & lead Home 3 Cartful Turnips- at Barns &c. The roads & traveling Bad and the Country feeling many Hardships accordingly &c.

Tuesdy 17th very strong Frost, and sun shine Th^r 23° – lofted a stack Oats from S Barn and put the Horse Corn Malt Barn loft – took in a stack of Oats large Stathill S Barn P.M. and at Sundries &c.

Wedy 18th Th^r 37° very strong Frost & some appearance of Change of weather – by noon Th^r 37° – Cloudy, but little fresh from S.E.- People at Barns Chaffing Oats and in some Hay – Gardner at Dunghill &c.

Thursdy 19th Th^r 32° Cold N.E. wind & no appearance of Fresh weather though expected at this Change of moon this day – People at the Mill of Kair – Brought Home 12B Meal on Horse Back by Kair House and Bridge. a Cow Bought at Crofts Cast Calf at 3 Months – People Cutting the roads to the Westw^d of this –

Fridy 20th Cold Frosty weather from the N.E.- forenoon rather fresh and Flirts of Sleet – P.M. Frost again – Brought Home all the Meal on Horse Back the first I ever Brought from a Mill in that Manner – Sundries.

Saturday 21st Th^r 34° Early this morning Came on Snow, Sleet & drift with High wind from N.E. which rendered traveling very Bad & Blew up the roads again, at the Barns dighting Oats for Montrose & lead 16 f^rt Turnips P.M.

Mondy 23d Th^r 38° Yesterday betwixt fresh & frost from the N.E. with Dark weather – this day same weather N.E. with drisling rain & moderate fresh – Sent One Cart to Bervie for 72 st English Coal & others digging and leading of Turnips – Above 100 People Employed Cutting the Snow from Lkirk Eastward.

Tuesdy 24th Th^r 41° moderate fresh & Calm dark weather – Sent all my Men to Cut the road from red Hall Bridge Eastw^d all day – with others am Informed quite Opened the road all the Way to Aberdeen – several Carters passed to the Westward this day for Montrose – at Barns &c Jobs.

Wedy 25th Moderate & pleasant weather but rather Frost than Fresh 3 carts to Montrose with 12B Oats and others at Barns dighting Oats for Montrose – P.M. took a stack of Bear W Lair N Barn.

Thursdy 26th Th^r 34° pretty Strong Breeze from the N.W. all last night and this day – with keen frost towards Evening – little of the Snow Gone – a General Fast this day –

Fridy 27th Th^r noon 30 degrees a very stormy night last with a Considerable deal of Snow & Blowing which Hath rendered the roads that were Cutt little better than before and almost Impassable – at Barns dighting of Oats and P.M. Lead 2 Cartful Turnips &c.

Saturday 28th very Strong Frost all night & this whole day. People Led Home some Whins & Broom from Thornylee for washing Brew Looms – Sent 2 Carts to Montrose at 4 A.M. three Horses on Each Cart with 8 Bolls in both Carts – No Coals, Chows, or Cunnle in it – The Fodder of Water lair Bear being so free Cutt and obliged to Feed all the Horse and most of the Cattle with Hay this day & tomorrow – The Storm is universally allowed to be the most Severe in the memory of Man. no 4 wheel Carriages Have been able to travel for 5 weeks by past & no Appearance when so Except for Meal every kind of people are Incommoded for one thing or other – and the longer it Continues will be Increasing – Therm^{r} 29° The Quantity of rain in the Gage not Ascertainable this two Months being Frozen & it doth not appear that it Can be Measured of Snow.

Diary for the Month of March 1795

Mondy 2d Yesterday very strong Frost & Sunshine – this day the same – Therm^{r} noon shade 28° – the Snow now Carries Men universaly on the Surface which I never saw before at any season, but not the Horses & Carts. All People at Barns dighting & sacking Oats for Montrose and some out of lofts – in Turnips, and Hay – Jobs –

Tuesdy 3d very strong frost last night & Dung Frozen in Byres 3 Carts with 8 Horses to Montrose with 14B Oats & 7B by Messrs Spark & Andrews Carts to D^{o} – Got only $\frac{1}{2}$ Boll Coals – Passed at Marykirk on the Ice both Going & Coming – at Barns dighting &c Oats – Snow Shower from 2 P.M. to past 5 from the Westward –

Wedy 4th Th^{r} 43° rather Fresh all day till 4 P.M.- frost again – wind S.W. People at Barns dighting & sacking Oats for Montrose tomorrow, and P.M. lead Home some Turnips.

Thursdy 5th Frost in the morning, but Turned fresh Hazy weather with drissling rain after noon & Heavier by 4 P.M.- Sent 3 Carts to Montrose with 14B Oats which went & Came by river Marykirk, roads very bad – digging of Turnips –

Fridy 6th Th^{r} 36° Fresh Thick weather & Mild with Small rain pretty heavy towards night at Northw^{d} – dighting & sacked 14 Bolls Oats for Montrose – and lead home 4 Cartful of Turnips - the Larks singing for first time &c –

Saturdy 7th a Good deal of rain & stormy weather from Northw^{d} – and Melted a good deal of the Snow & Brought down the rivers, the Ice Broke Early above this place but not below it all this day – frost again in ye Evening – People dighting of Rye Grass & Sundries triffling –

Mondy 9th Yesterday fair weather & fresh wind South^{y} – this day very Early a good deal of rain & fresh S.E. wind – the Ice Broke only as far as the Nether End of Bleaching Green Stopped by the New Bridge – at Barns Chaffing Bear

Tuesdy 10th Strong wind S.W.betwixt fresh & frost – taking in Bear S Barn, and lead home 4 Cartful Turnips, and dighting Bear N B – Th^{r} 42° – Some of the Snow Gone – Sundries

Wedy 11th almost Constant rain & stormy from ye East – and River Broke at the Bridge at 2 P.M. – at Barns dighting of Bear sold, Gathering dung &c about Houses – a good deal of the Snow gone on higher places – roads Impassable.

Thursdy 12th Sleet and Snow night & till 10 A.M.- Country White again day fair from the N.E. & rather drought – People taking in Hay, and P.M. turning Dunghill feed the Yield Cattle with Bear Chaff which agrees well –

Fridy 13th Frost & a Considerable fall of Snow above 2 Inches from E.N.E.- day frost turning a Dunghill Cottar Bank – and lead Home 4 Cartful Turnips P.M. The river is fallling in Considerably few traveling as yet –

Saturdy 14th Therm^r^ 34° very Strong Frost & Snow almost Carrying the roads better than for weeks by past Sent 14 Bolls of Oats to Montrose & returned by the river Marykirk with six firlots of Coals 8 P.M. – The Snow very little vanished & on the Braes as Closs as in the Season – In the Morning Frost uncommonly great & Cattles dung Frozen in some of the Byres not very frequent the season.

Mondy 16th Th^r^ 39° Yesterday very Cold weather & frost E.N.E. – Snow Carrying Foot People everywhere, this day stormy weather with snow & sleet whole day – no work without Doors – Chaffing Thornylee Bear, and at sundries rather Fresh & new Snow not Lying – Sacked Bear for Montrose – Hear they Have Ploughing on the Coast –

Tuesdy 17th Th^r^ 35° Very Early pretty strong frost and about $1\frac{1}{2}$ Inch of new Snow from N.E. – Sent three Carts to Montrose with 12 Bolls Bear at 18/6 – and to Bring out Coals & Porter in return – day Bleak weather & little Thaw – the Country nearly as Closs with Snow as at any preceeding time – there must be an Uncommon Scarcity of Fodder soon – Hear most of the Fatt Cattle which are very few Indeed are Sold about 6/ per stone – sold Two horses @ 3£10 S^tr^

Wedy 18th pretty strong frost & strong wind from N.E. – People Chaffing Bear and leading of Turnips 4 Cartful, this day rather fresh, but the Snow goes off but Insensibly Indeed – Sundries –

Thursdy 19th Strong winds variable N.W. and S.W. & Middling Thaw – the roads tolerable good – Sent three Carts to Montrose with Bear & to Bring out sundries, took up a Pitt Tartan Potatoes in good order & Housed them Saw Some Ploughs going redmyre.

Fridy 20th Th^r^43° This day fair & though no great Thaw the Snow wearing away gradualy – wind N.W. – at Barns dighting & sacking of Bear for Montrose and lead some Turnips P.M. – at sundries –

Saturday 21st Th^r^ 44° This day fair weather, drought and a tolerable thaw of the Snow, Gradualy – the Rivers Increased moderately – 3 Carts to Montrose with Bear & out $1\frac{1}{2}$ Boll Coals, and at Barns Chaffing Bear – Sundries.

Mondy 23d Yesterday Moderate Thaw Wind S.W. this day the same, the river out much Evening – dighting & sacking of Bear for Montrose, took stack Oats Morning, Sowed Garden Pease & digging for D° Potatoes &c Jobbs.

Tuesdy 24th Th^r^ 46° This day very fine dry weather from S.W. and the Snow mostly gone on the Fields – most People Ploughing Sent 3 Carts to Montrose with Bear 11B – Oats 1 – Brought 12B Coals – Transplanted Field Turnips into the Garden for Seed very thick, Vegetation very Slow Indeed.

Wedy 25th Very pleasant dry weather from S.W. – leading Turnips to Transplant

– and P.M. yoked two Ploughs on West side of Robs Hillock ley first time for above Nine weeks – took in pease S B – and at sundries.

Thursdy 26th Th^{r} 46° White frost in the morning & pretty Hard also, One Cart to Montrose with 4B Oats – a Calf & Two Kirn of Butter sold at 11^{ds}per lb – 3 ploughs at West side of Robs Hillock Ley – and at Garden Sundries.

Fridy 27th pretty strong frost last night & this morning – wind S.E. & very Cold – three ploughs at Robs Hillock ley – and Brought Home the Sheep Flakes – at Garden sowing small seeds – and dressing at Strawberries &c.

Saturdy 28th Cold Flanders Frost and drought from S.E. – Sent three Carts to Montrose with 16B Oats & Out Coals 12 Barrel – at Garden planting Early Cabbages, Sowing Small Seeds and Dunging N.E. Corner Robs Hillock Ley.

Mondy 30th Yesterday Cold drought from the Southward – and this day the same – about 2 P.M. Came on rain from the SSE, but not Heavy – three Ploughs finished out all Oat seed on Whiteriggs and One Cart leading of Turnips & Sundries – Sacked $8\frac{1}{2}$ Bolls Oats to sow – Planted Curleys and at digging of Onion & Carrot Grounds &c Jobs –

Tuesdy 31st last night Came on rain from the S.E. which Increased in the Night, and through the day pretty Constant & deluged the fields very much – 4 Ploughs at the Ley field Water Lair 3 years old for Oats, Excessive rich, till Dinner time – Sacking Oats P.M. – Some few Have Sown Oats Neighbourhood but it had been better not the Fodder will be Excessive Scanty – I Have Only three stacks – but a Good Sowe of Hay – Fatt Cattle very scarce but the Butchers will not give 6/ per Stone Including ye fifth quarter and Hide, tallow &c although is worth above 5/ per stone 4^{ds} to 5^{ds}per lb – Meal at 16 to 17/ Oats D^{o}- Bear 18/6 to 19/6 – the Pease 17 to 18/ – Wheat 27 to 30/ – all Sorts of Linnen Manufactures are very brisk – Canvass &c Sailors Enlisting without Impressing a New thing & very Political late Act of parliament.

Quantity of rain is 1 Inch

Diary for the Month of April 1795

Wedy 1st Dark Gloomy weather from the Southward & P.M. some light rain – the land not Wetter – 4 ploughs at the Waterlair ley from 8 to 2 P.M. – after dinner took in some Cart loads of Hay W Barn & Sundries –

Thursdy 2d Early this morning Came on very Heavy rain & a strong Gale from the S.E. – no work without doors – Chaffing Oats N Barn, and the river much out &c –

Fridy 3d Bleak Coldish weather from the N.E. but dry – 3 ploughs at Water Lair ley and One Cart leading of Turnips took into Barns last of Oats & Pease Stacks – dighting & lofting Oats, and at sundries – Garden pruning &c –

Saturdy 4th Th^{r} 46° Noon Frost slight day fair & drought from the Northw^{d} with sun shine – 2 Carts to Montrose with 10B Oats and in return with Coals Two Ploughs at Waterlair ley Planting Alders in the Bog planting & sundries –

Mondy 6th Th^{r} 46° Yesterday very pleasant drought from N.E. and sun shine - this day Cold N.E. and fair – 3 ploughs finished all Tillage for Oats of season & One Cart leading of Turnips – dighting of & sacking Pease for seed – Sold all my Fatt Cattle about 6/ per stone for Montrose.

Tuesdy 7th Cold Stormy weather from N.E. but fair till near noon when some rain A.M. begun Oat seed Water Lair Sowed 7½ Bolls & partly Harrowed it too raw – at turning Dunghills P.M. – at Garden – Bleak weather –

Wedy 8th Good deal of rain in the Night – day not fair, but three Harrows Broke in East side Thornylee for Pease & three ploughs seed furr for same, but very Middling Order as is season – about 1 P.M. Came on Increasing rain & storm from the N.E. which stopped work without & Brought down the rivers very High –

Thursdy 9th The Heavy rain Continued all night, and all this day - and the rivers were Out as much as for years by past – no work without till past 3 P.M. – at Turning Dunghill - Great Numbers of Cattle Going to the Southw^{d} this week very high prices from North.

Fridy 10th The rain Continued all yesterday & this day little or no drought from the S.E. – Saw some Ploughs going but ours Could not for Wetness – turning dunghills and P.M. lead Home 4 Cartful of Turnips & Sundries – Great Numbers above 400 Head of Cattle Have passed here for the Southward at very high prices Indeed –

Saturdy 11th Bleak weather from the N.E., but chiefly fair – the Grounds Wetter than this season before, people at levelling down an Old Dyke red Hall Park and P.M. driving dung to a Midding D^{o} park for Turnips – In some Hay, and at sundries &c. –

Mondy 13th Yesterday fair weather & Some drought from S.E. – this day same till about 9 P.M.Came on Heavy rain from S.E.- 4 Carts driving Dung to Midding for Turnips at Red Hall Park, and P.M. at sundries. Planting Cabbages and Varieties in Garden &c –

Tuesdy 14th a Considerable rain in the night – day light showers from the Southw^{d} – 4 Carts driving Dung to Red Hall Park & P.M. at Barns dighting Pease & Jobs. The Water this month large 3½ Inches Water Gauge. Above 600 Head of Good Cattle passed to the Southw^{d}. The Fields Excessive Wett & deluged almost –

Wedy 15th Stormy weather from the Southw^{d} & P.M. a Gale – 4 Carts driving Dung to Red Hall Park – and P.M. at Sundries – Sowed Carrots & small seeds.

Thursdy 16th Th^{r} 57° Strong Gale from SSW & drought – took in some Bear for seed – 4 Ploughs at Thornylee for Pease – about 3 P.M. a very Heavy shower for near an Hour – Sowed Onions & remainder of small Garden seeds – Everything latter than ever known – the Grass is rising – Good Bear now One Guinea per Boll & fatt Cattle 7/ per stone fine Quarters or 6/ Beef –

Fridy 17th Strong drought from the Westw^{d} – with some Hail about Noon – Yoked our Ploughs twice this day the latest by three weeks ever before – 4 Ploughs at seed fur for Pease Thornylee – at sundries –

Saturdy 18th about 6 A.M.Came on rain from Southw[d] which stopped sowing all day – Yoked 4 Ploughs to Steer Clean Fur the Turnip Ground for Bear, dry enough and at Barns Chaffing Oats for seed – & rain P.M. again but neither very Heavy – Very Bad weather Indeed.

Mondy 20th last night a Considerable deal of Closs rain & stormy foggy weather from the S.E. – this day fair & drought from the Eastw[d] & Mild – Ploughs still at steering of Cottar Bank – & P.M. sowed 3 Bolls Pease on the East side of Thornylee & Harrowed D[o] – 6 Harrows at the Water Lair Ley park – Severals begun Sowing.

Tuesdy 21st Constant rain all this morning & till about Noon which Deluged the Fields Immensely – 4 Ploughs at steering of Cottar Bank too Wett –

Wedy 22d Th[r] 53° Thick Hazy weather & fair till 4 P.M.- Heavy rain for above one Hour – Ploughs at steering Cottar Bank – small fur, but much too Wett – but other must ly Idle – lead Home 3 Cartful Turnips & Sundries –

Thursdy 23d Th[r] 56° Very thick fogg till 10 A.M. & then warm dry weather 4 ploughs at Steering Cottar Bank – and leading of Turnips nearly Shott, Observed several People sowing & Harrowing P.M. – at sundries &c –

Fridy 24th Fair & dry weather from the Southw[d] – Yesterday a good deal of Thunder N.W. – this day sowed 3 Bolls pease being all Intended & Harrowed D[o] midlingly – some men Harrowing to Waterlair near Breering & P.M. sowed 8B Oats West side Robs Hillock ley & Harrowing it.

Saturdy 25th fair weather till past 8 A.M. a Moderate rain warm till past 11 A.M. – Sowed 5B Oats E[t] side Robs Hillock 13B total on 13½ Acres – P.M. Sowed all Back Yards, long Burns & most of Goulds Hillock and Gave a double Tine of Harrows to it – rain again after 4 P.M.- but mild & warm & growthing Ground – Gooseberries flowred –

Mondy 27th last night before 8 P.M.Came on much rain & wind from the S.W. which lasted till very Early morning – this day drought strong – Early begun Sowing out the Goulds Hillock – & all Old Park & Gibsons D[o] Having Sown 20B this day & partly Harrowed the same which Finishes all for Oats & Pease of season – Have Sown on 49 Acres 53B 3f 1p or 17½ pecks per Acre.

Tuesdy 28th Th[r] Noon 54° Early in the morning Excessive rain, but Broke out drought again – mostly Harrowed out North Parks – about ½ past 1 P.M. Came on rain very Heavy from 2 to ½ past 3 P.M. & stopped all work without doors, dighting & lofting Oats &c Jobbs.

Wedy 29th Th[r] 51° very strong Gales from S.W. with showers of Hail & very Cold weather – 3 Ploughs A.M. steering at Cottar Bank and one Cart leading Turnips P.M. 9 Harrows did out N Parks & mostly that of Robs Hillock Ley – at Barns &c Jobs – this day lost a Colt rising a year old-of what Disease I know not-it was Ill but 16 or 18 hours &c –

Thursdy 30th Strong wind in the fore part of the day, but about 10 Came on rain which Increased much till near 2 P.M. & after that heavy showers at Intervals which wet the Grounds Immensely – Harrowed out Robs Hillock before y[e] rain good order and P.M. four Ploughs at steering Cottar Bank – a Good Many Cattle at Laurencekirk Markett quick sale & very high prices for Fatt & lean Ones – The Oat Seed not much above half done in many places & in real bad condition – The Fodder in General Consumed & very little pasture – Prices of Grain still rising daily – Bear 21/ &c – rain in the gage 5:5 Inches very great in seed season.

Diary for the Month of May 1795

Fridy 1st Therm^{r} Noon 51° a Considerable rain this morning till past 7 A.M. fair but some showers P.M. – Sent 2 Carts for Lime Shells to Mathers for Tiled Houses, and One Plough steering Bear Land P.M. – five Harrows Harrowed 3 pecks Lintseed & all of the Cow Wards & Cross D^{o} Goulds Hillock Oats – leading Turnips.

Saturday 2d Th^{r} 54° Some White Frost Early – day fair & drought – 9 Harrows did out Cowfolds & Golds Hillock finaly-and P.M. at the Back yards at D^{o} – Inn Hay, dighting seed Bear &c Jobs.

Mondy 4th Yesterday strong drought at S.W.- this day high wind D^{o} 9 Harrows did out Compleatly the Back yards & long Burns – & P.M. yoked 7 Harrows to Brake Cottar Bank for Bear and a Plough on West side of Thornylee for Potatoes – Sundries –

Tuesdy 5th Th^{r} 58° Strong drought from S.W. 6 Harrows at Cottar Bank and one Plough for Potatoes West side Thornylee – & a servant to Stonehaven for Hair to Lath plaister for the Houses, Brewhouse & at sundries –

Wedy 6th Th^{r} 59° very pleasant warm dry Clear weather from S.W.- Harrows at Cottar Bank – a plough drilling for Potatoes all day P.M. 3 Carts driving dung Thornylee for them all Hands at Planting of them – In Hay – Considerable growth –

Thursdy 7th very strong drought from Westw^{d} – Planted out A.M. about 3 Rood Ground with about near 7 firlots of Potatoes well Dunged – P.M.- Steering North End Cottar Bank & Chaffing Bear for seed – Sowed Haws &c Garden &c

Fridy 8th Th^{r} 42° Cold stormy weather from the North & frost in the morning day very Cold & high wind – Ploughs at steering N side of the Cottar Bank & P.M. yoked 4 ploughs for seed fur for Bear E^{t} side field, and at sundries –

Saturday 9th Some Snow Showers morning – day Cold & rain P.M. 4 Ploughs at Bear seed & P.M.sowed about 3 Bolls East side Cottar Bank & partly Harrowed D^{o} – The Oat seed is just about being finished & Severals begun that of Bear - Sundries

Mondy 11th Yesterday Cold dry weather, this morning & all day a very Strong Gale from the N.W., with Flying Showers of rain & Hail – 3 ploughs all day at Cottar Bank and three Harrows on D^{o} – Barns &c.

Tuesdy 12th Th^{r} 49° Cold & Stormy weather from the Northw^{d} and P.M. Showers of rain – 3 ploughs & Harrows at Bear seed Have sown about 10 Acres ready for Grass Seeds – Observe some People yet sowing Oats – pasture Grass pretty good a much greater Quantity of Cattle brought to ye South^{y} usual.

Wedy 13th Cold weather from the Northw^{d} & showers of Hail – Ploughs at Bear

& one Cart dunging fallow new Ground - Evening a number of Men Sowed about 10 Acres East side of the Cottar Bank with Grass Seeds pretty richly Considering the prices of Clovers is so very high – Sundries

Thursdy 14th Moderate weather but not warm from the Northward – all Ploughs and Harrows at Covering y[e] Grass Seeds and sowed about 3 Bolls Bear & partly Harrowed it – The Ground pretty good for seed & at sundries – Sowed many of the Annuals Garden &c –

Fridy 15th Th[r] 40° During whole night and all this day till 1 P.M. almost Constant Storm of Snow & Wind from the N.E.- and no work without doors till after Dinner when Braking in the Park at Waterlair for Bear with 8 Harrows – at sundries dighting seed Bear &c Vegetation Checkqued.

Saturdy 16th Th[r] 40° Ice on dubs Cold dry weather from the Northward all day – four Carts drove out about 130 Cartload of Dung to Cottar Bank Fallow which is more than I ever did in one day before – Middle day Spreading of D° – The present Stormy & frosty weather Have stopped the Growth of Vegetation almost totally – Bear Chester is now selling for 22/ & Oat Meal 17/ to 17/6 – The Fodder quite Consumed & Cattle obliged to shift except where Hay which is plenty here – Heard the Cuckoo y[e] 17th Inst.

Mondy 18th Yesterday Growing weather some slight rain South[y] and this day same with more D° – Carts dunged out the Cottar Bank & people spreading of D° P.M. – three Ploughs at same – upon Whole no Ill season at present –

Tuesdy 19th very Pleasant weather from S.W. and three Ploughs at Cottar Bank – Others Spreading of Dung and One Cart to Mathers for Lime shells to Houses repairing at at sundries fine Growth.

Wedy 20th Th[r] 58° very pleasant dry weather from S.W. all Ploughs at Cottar Bank Sowed out P.M. 15B 3f Bear total & chiefly Harrowed it for Grass seeds – Bought Two Bull Calves for rearing three of mine Having Died and in General great deaths amongst them – of Fluxes –

Thursdy 21st Th[r] 62° Strong drought from S.W. & very warm grow weather – all Harrows finished out all Bear seed Here for Grass seeds & P.M. 4 Carts driving Dung to Water Lair – 2 Men at Dunghill there –

Fridy 22d very strong drought from S.W. whole day Blowy 4 Carts at driving two draught of Dung to Waterlair and drove out all the Dunghill there on the land Wewmen spreading D° & Men – a Plough P.M. West side. The Oat Seed is not Generaly done yet in some places –

Saturdy 23d Dark weather & Some Showers of rain from the N.E.- Early Sowed the remainder of Cottar Bank with Grass Seeds & Harrowed the same – Water Furred D° – Ploughing Waterlair & Sowed about 6 firlots Bear & partly Harrowed it – intolerable Mould without steering – Sundries &c.

Mondy 25th Yesterday very Cold Bleak weather from the N.E.- this day the same – all Ploughs & Carts driving Dung at Water Lair & P.M. sowed part D° – it is now growing too dry and not so good seed furrow – at sundries &c.

Tuesdy 26th Cold stormy weather from the North with Gusts of Snow & Hail at Intervals – dunged out Water Lair & about finished ploughing sowing & Harrowing it, which Finishes all of season and amongst the first in the Neighbourhood Some as Arbuthnott Burnton have two or three weeks work yet –

Wedy 27th remarkably Cold & stormy weather from the North with some slight skiffs rain – Harrows at Water Lair now Excessive Hard & requiring rolling – Braking in P.M. Red Hall Park for Turnips & Rolling Cottar Bank rather late as the Bear & Grass Seeds on it first sown are well Breered – Every kind of Vegetation stopped –

Thursdy 28th Wind from S.W. and Milder Came on rain at 9 A.M. pretty Heavy from noon to 4 P.M.- much good. Harrows at Braking in Red Hall for Turnips rolling &c learn that at Glammiss Markett yesterday Cattle dull –

Fridy 29th Cold stormy weather from the Northward with flying light showers – Ploughs steering for Turnips & rolling Bear Land and at sundries – very disagreeable & no growth – put the Horses to Grass at 9 A.M. on Bank –

Saturdy 30th Frosty in the night – day fair – wind shifted to the Southward about 10 A.M.- and yet very Cold – Sowed the Water lair park with Grass Seeds Early & Harrowed D° – Others at steering red Hall Park for Turnips &c Jobbs.

Sundy 31st Some light showers in the night & morning – day dark foggy weather from South by E – The Bear seed will take all this week & more in several places – the Grass Pasture & for Hay the most Backwards for years – most of that for Hay eat up for want of pasture, and the Country in great strait of Beastials Meat – Cattle selling however very high & much demand to the Southw^{d}. The Fruit Trees not fully Blosomed nor Ash or Elm Foliated – as Backward nearly as 1782 – Quantity of rain this month is 1 & $\frac{1}{2}$ Inch – Seed Bear 22/ per Boll – Strawberries Flowring – Field plants a Month latter than usual –

Diary for the Month of June 1795

Mondy 1st Th^{r} 55° Showery weather & fine Growth from South – all the Ploughs at red Hall park – water furring & rolling of the Waterlair park – at sundries – the Bear looking very Brisk and though Backward looks well –

Tuesdy 2d Th^{r} 54° last night & this morning a very Heavy rain from S by E near $\frac{1}{2}$ Inch in Gauge – all Ploughs at red Hall Park & rolled out Waterlair – set up the Cattle Flakes on Dickies Haugh 4 year ley – and sundries.

Wedy 3d Pleasant weather Southy – with Flying showers at Intervals – all the Ploughs at red Hall Park nearly done and at sundries Mending Dykes &c – Towards Johnston Observed much Bear seed to – Grass bad –

Thursdy 4th Considerable rain night, day foggy Growing weather from Southw^{d} – all Ploughs did out red Hall park – Mending up Dykes and at sundries –

Fridy 5th Th^{r} 62° remarkable fine growing warm weather from the S by E – 4 Carts driving dung all day to Turnips – and Flited the Cattle Flakes P.M.- with some – a Considerable Growth of every vegetable –

Saturdy 6th pretty strong showers of rain at Intervals – 4 Carts driving dung for Turnips all day – turning Dunghill for Turnips rather late but rotten – The Breer of all kinds looking very well & Grass rising fast –

Mondy 8th Th^{r} 50° Yesterday morning very heavy rain from the Northw^{d}, after dry & very Cold – this day a storm from the N.W. with flying showers all day – 9 Harrows at the red Hall Park for Turnips & People Turning Dunghill for D^{o} – Observe the Bear seed not quite Finished as yet –

Tuesdy 9th Strong Drought & very Clear weather from the North, Harrowed out the Red Hall Park – turning dunghill on D^{o} nearly finished – Marking Furrows for Dunging of D^{o} – P.M. Harrows Braking in Thornylee for Turnips &c Jobs –

Wedy 10th Strong drought from the Eastw^{d} & sun shine – 4 Carts driving dung out upon the Turnip ground – and at Cammock Muir Markett sold two Milch Cows about 5/ per stone – Many buyers – Sundries –

Thursdy 11th Coldish Bleak weather from the Northw^{d} but fair – 4 Carts driving Dung morning – & P.M. Drilling E^{t} side of the Red Hall park for Turnips – P.M. Sent to Dava for 3 Cartful of Flag Cutt Stones to Pave the Kitchen with.

Fridy 12th very bright Dazling sun shine & Strong drought from the North – Carts driving stones from Dava & drilling Turnips all day – 2 Carts at driving Dung P.M. & others Spreading of same.

Saturdy 13th Th^{r} 54° Considerable white frost in the night, day fair & Bright Sun Shine, very Early Yoked at the Turnip field, drilling and Dunging so as to Make Half days Work from about 9 to 10 A.M. on account of a Marriage of one of the Serv^{ts} thereafter.

Mondy 15th Th^{r} 55° Yesterday very pleasant dry weather from the East – this day the same – 4 Carts driving Dung A.M. and P.M. sent three Carts to Dava for stones to Kitchen floor – spreading dung & drilling of Turnips – very foggy weather last night & morning – and strong sun shine –

Tuesdy 16th Closs daggy weather from the Eastw^{d} but little rain – Carts driving dung all day to red Hall Park – partly from the Town & Cottar D^{o} – & Drilling two Yoking – Observe no person yet Begun to Dung for Turnips hereabouts –

Wedy 17th Bleak Gray weather all day, & some Appearance of rain Evening – all Carts driving Dung to Red Hall Park & one Plough Drilling of D^{o} – Others Spreading of Dung & Gathering of Dung Court &c Jobbs –

Thursdy 18th Cold stormy weather from the Northw^{d} – with showers of rain but of no great Extent. Carts driving out Dung to finish whole of the red Hall park & drilling & rolling of D^{o} –

Fridy 19th Very Cold Stormy dry weather from the North with Skiffs not observable – One plough at red Hall Park drilling nearly finished – Carts driving dung Ploughing & Harrowing at West side of Thornylee for Yellow Turnips- and at Sundries.

Saturdy 20th fine pleasant dry weather – at preparing of the Dung for Thornylee

& Sundries – Observe scarce any person yet begun to Dung or sow of Turnips – The leaves of the Standar Apple Trees Pears & Plumbs Blasted – Caterpillars spreading –

Monndy 22d Yesterday Evening a Shower of rain – this day warm & Pleasant – Observed about the Town of Aberdeen the Cutting Grass better Hodged than with us – the Crop from Stonehaven to Dee Bridge Miserable-at Turnips &c.

Tuesdy 23d Bleak weather till about noon – much rain – People Employed about Thornylee Turnips – Preparing Dung, and at Sundry Jobbs –

Wedy 24th Bleak Coldish weather from the N.E. – at Turnips Thornylee – Observed there was few or none Turnips Sown about Aberdeen – the Town is so swarming with Soldiers they forget other things –

Thursdy 25th Thick Scots Mist & drissling rain whole day – Ploughs at Thornylee & Gathering or about Dung for them – Planted Collyflower the great Crop – Houghing Kale of Kinds &c.

Fridy 26th Drissling rain Continued till near noon when faired but Cold from the N.E. & little Growth of any kind of thing at present – three Ploughs at Thornylee – and at sundries.

Saturdy 27th very fine day wind SSW – Ploughs and Harrows at Thornylee – Steering NW Corner of it – Observed the Crop on Water lair looking well but the Hay will not be Heavy on it – Sundries.

Mondy 29th Yesterday some showers of rain very light in general moderate at S.E. – this day the same but more pleasant. Harrows did out the Thornylee and draining of D^{o} – P.M. Yoked 4 strongest Horses (after yoked six Horses on it) to Brake Ground new, never touched before, on Brae of Waterlair, it is Inclosing it is of a light frably Soil, mixed, with Immense Numbers of Stones of moderate sizes and a Rock Bottom about three feet under the surface – There is some Hun^{ds} Acres of this on the Estate, and at present not one Sixpence per Acre –

Tuesdy 30th Thick Hazy weather with very little rain from the N.E. - One Plough all day at Water Lair tolerable work – Others leading off stones for Sivers. The price of Bear Hath fallen from 27/ per Boll to – owing to the Legislature Stopping the Distilleries for a time – The Quantity of rain is In 1 : 75 – Everything as Backward nearly as 1782 – One of my year old Queys burst on the Grass this day by Hoving & lost by neglecting &c. Garden Pease Podding Rose Campion only Flowring.

Diary for the Month of July 1795

Wedy 1st This day thick dark, dirty weather from the Northw^{d} One Plough as yesterday at Waterlair & Broke her near the Evening – Carting & filling up Sivers on Thornylee & Covering of same – Everything very Bleak –

Thursdy 2d Bleak dark & dry weather from the Northw^{d} – preparing & turning Dunghill for Thornylee & P.M. 4 Carts at driving out of part of it – at sundries –

Fridy 3d Bleak dark Gloomy weather – little or no vegetation - 4 Carts drove out

Dung A.M. & drilling of Turnips P.M.- 6 Horses at Water Lair Brae again – Heard there was much Sale at Megrie fair yesterday – Sundries.

Saturdy 4th Bleak weather till P.M. Broke out a little sun shine and Milder – One Plough all day at Water Lair and one Cart driving some dung to Thornylee which spread, drilled & seeded & rolled P.M. – good order.

Mondy 6th Th^r 60° Yesterday Moderate weather & drought N.E. – this day warm & Sun Shine at Intervals – One Plough at the Water lair Brae & one cart leading Earth into dung Court – Others Thisling Oats – flited the Flakes –

Tuesdy 7th Th^r 60° Fine dew last night day very warm dry weather – wind East^y – One Plough at Waterlair – and one Cart leading of stones to Dyke at D^o – People Nearly Thistled out all the Oats on the Town – Looking well – Grass good –

Wedy 8th warm weather but wind still Northerly & drought – being the Sacrament fast all People at Church One of our largest Cows Hoved with Grass at 9 A.M. and Obliged to Slaughter her – to my Considerable loss I am Certain - Careless Herds Indeed – Grass too good –

Thursdy 9th very strong drought from the North – The Plough at Water lair – People Houghed the Potatoes at Thornylee looking very well, and leading Whins, & Sundries – little or no Vegetation being so Cold –

Fridy 10th Th^r 58° very Cold Stormy dry weather from the North - Sent 4 Men & two women to the White Water Moss to Cart Sods – with 3 spades & Barrows to Roll out to the dryer Grounds – Harrows at Thornylee &c –

Saturdy 11th very Boisterous wind from the North – One Plough at Waterlair – & One Cart Sundries – this storm is Blasting many things – yet the Crop keeps up Surprisingly well at least here –

Mondy 13th Th^r 58° Yesterday a severe storm of wind & very Cold from NNW Blasting vegetation – this day more moderate, but very strong drought from the Northw^d – towards Evening more moderate & appearance of Milder weather – In the Morning the Plough at Waterlair & afternoon flited the Cattle Flakes Haugh – Oblige to Water Garden Things – the Field Pease are nearly ruined by last weeks Cold & drought.

Tuesdy 14th Strong drought from Northw^d – & little dew in the night a Plough Water lair – & one Cart sundries – at the Sheep Markett Paldy fair – The Grieve Bot 19 Wedders & Ewes to Feed at 4/ each to feed to 8 or 9 lb per leg – the Weathers fully as Cheap as the Ewes

Wedy 15th Cold but dry weather from the North Chequing Growth – the Plough at Water lair & a Cart driving in Earth to Dung Court – Cutting up withered whin &c for kitchen Fuel and at sundries –

Thursdy 16th Th^r 59° Moderate weather the wind shifted to the S.W. & P.M. some light rain – the Water lair Plough as before – at Paldy fair not one fourth of the Cattle usual, and these sold the lean at from 5 to 6/ per stone when full Fatt in the spring so Could not Buy any – There never was such a Markett for the Sellers – The appearance of rain towards Evening &c –

Fridy 17th Th^r 57° a strong Gale from the North with drought – Broke the Sock Waterlair Plough morning – at driving Gravel to Publick roads P.M. –

Saturday 18th Moderate weather from E.S.E. but fair – the Plough at Waterlair – and one Cart at the roads – the Crop is suffering much for Want of rain and Discolouring on the Thin Soils &c – Flited Flakes –

Mondy 20th Yesterday Bleak weather from S.W.- this day same a shower from $\frac{1}{2}$ past Noon to $\frac{1}{2}$ past 1 P.M. The Plough at Waterlair and two Carts driving Gravel to Roads – the drought is Checking the Grain severely in many places – dark weather Coldish –

Tuesdy 21st Bleak dry weather from the SSW with very little rain the Plough at Waterlair and a Cart driving Gravel to Publick road – there is a Considerable Growth in the Grass & also of the Trees – &c Jobbs –

Wedy 22d Early this morning some Thunder & Lightning & some smart showers of rain from the Eastward the rain Continued P.M. – the Plough at Waterlair and laying to a Furr to Potatoes Thornylee &c Jobs.

Thursdy 23d a very Considerable deal of rain through the day which will do much good – the Plough at Waterlair morning done – P.M. at the new piece of Ground at Thornylee – at Garden & sundries –

Fridy 24th several showers in the night & morning – the Strong Plough at ploughing the road at Will & that at West side of red Hall park for Graveling – driving Earth to Dung Court and sundries – Planted Curleys & sowed Spring Seeds.

Saturdy 25th Th^r 62° very pleasant weather but rather Bleak from N.E. till P.M. sun shine – Plough at rough ground R Hall park done – and at driving Earth into Dung Court – Harrowing at new ground Thornylee – 2 Horses to Aberdeen with my second son William for the Navy – and at sundries - a few strawberries.

Sundy 26th Th^r 63° Noon very warm weather & Clear sun shine – the Cattle Startling & running with the Flys for first time –

Mondy 27th Closs Foggy weather till noon Came on pretty Closs & heavy rain from S.E. which stopped work without Doors – after Brakefast begun Cutting Hay Own Men 4 Blacklands - P.M. Dighting Oats sold to Dragoons Montrose at 20/ per Boll.

Tuesdy 28th Thick Hazy in the night day Clear & strong drought from W.S.W.a W – P.M.- people driving Earth to Dung Court – flited the Flakes & after Dinner Cutting Hay 4 of our Men – no Sheep to be got at Garvock rose 50 per C^t within this fortnight Owing to the Scarcity of Cattle –

Wedy 29th Very strong drought from West a S.W.- till P.M. several showers – 4 people Cutting Hay at Blacklands – One Plough all day taking away the Fur from first sown Turnips – rather backward.

Thursdy 30th Strong drought from the Westw^d – little skiffs P.M. Cutting Hay till Brakefast, & at Turnips – after every Body at Garvock Markett – Cattle higher than at Paldy fair – Servants very Extravagant in their demands & believe but few Engaged –

Fridy 31st fine dry weather wind Chiefly So^{y}- but drought great & Stinting ye Crop Bear Shooting, Pease Blooming, & few of new Oats Shooting – the Bear Crop looked upon Generally as thin & Scanty – 4 People Cutting Hay – a Plough at Turnips which are Scarcely ready yet for Thinning – The Quantity of rain 0.75 Inches – the Grass Failing or Withering rather – Oats up to 25 Sh Str per Boll – Bear 26/ Wheat 40/ Oat Meal 20/ per Boll Hay not Settled – Garden Pease & strawberries only just ready –

August Diary for 1795

Saturday 1st Mild weather from S.E. with no drought but little rain Cut out the Blackland's Hay before dinner and after Flited the Cattle Flakes Haugh – 2 Carts to Montrose with 10B Oats Sold at 20/ per Boll & Brought out Coals in return rather dear – and at sundries –

Mondy 3d Yesterday two very smart showers of rain at 2 & 4 P.M.- this day showers at Intervals wind Northerly – lead Home Whins & Horse & Hand Houghing of Turnips Red Hall Park – Bad for the Hay – roads dirty –

Tuesdy 4th Therm^{r} noon 59° late last night & Early this morning very Heavy rain for Hours – day fair – a Plough all day at Turnips & Men at Hand Houghing first sown D^{o} – Some Bear Layed – This will Hurt the Hay if it Continues &c –

Wedy 5th Th^{r} 62° Strong Hoary dew in the night but day very warm & sunshine – People at Turnips morning – after at Gathering & Coling Hay all day in good order, no worse of ye rain – Bought 22 fir Trees at the Garrol Wood about 17^{ds}each reasonable – Oats shooting & looking well.

Thursdy 6th Th^{r} 62° very fine day after a thick Frosty Fogg – Twelve People Gathering & Coling of Hay all day in good Condition – P.M. into large tramp Coles – the cattle Startling & Obliged to be Brought to the Dung Court.

Fridy 7th Th^{r} 60° Strong Breeze from S.E. & drought – 10 people at the Gathering & Coling of Hay Tramp Coles all day till 5 P.M. when nearly all Smithy park was done stopped by wind & rain – at Turnips & dighting & sacking Oats for Montrose.

Saturday 8th Th^{r} 63° Yesterday Evening Excessive Heavy showers of rain and this day several lighter ones – 4 Carts to Montrose with 22B Oats & three of Pease & to Bring Out Coals – at Sundry Jobbs on Farm &c.

Mondy 10th Yesterday showery – this day same – Sent 3 Carts Twice a day to White Water Moss for Sods in good order – Others Horse & Hand Houghing of Turnips red Hall Park – and at Sundries

Tuesdy 11th Very Pleasant dry weather from the Southw^{d} – got three Cartful sods from the Moss & Built them – after Brakefast 4 Own people at Cutting Hay Waterlair a good Crop and others Horse & Hand Houghing Turnips –

Wedy 12th Th^{r} 71° very Pleasant & warm weather from S.W.- 4 Hay Cutters as yesterday – and others at Turnips much in need of dressing but People Cannot be got to Hire being so Scarce – Cattle Startling & Brought or run Home before Brakefast – Sundries –

Thursdy 13th Thick Closs Foggy weather, Every thing Wet – 4 people at Water lair Hay Cutting & three people at the Turnips thinning above Half through – and finished taking away first fur from D°- and took away a second from those first done being fine and P.M. Flited the Cattle Flakes Haugh – Gooseberries ready –

Fridy 14th last night and this morning very Heavy rain and Squally weather from E.S.E. which brought down the Rivers a good deal – day thick Fogg & Growing weather – after Brakefast 4 Men finished cutting the SW park on Water lair – Horse Houghing Thornylee Turnips very fine but no thinning of any this day Ground so Wett –

Saturdy 15th about 8 this morning Came on Heavy rain which lasted more or less all day – Thinned & dressed However the Turnips on Thornylee very Luxuriant and after at red Hall Park some of the rankest of the Oats & Bear swayed, Pease field Podded, the Grounds & the roads very wett & dirty – sundries –

Mondy 17th Yesterday morning very Heavy rain & fields very wet this day frequent showers & Thunder at 4 P.M.- people and Hired at R H Turnips very bad season – the roads Bad, and the rivers out P.M.- Some Oats Lodged –

Tuesdy 18th Th^r 62° Fair weather and some drought with sunshine from W^y All people & Hired ones at the Turnips red H Park which finished first dressing of – The Oats & Bear mostly raised again – The Oats appear the richest Crop for many years by past in General but late being not near shott in any place almost –

Wedy 19th Fine pleasant weather from S.W.till 2 P.M. some rain which Stopped the Hay Coling at Waterlair & then at Turnips Thornylee – The Hay in Small Coles is in very bad Condition & much Spoilt mine in large ones.

Thursdy 20th very pleasant day & strong drought from S.W.-Ten people at Waterlair whole day at Coling of Hay into Tramp Coles in good order – a vast of seed lost by ripeness and late rain but great Quantity remains yet, sundries.

Fridy 21st very fine dry weather frosty this morning – Coled out Waterlair 13 Tramp & 29 Hand Coles full 2000 St on $6\frac{3}{4}$ Acres – P.M. 5 people own Cutting 2^d park Waterlair not near so weighty Crop – People at Sundries P.M.

Saturdy 22d a Considerable rain last night & this morning & some showers in the day – 4 people own Cutting of Hay all day Waterlair – and a Cart to Inglishmaldie for wood to the wright Flited the Cattle Flakes &c.

Mondy 24th Yesterday dry weather – and this same from S.W. with warmth & sunshine – Cutt out all the Hay of season, at the Turnips much Injured by ye rains and will require good dry weather to recover them.

Tuesdy 25th very strong drought & warm weather from Southw^d – all the people Coling of Hay at Waterlair, and at the Sheep Markett Laurence fair Bo^t 40 Ewes to feed to about 7 to 8 lb per Quarter well Conditioned at 4/6 each.

Wedy 26th Blowing weather all night & first Half of this day – Coled out all Hay of Season – in fine order – and after Cutting Rushes &c for Covering it at Floors – some little drops of rain P.M. – warm weather –

Thursdy 27th Frosty dew in the night day strong drought S.W. People Horse & Hand Houghing of red Hall park the 2[d] time- now growing at L fair an unusual Number of Cattle very fine ones as high as 19£ Str few Buyers & little Sale- my three three year old Stotts Fetched £12-12 Str each-

Fridy 28th Pleasant dry & warm weather from various – all People at Turnips, Hand & Horse Houghing all day, Turnips Seed Garden-46 lb Ams[d] – Flited y Flakes –

Saturdy 29th very fine dry weather from the Southward, Sent 4 Carts to Garrol Wood for 22 fir Trees – others at Turnips dressing now growing very well – the Bear Ripening apace, but the Oats very little Alteration in them –

Sundy 30th Strong drought from the Southward and some appearance of rain but went of – with wind and very growing weather –

Mondy 31st pretty strong wind at South which prevented us leading of the Hay this day as Intended – all People Horse & Hand Houghing the Second time the Turnips red Hall Park y[e] most part done –

The Turnips Have Failed mostly on strong grounds or where late, but here good – the Bear will take two to three weeks yet to be ripe – and the Oats although they look universally Green as Yet are filling apace, as is the Pease, and the price of Grain falling much to the South – Quantity of Rain in August is three Inches – which fell Chiefly the first 12 days – Grain selling in the Look of it about 20/ per Boll Conjectured – and after Turned out twice or thrice as much. J.B.

Foals – Corn & Fodder

Diary for the Month of September 1795

Tuesdy 1st very strong drought & wind from South – Could not Lead the Hay for it – all people at Turnips mostly dressed out – flitted y Flakes –

Wedy 2d Thick hazy weather but dry-from S.E. – Early Yoked 4 Carts and lead Home all Blacklands Hay into a Sowe 46 feet by 12 – Twenty five Cartload – and want about Six Cartful Waterlair to Finish it –

Thursdy 3d Thick foggy warm weather from the Eastw[d] – Brought Home 5 Cartful Hay from Waterlair finished the Hay & Thatched it Immediately in fine order &c –

Fridy 4th Thick Hazy weather from East – with about 20 people Early yoked three Carts & lead all the Waterlair Hay 14 Acres into a Sowe 15-by 60-in very fine Order which finished after it was dark –

Saturdy 5th Blowing dark weather from S.E.- and some rain in the Forenoon but Cleared up a little afternoon – Own People & Hired Cut Rushes & Thatched the Hay Sowe on Waterlair finished it P.M. – Stormy again Evening – the others at Cutting rushes P.M. –

Mondy 7th Th^r Noon shade 65° Yesterday fine drought from Southw^d – this day the same - lead home 4 Cartload Whins &c and flited the Cattle Flakes – after giving last 2^d dressing to the Thornylee Turnips, and at sundries – Observe some Inequality of the Bear ripening –

Tuesdy 8th Variable weather – wind N.E.a S.W. – about $\frac{1}{2}$ past 10 A.M. an Excessive Heavy shower for 15 Minutes – after fair – Three Carts to Stonehaven for 240 stones English Coals which returned after 12 hours absence – Second dressing of Thornylee Turnips which are good – Sundries

Wedy 9th Frosty dew in the night, with showers through the day – all People at Hand Houghing of the Turnips red Hall Park, and at Sundries.

Thursdy 10th very strong White Frost morning wind S.W. – the Potatoe Stems nor Indian Cress touched – all People at Turnips- and One Cart to Johnshaven for Great Coal – road Bad through Garvock.

Fridy 11th Frosty Dews in the night day warm & sunshine from S.W. - all People nearly third dressing of the Red Hall Turnips – & P.M.- laying to a furr to the first of them ready for it – & Flitted y Flakes – Observe a few spotts Bear Cutting Neighbourhood –

Saturdy 12th very strong drought from S.W. – Sent three Carts for Coals – Two Ploughs laying to a furr to Turnips Thornylee which finished – Others Houghing Red H Park – the Frosty dews Have not Hurt Pease or Potatoes.

Mondy 14th Yesterday drought Chiefly fine weather – this day the same filling the Corns very well – Sent three Carts own & 2 Borrowed to Johnshaven for Great Coal-& two Ploughs laying to a fur to Red H Turnips now very Clean, and as things rather Backward I should Suppose would be good.

Tuesdy 15th Th^r at Noon 58° White Frost morning day very Hot and strong Sun – 3 Carts lead two Cart load Turfs from the Garvock Hill & Built them – and two Ploughs laying to a Furr to Turnips as yesterday – P.M. Cows & Horses on Kavels as Good Grass as if it was in the end of June being Hained two Months.

Wedy 16th Foggy weather in the morning, at $\frac{1}{2}$ past 11 A.M. Broke out very warm Sun & drought – three Carts Brought Turfs from Garvock & two Ploughs at Turnips – P.M. Pulled the Lint – a very poor Crop owing to Bad Seed – Observed some of it stocked quite uncommon & shows strength of soil – Observe severals Cutting of Bear & Pease –

Thursdy 17th Therm^r 58° Thick Foggy weather till Brakefast time – then very warm or rather Hot – till 11 A.M. when Clear & fresh Breeze from Southward – Begun HARVEST with twelve shearers Half Men – on long Haugh Bear – fully ripe as is the whole on this farm –

Fridy 18th Th^r 57° last night & this morning some light rain & stormy SE - day Cloudy & fair – about 10 A.M. begun Cutting Bear as before all day – morning three ploughs laying to a fur to Turnips nearly finished – In Whins &c fireing – and at sundries –

Saturdy 19th Early Yoked Shearing as yesterday, but stopped by light rain & rather stormy at 2 P.M. – after Flitted the Cattle flakes, & did out Turnips –

Mondy 21st Yesterday Closs Foggy weather from the Eastward – and this day same – Early all shearers at Cottar Bank Bear, and rather speedy with it although not hard work – it Lyes well & in order – Sundries –

Tuesdy 22d very fine weather and drought from S.W.- finished all Bear Cutting here 452 Th^{r} about 22 Threaves per Acre and after 4 P.M. at Waterlair Bear fully ripe also and Exceeding Great Grain as I ever Observed –

Wedy 23d very fine weather & drought from the Westward, People finished out Cutting of all Bear of Crop – and in fine weather – a Harvest now General

Thursdy 24th fair weather till about 8 A.M. Came on Smart rain which lasted till near noon when fair & strong drought from S.W.- People at the Pease Thornylee tolerably well Filled but moderate Crop in the General they & Beans Have failed much –

Fridy 25th Thr 59° Mixed weather but fair & no Great drought from S.W. – People Finished Cutting all Pease before noon – and three ploughs at giving Furr to Thornylee Fallow Angle wise very fine – Cutting Rushes –

Saturdy 26th The weather is now much Cooler than before a N.W. – People Cutting Rushes for Thatch and others took out of Water 10 nights the Lint and spread it Head Nether Burns – Sent two Carts to Inglishmaldie for Birch Wood, Moderate priced.

Mondy 28th Yesterday morning all the Potatoes stems and Indian Cress Frosted Entirely – this day Cold & no great drought from S.W. – Sent three Carts for English Coals to Johnshaven – and at sundries – Observe some leading Bear but not good.

Tuesdy 29th Fair weather wind Easterly and Appearance of RAIN - 3 Ploughs at Thornylee – and P.M. Lead Home Whins &c Observed severals leading Bear &c but ours was in Condition Came on rain but not Heavy about 6 P.M. – at sundries –

Wedy 30th Therm^{r} Noon 58° Moderate weather from S.E. but very little drought - three Ploughs at giving a Furr to Thornylee fallow nearly finished – and others at sundries – Did not Judge it right to Lead the Bear although severals Have done so, not so well Ripened or long Cutt, but their Stacks Smaller than ours are usually – The Oats will be now ready for Cutting being full ripe on some Parts of the Fields Only – Very Foggy and warm Evening no Encouragement to lead – The Oats few of them ripe & very green as yet in many places – The Wheat, Pease & Beans & the Bear will not turn out Good either in Quantity or Quality – the Bear not whole meal and in the South it is said to be the same Quantity of rain this month One Inch mostly at the beginning – Cattle Sale now duller, but yet prices so High as not to be Bought to Keepe –

Diary for the Month of October 1795

Thursdy 1st Therm^{r} Noon 57° Foggy mild weather morning, but Broke out fair

& Drought from S.W.- after Brakefast yoked 4 Carts & lead Home about best Half of Cottar Bank BEAR – Women Shearing of OATS Back Yards fully ripe, but not from End to End Ridges

Fridy 2d pretty Early morning Came on rain from Northw[d] which Stopped all Harvest work-whole day – People at Barns, Thatched 2 Stacks Bear & Ploughs did out the Fallow on Thornylee – and at sundry Jobs –

Saturdy 3d about 10 last night Came on very Heavy rain which lasted till two or three morning, which However did not go deep into the stooks, day fair & some drought S.E.- Carts driving Dung to Dickies Haugh Ley & spreading D°-

Mondy 5th Yesterday moderate drought at Southw[d] – Heavy dew & Corns Wett, but begun Early to Cutt up the B Yards Oats Gayte & then stooked when dry by 11 o'clock and stopped by $\frac{1}{2}$ past 1 P.M. with Heavy rain from the South – after our people at Barns at Sundries –

Tuesdy 6th a Great deal of rain & after 9 last night a strong Gale & Squally from North – day fair & shearing Oats dry all day – Set up Hood Sheaves of Bear to dry – Observed some few leading Bear – flited y[e] Flakes –

Wedy 7th a very Strong Frost this morning – Early Cutting of Oats & Gayting – after 10 A.M. yoked all Carts and Brought Home all the Bear here & Waterlair – in good order – Women shearing and fair day wind Sy.

Thursdy 8th dry Bleak weather from the Eastw[d] all day – Shearers at Oats all day – & turned the Pease will win – at Drumlithie few Cattle & little Sale but still the Prices double of what Feeding Can afford &c –

Fridy 9th last night about 8 Came on Heavy rain & Storm from E.S.E.- which stopped by 4 this morning – day rain after 8 & no Shearing rest of the day – at Barns & at sundries making stathells – rain & storm again 8 P.M.

Saturdy 10th The rain & storm brought down the rivers as high as for some years by past & alongst with it some grain from the upper Farms on this river – our People shearing at Robs Hillock Oats all day, there is much dammage the Corns being Broke down, and almost Impossible to Shear them without loss of Grain.

Sundy 11th fair weather at East till $\frac{1}{2}$ past 9 to 11 o'clock P.M. which Has wetted the Grain & washed the roads Clear the Crop to Johnston very Greenish,much Bear out & to Shear

Mondy 12th Wind N.E. Fair Bleak weather all this day – Shearing & Gateing R Hillock till noon when Stooked – then at Gibsons Park totaly Broke down & spoilt by the rain & wind in many places, especially where thinnest –

Tuesdy 13th Blowing weather from the East but dry till 10 A.M. when Came on a severe Gale of wind & Rain from same Quarter- dispersed the shearers – it Continued till near 5 P.M. & Brought down the Rivers &c very much – at Barns & sundries Flakes

Wedy 14th Thick Hazy weather from the East – Observed the first Cutt Oats beginning to Grow the Hood Sheaves & some of the Gayte – Shearers after Brakefast begun Cutting Oats B Yards in Gayte being wet all the day – dunging Haugh ley morning – Observed the Oats at Johnston Growing & Hear much Bear is so –

Thursdy 15th Closs Foggy weather till 10 A.M. when Came on severe rain & wind which stopped shearing & deluged the Fields Observed the Oats Growing that were Cut about a week ago & others uncut sadly smashed – People at sundries – Abated by Dinner time.

Fridy 16th About Six this morning Came on Heavy rain for two Hours from South – at Noon Cleared up & begun shearing of Auld Town park Oats Greatly broke down by the rains & winds & Impossible to shear Clean & much Shedding by being over ripe in the Cutting – Gayting – Heavy showers again after 3 P.M. – Observe all the Oats that are Wett Springing fast, not six days Cutt – and our Neighbours Bear the same – those who Have not Hooded their stooks will suffer severely – The Pease will be lost in many places Especially where they are mixed with weeds – mine is Clean of them – rain in the Water Gage is 5 Inches of rain already.

Saturdy 17th Therm^{r} Noon 56° This day mostly fair weather but little drought S.W. Early all people at shearing Auld T park Oats Gayte till after Dinner then stooked, greatly Broke. The Oats in stooks & Gayte so Immensely Grown & spoiling and if not Checkqued soon will Greatly Injure ye Crop.

Mondy 19th Therm^{r} 56° Yesterday about 8 A.M. Came on Heavy rain which lasted till 4 P.M. – this day fair till 3 P.M.-Heavy rain all the rest of the day & Evening at S.S.E.-Stopped Shearing Gibsons Park Oats – Observe the Oats growing that are not cutt a very rare Occurrence though not near the Ground – Flited y^{e} Flakes.

Tuesdy 20th rain in Gage $6\frac{1}{4}$ In. An Immense rain last night & rivers very much out but day drought & P.M. strong wind S.W. – finished out Gibsons & Auld Toun parks before 11 and after at R Hillock pretty dry & stooked it – Shaking a good deal afternoon & with loss of shearing very Considerable Indeed – Observe the Pease much Swelled, opening &c

Wedy 21st very strong drought & Blowing weather from S.W. till 8 A.M. more moderate-but dry – Cut up all R Hillock Oats stooked – then the N Burns being all in this Town and after Turned the Pease & after at Waterlair Oats – In the Evening by $\frac{1}{2}$ past Seven rain Came on again.

Thursdy 22d This morning Early rain from the Eastw^{d} – day fair but little drought – after Brakefast begun & finished out all the Oats at Waterlair being whole Crop in 20 nights supper Meal – Observe severals leading, wyling &c

Fridy 23d Frosty Dew till noon & no drought – Cutting Rushes &c for Thatch and after Dinner Begun leading of the last Cut of R Hillock Oats in Stooks & some Gayte in tolerable good Condition for Keeping into three stacks.

Saturdy 24th This morning Early Came on rain a South with very stormy weather which faired before 11 A.M. and after Miday Increased into a very strong Gale from S.W. which drove the stooks & Gayte & must Have shaken much of that unshorn – layed & Graveled several Stathels for stacks &c.

Mondy 26th Yesterday and this fine drought from S.W. – at 5 A.M. yoked all Carts, and lead in all North Parks Oats Others Binding the Gayte of them & Robs Hillock & part of R Hillock stooks West side of them into 7 stacks I believe in good order & a great days work indeed –

Tuesdy 27th Fair weather till 8 A.M. when Came on rain a S.E. which stopped the leading Golds Hillock – Set up all Wyled Oats & Gayte & Hooded those at Waterlair & set up Gayte Flited the Cattle Flakes before dinner – perfectly stormy & rain all day –

Wedy 28th Th^{r} 50° Fair weather till about 8 A.M. Came on rain from Southw^{d} which did not last long – dismissed the Shearers till Leading again yoked 4 Carts and drove out Dung to Haugh Ley – and sett up all Gayte & stooks blown down – and at sundries –

Thursdy 29th Th^{r} 50° a Heavy Shower & Boisterous wind at S.E. last night, but faired up & strong drought & Stormy all day a S to S.W.- leading Oats all day – Wyling stooks, Binding Gayte and Opening & by Spreading all those most Growing In some stooks near the Garden – Only 3 Sheaves not Growing others one Half free & some more – Many leading & Hutting on the Grounds – & some taking it Home without Picking –

Fridy 30th very pleasant weather & Good Harvest from S.W.- Lead Home all of Waterlair Oats very light Crop & after at leading from different fields here Picked – then lead Home all the Pease 14 Cartful into three tolerable stacks, and in no bad order Considering not more dammaged than the Oats.

Saturdy 31st very strong drought & stormy S.W. till 8 A.M. some rain, Had Got in all the Crop but about 4 cartful of them out – after Much rain till Dinner time – Thatching stacks and Cutting of Rushes & S Bog Planting – Flited the Cattle Flakes P.M. & at sundries – Quantity of rain Immense for Principal harvest is Seven Inches & three Quarter Inches. The Most of my Crop which Bulks well and as far as I know Hath few Equals in Condition for Many Miles around – I Have Had it often longer 25 supper meals and in far worse Condition – many of the principal Farms Have not yet Got in much of their Bear & little Oats and a Great proportion of the latter to Cut up – Pease Out &c. On the whole this is One of the most Destroyed Crops already & will be more so yet by ill keeping since 1782 or Even it. J.B.

Diary for the Month of November 1795

Mondy 2d Therm^{r} 38° Yesterday a very Strong Gale from S.W. & drought about 5 P.M. wind shifted to the N.E. with stormy rain - this day same with showers of sleet & very Cold weather – People Thatching of stacks, and set up Gayte, and the gardner & others Digging of Potatoes, and at sundries –

Tuesdy 3d Strong Frost & Considerable Snow night & first Half of the day which lay on the Braes & Hills all day & sorryful to see it Amongst the Growing Grain – Thatching out all the Stacks, and Digging of the Potatoes & at Sundries.

Wedy 4th Gave ye Feast a very strong Frost last night & this morning the Ice almost Carrying and betwixt 2 & 3 P.M. Came on Heavy RAIN and stormy at South – Brought in the few remaining Cartfuls of the worst pickings of the Crop & Built them Corn Yard – and at Digging & Pitting Potatoes all day - Flited Cattle Flakes Haugh. Observed many for Days by past leading off the Hook.

Thursdy 5th Fair weather all this day but little drought S.W., all the People Digging of Potatoes, Midling Crop & Earthing of them Back Yards – few Cattle at L kirk Markett little or no Sale but dear – Much Corn & Bear to shear & out as yet everywhere in this Neighbourhood &c.

Fridy 6th fine dry weather from N.W. all day – People at Digging up the whole of the Potatoes which Secured by Covering into 3 Tumuli & three pitts S side Garden Backyards - about six or 7 Bolls – and observed Many leading shearing &c.

Saturday 7th Th^{r} 53° Some rain last night & day very strong drought and P.M. Strong Gale at W.N.W.- Sent 3 Carts to Stonehaven for 240 st English Coal- and at sundries – good for stacks.

Mondy 9th Th^{r} 44° Yesterday very pleasant weather from the Westw^{d}- and this day uncommonly so – 4 Carts driving Dung to the Haugh ley from Dung Court & Flited ye Flakes P.M. Observe severals leading stooks & Hutts in Neighbourhood.

Tuesdy 10th Strong Frost White & Ice pretty thick all day – 4 Carts at Driving Dung to Haugh all day – put in House Coals P.M. and at sundries – Observed Ld Arbuthnot & Waterlair &c leading of their Pease in bad Order – Gardner at Hedges Cutting &c.

Wedy 11th very strong frost & white scarcely fit for Pasture before $\frac{1}{2}$ past 10 A.M. – 4 Carts driving out Dung to Haugh and sundries – Observed a Field Oats at Allerdyce belonging to Mr Barclay of Urie Proprietor to shear and the Crop not Settled quite & very Indifferent to Bervie

Thursdy 12th Fine fresh dry weather from the West – 4 Carts driving Dung from D Yard to E^{t} Haugh & spreading of D° – Cleaning of Hedges & Cutting same – Observe severals at the roads towards L kirk – The Foggage quite spent in most places.

Fridy 13th Therm^{r} 48° The weather Altered to stormy from ye Northw^{d} with Flying showers of rain at Intervals – 4 Carts driving Dung all day to Haugh & spreading of D° very thick – Hear every day of the Bad Accounts of the Crop – Oats as low as a Firlot of meal per Boll & Bear at Eleven pecks D°- this at Kineff & Knoxes – There is much out & to Cut at Auchwhirry, Morphie & other places – Grounds now dryer –

Saturday 14th a very stormy Blowing Night but dry Except some skiffs of snow from North – 4 ploughs begun a steering Furr Anglewise to the Oat field on Waterlair Clean furr rather than rib on such strong Ground which if drought at seed time turns up lumpy & requires more ploughing – P.M. Flited the Cattle & Sheep Flakes Haugh & Spreading Dung on D° –

Sundy 15th Therm^{r} 43° Strong drought & Black Frost in the morning – wind S.W.– the Grounds Considerably dryer & some few I observe ploughing down their

Lime on Fallow – and others sowing Wheat on D° – Horses on Grass Waterlair which is very fine Hay Foggage Hained since Cutting –

Mondy 16th uncommon fine pleasant & dry weather from S.W. – four ploughs at Waterlair – and spreading of Dung being done on Haugh ley – at reding up of Byres &c in Case of Occasion to take in the whole Cattle – Hear some Drovers stately Williamsons Have not sold their Cattle in the South Country Marketts but Obliged to Hire Meat for them through Winter in Lothians – am Informed the whole Crop of Barley & Bear at Luoman Angus there is Sold for 25/ per Boll ready money on Delivery at Montrose betwixt this & April next – Meal 15 to 16^ds^ per peck and Beef & Mutton from 5 to 6/ per stone –

Tuesdy 17th Strong wind from the S.W. & dark weather – 3 ploughs at Waterlair – and one Cart at sundries – pasture no good.

Wedy 18th about 7 A.M. Came on a very Heavy fall of SNOW which with a strong Gale from the N.E. stopped all work without doors whole day & obliged to House all the Cattle & the Sheep in the Evening into Shades – the Snow several Inches deep & Country appears as in the dead of winter – People Threshing Oats S B for Fodder –

Thursdy 19th Strong Frost all day & the Snow not vanishing in the least at Noon – three Ploughs at Waterlair, no Hindrance by Frost, but too much snow – However not Amiss For Sterage Fur for Bear – begun to Lead first of the Turnips of Season Thornylee, and never before at first Obliged to Digg them – at Barns Threshing Oats & Sundries.

Fridy 20th Therm^r^ 30 deg^r^ very Hard Frost & Clear weather from N.W. – 4 Carts driving Gravel to the red Hall Park road before formed and sundries – Hear the road to Stonehaven is Stopped by Snow on Wednesday last very Early Indeed, and suppose never known to be so Early by any person living

Saturdy 21st Th^r^ 29° Very Strong frost till noon when a little slack from S.W. – 3 Carts driving Gravel as Yesterday & Home some Turnips Thornylee – a Cart to Montrose for Sundries.

Sundy 22d Dry weather from the S.W. but little or no Freshness - about 6 P.M. Came on a very strong Gale from W a WNW but did no Material dammage – abated by midnight –

Mondy 23d Betwixt Fresh & frost – a good deal nearly whole of the snow gone here, but lying on the rising Grounds to the Northw^d^ – 3 ploughs at Waterlair tolerable season and One Cart leading of Turnips Red Hall Park – took in 2^d^ stack Oats N N Barn and at Sundries –

Tuesdy 24th variable weather fair & rain at Intervals – 3 ploughs at Waterlair and one at Thornylee Pease land – at Barns dighting Oats for the Mill, and at sundries –

Wedy 25th Therm^r^ 30° very Keen Frost & forenoon some Snow for 2 Hours at NNW 3 Ploughs finished out Waterlair, & at Thornylee after Pease & one Cart with $7\frac{1}{2}$ B Oats to Mill of Kair, and leading of Turnips – P.M. Pulling & Heaping of D° on Red Hall Park – Gardner at sundries Hedges Cutting &c.

Thursdy 26th very strong Frost to stop all Ploughing – 4 Carts driving Gravel to red Hall Park road & Brought Home Turnips last draught – Evening sleet & snow at N.W. – Sundries.

Fridy 27th Strong Frost Continues – the road a sheet of Ice no Horse work – Digging Earth for Miding Layer Back Yards- at Auchen Blae Crofts Markett Bought a Farry Cow about 4/ per stone when Fatt –

Saturdy 28th Th^{r} 30 Strong Frost & High Wind morning from N.W. took in a Stack Oats S B – Bottom Waterlair & of other fields – in fine Order – Men & Carts digging & driving out Earth from Ditch E^{t} side Garden and at sundries.

Sundy 29th Strong Frost Continues all day but towards late P.M. Came on rain & fresh weather from the N.E. – left the Sheep out East Haugh all the night to Pasture at Freedom –

Mondy 30th All the first part of this day pretty Closs rain & foggy weather – Waters Broke & a little Out – lead Home 2 Cartful of Turnips – at Ditch Scouring, and at N Barn Chaffing of Oats.

The Cattle Eat much Fodder which is good. The Quantity of rain a little Uncertain Owing to the Water being Frozen, but Believes it about Two Inches & one Quarter- no Certain prices yet of Grain the Distilleries being stopt by the Parliament prevents the Sale of Barleys – Hear of some sold from 19/ to 25/ but its Quality is totally different some of it being double value of other – Oat Meal about 20/ per Boll & Scarcely to be Had – Manufactures of all kinds very Brisk – Beef 4^{ds} per lb & fatt Cattle from 5 to 6/ – all kinds of Cattle High priced. Oat Meal from 8 to 12 pecks per Boll & Bear from 3 firlots to five D^{o} –

Diary for the Month of December 1795

Tuesdy 1st High Wind from 6 to 10 last night – this morning a Considerable rain – three Carts leading Turnips & some people at the Mill of Kair – at Barns &c Jobs – The rivers Out & Grounds very wett frost not Out –

Wedy 2d Slight Frost morning, day very little showers at W.S.W. 4 ploughs at Thornylee Pease land ribbing Anglewise rather wett above & soft – at Sundries on the Farm.

Thursdy 3d Th^{r} 39° Very strong Gales from the West with slight frost that Hath Dryed the Grounds much – nearly finished the steering at Thornylee, and at Barns dighting Oats for the Mill – Observe the Ploughs in general going this day –

Fridy 4th Strong Gales from the Westw^{d} – 3 Ploughs begun Ribbing the Upper End of Robs Hillock across for Bear and one D^{o} finished all on Thornylee – Sacked $13\frac{1}{2}$ Bolls Oats for the Mill of Kair P.M. and at Sundries-

Saturdy 5th Th^{r} 44° very pleasant day from S.W., but Had been a great deal of rain in the Night, & fields wet this day – 3 Ploughs at Ribbing Robs Hillock – One Cart to Mill with $13\frac{1}{2}$ Bolls of Oats – Gardner dressing Hedge around Bog planting &c.

Mondy 7th Yesterday frosty weather – this day same & Foggy fresh - 2 Ploughs at Golds Hillock ribbing for Turnips – and one Cart leading Turnips from Thornylee – & a Cart to the Mill with six Bolls Oats – and at Sundries –

Tuesdy 8th Fair weather betwixt frost & fresh-3 Ploughs at ribbing of Golds Hillock dry and One Cart with Oats to ye Mill others dighting in the South Barn the Oats – Sundries.

Wedy 9th very Pleasant fair weather & Calm – 4 ploughs at the ribbing for Turnips – and P.M. at sundries – weighed of our Bear 18 st 4 lb per Boll Montrose Measure – Observe Grounds in general very wett –

Thursdy 10th Closs Foggy weather & warm – 3 ploughs at Golds Hillock and one Cart leading of Turnips from Thornylee and at sundries – the ground now much dryer –

Fridy 11th a Good deal of rain morning & at Intervals all the day - 4 ploughs at Ribbing Back Yards &c and P.M. at Scouring the Ditch East side Garden for Dunghill layer Observed severals could not Plough this day –

Saturdy 12th Constant rain from N.N.E. all night and this day till 11 A.M. when Faired – Lead Home some Turnips from Thornylee, and Chaffed a Stack of Bear N Barn with the Fanners, Gardner Planting Ashes at Loup Yards where the Oaks had Failed planted two years ago at least one Half of them, and at Sundries –

Sundy 13th Closs & Heavy rain all last night & whole of this day from N.E. & Stormy – In the Evening the Rivers Out as high as for Several years & fields deluged with water.

Mondy 14th The river last night at 7 was Higher for a Dozen of years by past – day fair till Evening rain again – 4 ploughs ribbing Back Yards wett – at Barns and at sundries – the waters & roads particularly the last surprising good & washed – Gardner Pitting –

Tuesdy 15th after the rain which was Heavy & High wind at N. Cleared Early this morning frost – Yoked three Ploughs to Rib Back Yards the lower End of it, wet work dry Enough – the People at the Mill of Kair with 27B Oats – Meal – at Barns &c – leading of Turnips from Thornylee and Sundries – Rain in the Gauge in this month already 4 Inches –

Wedy 16th Fair mostly all this day & Cloudy Closs weather Southy 2 Ploughs at Back Yards ribbing & one Cart leading of Turnips Thornylee – One Cart to the Mill for the Meal & Dust &c – Observe most Farmers at Ley Ploughing stubble Ground being too wett – The Crop Hath Every Appearance of being very poor Indeed Quantity & Quality – and the more it is tried will be found so – In Angus & Strathmore worse than here – little or no sale of Bear from 18/ to 21/ but owing to the stop of the Distilleries very shy about it – this year will Turn Out a Miserable one for most Farmers –

Thursdy 17th rather Stormy & rain from the Southward, Obliged to Loose the Ploughs at 10 A.M.- then dighting & sacking of Bear sold to Andrew Young at Mill of Conveth at One Guinea per Boll ready money – &c Jobs.

Fridy 18th Dark weather & slight rain through the day – People at Barns Sent 17B Bear to Mill of Conveth – lead some Turnips Thornylee – Planting Ash as before.

Saturdy 19th Therm^r 43° Early this morning Came on Closs stormy RAIN from S.E. which stopped work without doors till Noon when yoked three Ploughs – did out Back Yards and Begun Ploughing E^t side Haugh Ley and One Cart leading of Turnips Thornylee &c Jobs –

Mondy 21st Yesterday fair weather & Strong Frost Evening – this morning Came on Heavy rain & storm from S.E. which stopped all work without doors – at Barns Chaffing & Dighting Oats S B – Sundries.

Tuesdy 22d Fair weather in general but some Heavy showers in the Forenoon – three Ploughs at E^t Haugh Ley & one Cart leading of Turnips Thornylee – at Barns Chaffing of Oats and at sundries.

Wedy 23d Strong Breeze of wind at West & fresh weather-3 Ploughs at E^t End Haugh & One Cart Leading of Turnips Gardner Planting Bog Planting Ash & other Trees –

Thursdy 24th Blowing Soft weather from S.W.- about 8 morning a smart shower of rain for an hour – day fair, three Ploughs at Ley as before – One Cart leading of Dung to D^o – and some Turnips – Barns &c &c –

Fridy 25th Th^r 42° Fresh weather & strong drought from the Westward 3 ploughs at Ley – and one Cart to Mill of Mondynes with rough Flax – and at Sundries.

Saturdy 26th pretty strong Frost & fine weather at S.W. – three Ploughs at Haugh Ley and one South End of Robs Hillock and P.M. at Barns & Sundries – Some appearance of rain Evening – the grounds now much dryer –

Mondy 28th Yesterday & last night a very strong Gale at S.W.- and this Moderate & fresh D^o – 2 Ploughs at Haugh Ley & one Cart Carrying Oats to the Mill of Kair and one D^o leading Turnips & at sundries – Planting Hard wood in Bog Planting – Chaffing Oats N Barns &c.

Tuesdy 29th last night about 8 Came on Exceeding Strong Gale of wind from S.W. veering to South with Heavy rain which lasted to Early this morning, damaging Houses, Stacks &c – this day fair & strong drought at W.S.W. – three Ploughs at Robs Hillock Ribbing & one Cart leading of Turnips from Thornylee – at Barns &c –

Wedy 30th Slight Frost morning & strong drought S.W. – 4 Ploughs Ribbing R Hillock for Bear now dry – and at Barns Chaffing & Dighting of Oats for the Mill of Kair, and at sundries planting &c &c –

Thursdy 31st Variable winds & weather Chiefly S.W. & Moderate with some showers – 2 ploughs at Robs Hillock which about Finishes all Ribbing of Season – One Cart with Oats twice to the Mill & One D^o Leading of Turnips – at Barns & Sundries – The Quantity of rain this month five Inches – and the Fields are now very Green & tolerably Dry but labouring in general never more Backward – The Crop is Turning out the longer the worse – the Oats want one Third or more in the Yield, and as much in the Meal which must produce an Immense Scarcity at last – yet Meal is 15$^{ds}\frac{1}{2}$ per peck or 20/6 per Boll – Bear 19/ to 21/ – Beef & Mutton about 4^{ds}per pound, and best Fatt Cattle not better than 6/ per stone – the Fodder will be very Scarce & dear – Manufactures and Trade was never Known Brisker –

Diary for the Month of January 1796

Fridy 1st Therm^{r} Noon 37° Squally weather from S.W. with showers of Snow which did not Ly here, but did on the Higher Grounds – One Plough Finished all ribbing of season and three Ploughs at Haugh Ley – at Barn Yard & sundries –

Saturdy 2d Some light snow in the night which did not Ly – Two Ploughs at Ley – and two Carts leading Turnips from Red Hall Park but very light Crop – Wind S.W.

Mondy 4th Th^{r} 43° Yesterday moderate weather – Slight Frost – S.W. wind – this day very Mild & warm – till Miday when strong wind S.W.- 3 Ploughs at Haugh Ley & One Cart leading of Turnips from Red Hall Park – Chaffing Oats Barns –

Tuesdy 5th Blowing weather from the S.W. & fair – Chaffing Oats S Barn – being Christmass O S all our People visiting their Friends – Dirty roads &c.

Wedy 6th Fine pleasant dry weather from S.W.- 3 ploughs at Haugh and one Cart leading of Turnips – at Barns dighting & lofting Oats for the Mill – Sundries – Grounds drying.

Thursdy 7th White Frost in the morning – day fair from S.W.- three Ploughs at Haugh Ley, and one at steering the Turnip Ground Thornylee tolerable order – at Barns – Taking in a stack Oats – and at sundries –

Fridy 8th Cold dry weather from S.W.- three Ploughs at Ley Haugh and one Cart lead Home all the Cattle Flakes to Make room for Ploughing – People at the Mill of Kair with 26 Bolls Oats – Meal 18-2-3-3 – P.M. at sundries –

Saturdy 9th Closs Foggy Mild weather from the Southward – 2 Ploughs at the Haugh Ley and One Cart Carrying Oats to the Mill and Brought Home the Meal – One Cart leading of Turnips – and at sundries about the farm – Spring Plants Appearing.

Sundy 10th Closs hazy weather from the Southw^{d} & warm – Evening rain but light – The Cattle not quite so Greedy on the Fodder as before & well up –

Mondy 11th Blowing weather from the S.E. & Cloudy with some rain afternoon – Three Ploughs going and one Cart leading of Turnips – Wewmen Chaffing Oats for the Mill –

Tuesdy 12th Squally weather & some smart showers from S.E.- till noon when strong wind & drought from S.W. – Three Ploughs at Haugh ley and one at Thornylee Ribbing after Turnips for Bear – at Barns dighting and sacking of Oats for the Mill of Kair –

Wedy 13th variable weather from the Southw^{d} & fresh – 2 ploughs at Haugh ley and one Cart with Oats to the Mill & one D^{o} Leading of Turnips – Gardner Planting Larch Bog Atry, and at sundries &c –

Thursdy 14th Blowing weather in the morning and showery through the day but not Heavy a S by SW – three Ploughs at Haugh Ley – and one Ribbing at Thornylee after Potatoes – the roads very wet & Dirty – and unless dry ground the Fields Similar –

Fridy 15th very Blowing & Squally weather from S.W. with great drought & fields new ploughed as our Haughs Seed Coloured – Two Ploughs at Haugh ley, one at late Turnip Ground – and one Cart at Leading of Turnips R H Park – and P.M. at Chaffing Oats S Barn – and at sundries &c.

Saturdy 16th a Considerable deal of rain & Closs Foggy weather – 2 Ploughs at Haugh ley – and one ribbed out all Thornylee and one Cart leading of Turnips – the rivers some out & Grounds very wet again – Gardner Planting Larches Bog Atry & at the S Barn Chaffing Oats for the Mill of Kair

Sundy 17th warm & Milder weather than in June last – the Bushes & Flowring Shrubs Budded & artichoakes growing – Hear of some Turnips on the Coast of Kineff being Bought just now at 6£ per Acre for Cattle Feeders – Fatt Cattle are not worth 6/ per stone, nor Fatt Sheep worth 3$^{ds}\frac{1}{2}$ per pound – Hear of Bear up to 24/ per Boll – Meal 15 to 16^{ds}per Peck – Cattle a little Backsett with the Fodder –

Mondy 18th Blowing weather from S.W. with some showers of rain - three Ploughs at Haugh Ley – and one Cart at Carting out Gutturs &c from before Byre Doors for Intended Turnips – at Barns dighting Oats for Mill – Gardner Planting Larches daily Bog Atry &c.

Tuesdy 19th last night Blowing weather and some smart showers of rain, day Bleak but mild & open weather a SSW – three Ploughs at Haugh ley & Black Water D^{o} and one Cart Carrying Oat to Mill of Kair and after leading of Turnips – at Barns dighting Oats &c.

Wedy 20th Fine fresh weather from S.W. and free of all Frosts for Many weeks by past and the Fields as Green as in April- three Ploughs at Back Waters Ley and one Cart Carrying Oats to the Mill &c Jobbs.

Thursdy 21st very Blowing weather with some showers of rain from S.W.- fresh – 3 Ploughs Back Waters ley and one Cart leading Turnips & at Sundries- Braking down an old Dyke Haugh &c Jobs the Gardner still Planting Larch Bog Atry

Fridy 22d Blowing weather from the S.W. – Ploughs as before, leading Turnips – Observed towards Aberdeen much Ploughed Ley, & stuble Ground seeming very wett & fields very Green the whole way – Turnips &c shott.

Saturday 23d Stormy weather from the South & P.M. Came on a severe storm of wind & Heavy rain which Brought down the Dee very high not Half the rain to the Southw^{d} of Stonehaven – Ploughs &c as Yesterday –

Sundy 24th Blowing weather from S.W.- but dry – rode from Aberdeen to this Place in four Hours with no difficulty – Salmon 1/ per Pound –

Mondy 25th During the night & morning very stormy weather and Heavy rain from the S.W. which brought down the rivers Considerably – yoked the Ploughs on the ley but Obliged to Loose for Wetness – P.M. In Stack Oats –

Tuesdy 26th Cold Stormy weather and Heavy rain after 1 P.M. – three Ploughs at Back Waters ley nearly finished & one Cart leading of Turnips red Hall Park & Barns.

Wedy 27th Fresh weather variable from S.W. with showers P.M. at Taunton fair a vast Concourse of People and servants of all kinds Seeking & Getting High wages and the women for Shearing not Engageable –

Thursdy 28th Slight Frost this morning & day fair & drought – 2 Ploughs nearly finished Back waters Ley & one Cart leading of Turnips – People at the Mill of Kair 32B 2f 1p Meal 23B 2p about 11 pecks.

Fridy 29th Strong Frost till 10 A.M. when Fresh and soon Came on rain & stormy weather at E^{t} – 2 ploughs at the North or Auld Toun Park for Pease and leading Home the Meal &c & Turnips – at Barns &c Jobs.

Saturdy 30th Slight Frost in the morning day fair – 2 Ploughs at Auld Town Park for Pease, too wett – Obliged to go Kavels Ley – at Barns dighting seed Oats.

Sundy 31st Showery morning day fair a South – Quantity of rain this month three Inches – The Fields as Green as in April or May & Fields Excessive Wett & although much Corns Shade in Harvest very little of it yet Breered no good sign of the Grain – The Mezerion, Heleborus Heymalis & Bulbs Flowred – Hear of no Farms sold as yet Oat Meal 21 to 22/ Bear 22 to 23/ & Hay 7^{ds}per stone Fatt Cattle 6/ Vincas & snowdrops appearing.

Diary for the Month of February 1796

Mondy 1st a Great deal of rain this morning & Rivers very much Swelled – three Ploughs at Kavels Ley & one Cart leading of Turnips – at Barns & lofting Oats for seed, Fields Excessive wet in General &c –

Tuesdy 2d Cold Bleak weather from South till 3 P.M. Came on rain pretty Closs - three Ploughs at Kavels Ley and one Cart leading of Turnips &c-at Barns Chaffing Oats S Barn &c

Wedy 3d Frost slight morning-day fair till $\frac{1}{2}$ past 1 P.M. Came on a smart shower of snow & wind West till $\frac{1}{2}$ past 2 P.M.- three Ploughs at Ley and one Cart leading Turnips - at Barns P.M. –

Thursdy 4th Fresh & dry moderate weather from S.W.- three Ploughs at Kavels ley and one Cart leading of Turnips – at Barns dighting of Oats for y^{e} Mill – at a Roup of Ash &c Fordun House it sold very brisk.

Fridy 5th very Early this morning Came on a very Severe Gale of wind & Heavy rain from S.E. which deluged the Fields & Brought down the rivers high – it lasted till past Miday – at Barns Chaffing Oats, no work without Doors.

Saturdy 6th Coldish drought from N.E.- three Ploughs at the Kavels Ley, and one

Cart leading of Turnips – People dighting Oats for the Mill NB & planting Ash, Elm & Larix East side of Plantation behind House.

Mondy 8th Yesterday Cold raw weather from E.N.E. with some smart showers of Hail & rain – this day same weather – three Ploughs at Ley-and one Cart leading of Turnips – In a stack Waterlair Oats for Seed SB & one Pease N Barn –

Tuesdy 9th Cold Stormy weather from the N.E. with Heavy rain and sleet from D°- rivers out – Sent 15B Oats to Mill of Kair and lead one Cartful Turnips – the Snow lying on the Hills, but not here Evening –

Wedy 10th Fine pleasant dry weather from S.W.- three Ploughs at Kavels Ley – One Cart leading Turnips – the Gardner Filling up Planting Shepherds Haugh Brae planted three years Ago – Bringing Home roof Cottar House from Waterlair – Observe several thatched Houses Fallen –

Thursdy 11th Slight frost morning, rest of the day rainy & stormy from Southward – three ploughs Kavels ley – One Cart leading Turnips – Observed towards Mill of Halkerton the Ploughed Grounds a perfect Guttur – Sundries –

Fridy 12th Blowing weather & high wind Westerly – three Ploughs at Ley – and one Cart driving Dung to Ley – P.M. at Barns dighting Pease Lofted $3\frac{1}{2}$B – not a Plough to be seen going this two days past for wetness – Strong growth on the Grass

Saturday 13th Blowing stormy weather – but Clear till 2 P.M.- Heavy Squalls & rain from N.W. at Intervals – three Ploughs at Kavels Ley – and one Cart leading Turnips – at Barns took 2^d stack Pease NB – Sundries –

Sundy 14th Blowing weather with some showers of rain & Snow from N.W. – The Publick road is not Bad, but the Fields in general very wett –

Mondy 15th Slight Frost in the night, day pleasant till Midday & some appearance of Spring – P.M.Coldish a N.W.- three Ploughs at Kavels Ley – One Cart driving Dung to East ridge formerly a road of said Field – Spreading D° P.M. and at sundries – Cattle Greedy of the Fodder –

Tuesdy 16th Closs Foggy weather & P.M. a strong Breeze a S.W. & soft weather – three Ploughs at Kavels Ley & one Cart leading of Turnips – Gardner sowed some Early PEASE and Pruning Fruit Trees D° – The Tillage well advanced in general although in wett Condition – The Prices of Grain precarious at present – Meal scarce & scarcely to be got & threatned Mobbings &c.

Wedy 17th very pleasant spring weather from the West – three Ploughs at Kavels Ley – one Cart leading Dung – P.M. dighting Seed Pease & lofting D° – levelling down an Old Dyke North side Haughs to be Ploughed in with Old Road for Oats.

Thursdy 18th Fine pleasant warm weather from S.W.- Two Ploughs at Kavels nearly done, one Cart Carrying Oats to the Mill of Kair, and one Cart leading of Turnips – Took in 3 & last stack of Pease N Barn.

Fridy 19th Thick & Cloudy weather with rain slight from ENE to SE & SW – 2 ploughs at Ley and one at Auld Toun Park for Pease – pretty dry – a Cart to

Woodstock with a sow & for a Trial of their Sale Ale P.M. – People Braking down an Old Dyke Haughs –

Saturdy 20th Fair weather & after 1 P.M. Broke out fine sunshine from S.W. – three Ploughs at the Auld Toun park N Side for Pease – Considerable Growth in the Ground – roads very Bad &c.

Sundy 21st Fine pleasant Clear weather from S.W.- there is a new Growth in the Grass the Cattle are picking up & particularly the Sheep – Therm^{r} Noon Shade is 46 degrees, but pretty warm in the SunShine.

Mondy 22d Moderate weather from Southward & appearance of rain - three Ploughs at Auld Toun park for Pease and one Cart leading of Turnips – P.M. at Barns Chaffing Oats, and Gardner pruning Fruit Trees.

Tuesdy 23d Dark weather & Coldish from the S by E with drought – 4 Ploughs at Auld Toun park for Pease and P.M. dighting seed Pease, and Braking down the South Dyke of Old Road through Haughs E^{t} End.

Wedy 24th Therm^{r} Noon shade 44° Beautiful day, free of Frost – 4 ploughs at the Auld Toun park as yesterday, and P.M. at the Dyke, at Barns dighting seed Pease yoked 8 A.M. – Gardner dressing Espalier Trees –

Thursdy 25th Dark weather till 2 P.M. when Clear & drought from ENE- 4 ploughs at Gibsons Park for Pease and P.M. Braking down the Haugh Dyke. The People at Kair Mill 24B 2f – Meal 18B 1f – Begun to in Hay –

Fridy 26th Flying showers of rain at Intervals from N.E.-2 Ploughs at Gibsons Park for Pease, and one Cart Brought Home Meal & one D^{o} Brought Home all the Turnips on red Hall Park, very Bad Crop Indeed. Women &c at Barns Chaffing Oats for seed & Sundries –

Saturdy 27th Cold Bleak dry weather from the N.E. Checking of vegetation & Browning the Grass – 4 ploughs at the Gibsons Park as before, and P.M. at Dyke and at Barns women Chaffing Seed Oats all the day – the Ground is dryer but roads Bad –

Sundy 28th Dark weather & sometimes Sun Shine, but Cold and wind Moderate a N.N.W. – fields drying &c.

Mondy 29th Therm^{r} 44° Crocus opening In the morning some showers of snow from Northward which lay on the Hills & Braes but not the low Country – 4 ploughs at Gibsons Park as before and P.M. at Sundries-Quantity of rain is three Inches four tenths. There is a Great Alarm for a Scarcity in the Ensuing Summer & a County Meeting of Heritors, Farmers & Ministers Consulted on the Business – Arbuthnots Farms are sold it is said for 21/ & resold for 23/6 – Fatt Cattle very dull & the Turnips almost totally Expended.

Diary for the Month of March 1796

Tuesdy 1st Therm^{r} 39° Cold weather from North with Showers after 1 P.M. – 4

Ploughs finished out near twelve Acres all for Pease, after Dinner at S Barn dighting seed Oats – Hear several Cairnton &c Have Sown Pease &c –

Wedy 2d Black Frost in the morning day fair & Coldish N.N.W. Ploughs begun Ploughing Red Hall Park after Turnips for Oats – One Cart leading off stones from D° and 2 Cartful of Turnips from Thornylee – at Barns –

Thursdy 3d Therm^r 32° last night about ½ past 10 Came on Snow and soon after a strong Gale from North with Gusts of Snow which Made all the Country White & lay all the day with Frost - 4 Carts driving Dung and P.M. at Barns & levelling an old Dyke Haugh &c Jobs.

Fridy 4th Strong Frost night – day fair Clear a NNW – 4 Carts driving Dung to Golds Hillock for Turnips – and in Hay West Barn – and at sundries –

Saturday 5th Th^r 35° Strong Frost all Night & day behind the Sun with a Breeze at S.W.- 4 Carts driving of Dung – and P.M. at Barns dighting Waterlair Oats – Gardner placed his Hot Bed in Garden – & Sundries –

Mondy 7th Therm^r 348 Yesterday Cold weather & frost from Southw^d- this day same & Frost quite Hard to stop ploughing all day – 4 Carts driving dung all day – and at Barns Dighting & lofting Waterlair Oats for Seed – at a Roup of Corn & Fodder Mill of Kair Oats 20/ and Bear with Fodder 22/ per Boll very dull Markett –

Tuesdy 8th Coldish weather & less Frost than before from E.N.E. – 4 Carts driving Dung to Back Yards for Turnips & Home some Yellow Turnips Thornylee – In Bear S B and P.M. levelling down Dyke Haugh &c Jobs.

Wedy 9th Some Snow on the Hills & Braes all day wind N.E.- 4 ploughs at Red Hall Park for Oats – and at the Barns Chaffing Oats for Seed; Gardner at sundries laying Coleworts for seed – Planted some Trees Bog Atry &c.

Thursdy 10th Stormy weather from the N.E. with very frequent showers of rain & snow on the Hills – Being a General Fast no work this day – Vegetation quite Checked –

Fridy 11th Cold but dry weather at E.N.E. – 4 ploughs at red Hall Park – P.M. lead in Hay West Barn – Gardner sowed small seeds, but not onions &c – The Turnips were Shot long ago, & now nearly finished –

Saturday 12th Cold Bleak weather from the East with flying showers of rain – 4 ploughs at Red Hall Park – P.M. at Barns, and Haugh Dyke, vegetation very dull Indeed – Sundries –

Mondy 14th Yesterday Dark thick weather from East & some showers – this day same & stormy from S.E. with some Heavy showers – 4 ploughs at the Red Hall park & one of them at Haugh Ley – P.M. at Haugh Dyke & at sundries &c –

Tuesdy 15th Fresh weather from S.S.W.- three ploughs as before and one Cart lead one Cartful of Yellow Turnips from Thornylee – after Stoneing of N.W. Corner of Red Hall Park on some new fallowed Ground, In Hay-at Dyke &c Jobs

Wedy 16th Therm^r Noon 46° In the morning Early a Considerable Fall of rain & stormy – day fair & drought – at Barns dighting seed Oats till dinner – One Plough at Haugh Ley – and 2 Carts driving on Earth on Stannery part by East stone Bridge, ploughed Ley and at sundries – river out – vegetation but very dull – no sowing –

Thursdy 17th Moderate weather from N.W. – One Plough at Haugh and three Carts driving Dung to West side Red Hall Park worst part of it – P.M. at Barns dighting stacking of seed Oats – Bought 12 fir Trees at Garol Wood @ 19/5 –

Fridy 18th Th^{r} 44° Frosty dew morning – day fair & pleasant N.E.- Two Carts driving dung Cottar Toun to R H park, one D^{o} Earth to Haugh ley and one plough at Old road through E^{t} Haugh for Oats – P.M. taking in Bear N B – Planted Potatoes Early kind Nursery D^{o}D^{o} Cabbages.

Saturdy 19th Fine dry weather from N.E. but no great drought-2 Carts to Montrose for Coals, Porter &c Sundries and two Ploughs at Red H Park – Women spreading Dung on it – and at Garden &c Jobbs –

Mondy 21st Yesterday Cold drought from North-this day the same & more Blowing – at 7 Begun sowing of seed by Pease on Auld Toun park, tolerable Order & Harrow on it very well-11B 2p near One Boll per Acre. Observed towards Bervie Self Corn Yards pittyfull.

Tuesdy 22d Blowing weather from N.W. & strong drought – Seven Harrows did out the Pease A.M.- and P.M. Sowed 4B Oats all Back Waters partly Harrowed it – at Barns Chaffing Barley &c Jobbs –

Wedy 23d White Frost in the morning, day Cloudy weather at West - Harrows on Haugh ley – P.M. sowed 3 Bolls Oats E^{t} End Haugh & partly Harrowed them – at Barns and sundries - Gardner at Onion Ground &c.

Thursdy 24th Th^{r} 45° Cold weather from N.W. till 4 P.M. when Came on pretty severe Snow & drift from North which stopped ye Harrows at 5 – Sowed on Haugh this day 9B Oats and partly Harrowed same – at Sundries &c &c.

Fridy 25th a Considerable deal of snow yesterday Evening & very hard Frost during the night – Frosted some Potatoes unluckily Exposed to it – driving dung A.M.- and P.M. Sowed 7B 3f Waterlair Oats & partly Harrowed whole - being 23B 3f on 22 Acres or better than 17 pecks per acre or above 15 pecks Meal per D^{o}. Snow gone by 10.

Saturdy 26th Frosty Cold stormy weather from N.W. & Snow Evening – 9 Harrows all day at Haugh Crossing it – at Barns dighting Barley & Sacking seed Oats from the Long loft dighting before doing so, Planting Cabbages &c.

Mondy 28th Th^{r} 32° Yesterday a severe storm Frost & Snow a N.W. & A.M. the same at Intervals – Carts driving Dung A.M. Back Yards and P.M. 9 Harrows did out the whole of Haugh ley – the Country White & wreaths of Snow on the roads, In Hay & Sundries.

Wedy 30th Moderate weather from North-Sowed all Kavels 15B 1 peck Oats & partly Harrowed same, Yoked a three year old Filley in Harrows quite Handy – Barns & Sundries.

Thursdy 31st Fine dry weather from the N.E.- 9 Harrows finished out all Kavels field – at Barns dighting Bear and dighting & sacking Rye Grass for red Hall Park – Quantity of rain this month is 1-4 Inches – Prices of Grain unsettled Seed Oats 25/ – Bear about 23/ – Pease 21/ to 24 – Hay about 7 to 8^{ds} – The Oat seed well begun – mine will be finished tomorrow if fair – vegetation is almost totally stopped – Sowed onions and almost the whole of small seed

Diary for the Month of April 1796

Fridy 1st Thick Hazy weather from the East but Moderate-4 ploughs finished Out whole of Oat seed A.M.- P.M. Finished Sowing of all ley red Hall Park on it 9B 3f 3p^{ks} total Sowed on 46 Acres 49B:1f in very fine Heart & Condition & should be a very great Crop –

Saturdy 2d Therm^{r} Noon 55° Pleasant & warm weather with some Growth in vegetables- three Harrows at Cross Harrowing red Hall Park – and three Carts to Garol Wood for Fir wood for Flakes – Observe the Cattle Backsetting with straw &c –

Mondy 4th Yesterday Closs Foggy growing weather from the Eastw^{d}- this morning a light shower of rain – Sowed all the Red Hall park with 5B Rye Grass rich in Plantain and from 4 to 5lb red Clover per Acre it being so dear as fifteen pence per lb – Harrowed same – Braking in R Hillock for Bear.

Tuesdy 5th a Smart rain Early morning, day Cloudy & growing weather at E.- Harrows did out R H park and P.M. at Thornylee D^{o} both fine & Clean – at Barns dighting & lofting Bear & Sundries –

Wedy 6th Fine pleasant dry weather from Southw^{d} – Harrows did out Thornylee and P.M. at Braking in of Waterlair for Bear – at Barns & other Jobs –

Thursdy 7th very fine dry weather from N.E.-Harrows did out Braking in Waterlair A.M.- and P.M. driving Earth to Backyards and Preparing a patch of Ground Back Yards for Potatoes & Dunging it-at rolling red Hall park after Grass Seeds

Fridy 8th fair weather & drought from N.E.- Carts driving out Earth and dung to Backyards – Planting Potatoes D^{o}- Observed to Montrose by way of Garvock and Home self by Marykirk fields very Bleak & Not looking well Sold all the Hay I Can spare to the Dragoons at 8^{ds}per stone.

Saturdy 9th Stormy weather & flying light skiffs from N.W.- Carts at Gathering of Stone from Thornylee to Bank the River with & Planting of Potatoes &c Jobbs –

Mondy 11th Therm^{r} 45° Yesterday Cold Bleak weather from NNE – this day the same at N.W.- Getting Carts & Horses ready for driving of Hay to Montrose – After dinner Built 4 Cartful of it & sent it down to Polburn on the road at Planting Potatoes Back Yards &c Jobbs –

Tuesdy 12th Fine pleasant & warm weather from the Southw^{d} – 4 Carts to Montrose with Hay for Dragoons & Brought out one Boll Coals at 8/7 returned by $\frac{1}{2}$ past 8.

Wedy 13th Th^{r} 52° Very fine dry weather & strong drought from S.W.-people at planting Potatoes Back Yards and taking in Hay and at Sundries on the Farm – Ground drying fast Observed some of the Field Pease through Ground –

Thursdy 14th Th^{r} 53 Fine pleasant dry weather from the Westward – two Carts all day Carrying of Stones from Thornylee to Bank the river with below the Bridge and Ploughed Harrowed, sowed & rolled 4 pecks of Lintseed at Nether part of Robs Hillock in fine order and pretty thick as the seed was but moderate –

Fridy 15th Remarkable pleasant & dry weather – Wind a S.E.- loaded 4 Carts with Hay for Dragoons Montrose of own and three Borrowed ones of Neighbours – & sent them down from Waterlair to Polburn for tomorrow morning. People at Barns dighting seed Bear – &c Sundries

Saturday 16th Therm^{r} Noon 57° Very pleasant fair weather from N.W. – at 4 A.M. sent my own four & three Borrowed Carts to Montrose with Hay to Dragoons, Sent on them per receipt 416 stones – two of mine from 70 to 80 stones each – The Rick will turn out very light Indeed – like the rest of the last Crop & Hay of other persons – The Oat Seed is now mostly finished Neighbourhood and in good order – the prices of grain falling in the South Greatly – Fatt Cattle no great demand. The Carts return only past 9 P.M. or 17 Hours absent – Imperial Crown Flowered.

Mondy 18th Th^{r} 57° Yesterday & this Exceeding pleasant & fine weather from N.W.- this day loaded 4 Carts of mine with Hay for Dragoons – and one from Kair with D^{o} and at Barns &c Jobs – Some small rain P.M. –

Tuesdy 19th Th^{r} 54 White frost in the night but pleasant & warm weather from S.E.- Considerable degree of vegetation – and 4 Carts own & one Borrowed with Hay to Montrose – Several people Sowed Bear last week – Oat Breer looking pretty well, although not very thick &c.

Wedy 20th Th^{r} 58° Frosty dew in the night, day Clear & pleasant weather with strong drought a S.E – at dighting rye Grass for self & others, P.M.two Harrows Waterlair Braking in - and at Sundries –

Thursdy 21st Th^{r} 53° very strong drought from S.E.& Clear Sun – Begun Ploughing & Sowing of Bear Seed at Waterlair – the Land is stiff and rather Moist, Sappy below – gave a double tine of Harrows before Sowing & as much after it will require much of it – Put Horses on Grass on a Park Waterlair Grass of three years very good Middle of day.

Fridy 22d Th^{r} 57° very strong drought from S.E.& warm weather – 3 Ploughs at Waterlair for Bear & three Harrows before & after sowing it, when ploughed soft & Faughy and Harrows but Indifferently – at same time if steered would be soon too dry & Hard without rain again – Sundries.

Saturdy 23d very strong drought from S.E. but some light showers about Noon- two Carts to Mathers for Some Shells to the Brae Thornylee-Others ploughing, sowing & Harrowing at Waterlair Bear – Sacking seed Bear S Barn &c.

Mondy 25th Yesterday at $\frac{1}{2}$ past 11 A.M. Came on some Heavy showers of rain &

Hail till past 2 P.M.- this day fair & drought a S.E.- Ploughs and Harrows at Waterlair better season than before – at Sundries –

Tuesdy 26th Th^{r} 52° Strong drought a S.W to S.E.- Ploughs & Harrows at Waterlair as before – Observe severals begun to sow but very few – Others steering &c the Grass is rising very slowley – by the Frosts every night &c &c.

Wedy 27th very strong drought a S.W. & Ground very Hard, all Ploughs & Harrows at Waterlair which finished with 5B 1f 2p Bear in tolerable Order for the Season – the first sown well rooted – dighting & Measuring of Rye Grass Sold to Sundries at 15 to 16/ per Boll.

Thursdy 28th Fair weather from N.E. till past noon when Came on rain moderate & lasted all the rest of the day – 2 Carts Brought Home two very large Cartful of Hay from Waterlair – 2 ploughs & Harrows at E^{t} side Thornylee. at Laurencekirk Markett a Good many Cattle but few sold but Excessive high priced 6/ per stone &c –

Fridy 29th Th^{r} 55° Fair but little drought & towards 5 P.M. Came on rain from the Eastward – Built own 4 Carts with Hay for Montrose & P.M. loaded 4 others Borrowed Carts for D^{o}- being in all 30 Cartload & stones 1402 -and at Sundries

Saturdy 30th Dark Hazy weather from the Eastw^{d} and little drought, but little of the Bear seed done towards Stonehaven – an Uncommon large Meeting of the County about the New Turnpike & Commutation Bill & Electing a Joint Collector of Cess Observe Quantity of rain this month Nine Tenths of an Inch, in General it Hath been a fine Month but Vegetation is not Early for all that. Prices of Grain at a Stagnation – Meal from 16 to 18^{ds}per peck it Hath rizen 2^{d} per peck & fallen in the South in one week – no sale of Barley or of Oats – Fatt Beastials are not higher priced than lean ones of same size which is Nonsense – Grass seeds very high, Rye Grass 16/ & Clover 15 to 16 per lb – Bear Meal 10 to 11^{ds}per peck - Sundries – Sowed the Annual Flowers Garden.

Diary for the Month of May 1796

Mondy 2d Th^{r} 45° Yesterday Closs & Cold weather a N.E. with rain at night, this day same – Ploughs E^{t} side Thornylee and P.M. five Harrows finished Waterlair, tolerably fine Mould –

Tuesdy 3d Frost during the night – day warm from the Southward – all Ploughs & Harrows at Thornylee tolerable Order – P.M. Sowed Two Bolls Bear E^{t} side & partly Harrowed the same. There is yet but little growth in the Grass – &c - Barns –

Wedy 4th Th^{r} 46° Cold Bleak weather & some showers of rain & Hail from the North – all Ploughs & Harrows at Thornylee – Yoked this morning at 5 A.M. for first time and sent the Horses to Waterlair Grass – rather Distant.

Thursdy 5th Bleak Cold weather from the North till Evening rain & sleet – all Ploughs & Harrows at Thornylee – at Mixing up Grass seeds for above field – & Sundries –

Fridy 6th Th^r 40° last night about 10 Came on Heavy SNOW which by morning was three or four Inches Depth Over all yet by afternoon was mostly gone – P.M. 4 ploughs at the Thornylee & sowed what was ploughed of it – still very Cold from the North the whole day – took last of Oats –

Saturdy 7th Cold dry weather from the North – Ploughs at Thornylee and P.M. sowed nearly East Half of said field with Grass seeds & Harrowed the same – a Cart to Montrose with Sundries & Brought out Coals &c.

Mondy 9th Th^r Noon 44° Yesterday Cold Bleak weather from North, this morning a strong frost which Blasted Some Early Potatoes and day fair & Cold at N.E.- all Ploughs & Harrows at Thornylee a very fine Season on it Indeed – most people not yet one Third done of Bear seed.

Tuesdy 10th Th^r 45° Dark & Cold Gloomy weather from ESE and appearance of rain – all Ploughs & Harrows at Thornylee nearly done - and Sundries – the Pasture wretched & other grass turning Black & decaying – Cattle Eat fodder Briskly.

Wedy 11th Th^r 54° last night some showers of rain, this day fair till 1 P.M.- More rain a South & at 5 P.M. Heavy D°- all Force, dunging, Liming, Ploughing &c. at Thornylee West side nearly finished –

Thursdy 12th Planted Collyflower Showery weather from South but P.M. Heavy rain at Intervals – Ploughed out & sowed Thornylee very tedious work – Sowed 11B 6 pecks Grass seeds $7\frac{1}{2}$ Bolls of very fine Rye Grass & Rib plantain 2 stone weight per Pease Firlot & about gold red & white Clover very dear.

Fridy 13th Therm^r 46° last night & this morning Heavy rain & stormy weather from NNW which brought down rivers – only yoked at 7 A.M. West side Robs Hillock in good Order and Water furred E^t Half Thornylee Sown with Grass Seeds last week – at Bervie Markett yesterday Cattle very high priced particularly Milch cows.

Saturdy 14th Th^r 46° Cold Stormy weather but more Moderate at N.N.W. – 4 ploughs all day at Robs Hillock in fine order and at Sundries, some vegetation but as far Backward as for some years by past –

Mondy 16th Therm^r 46° Shade at Noon White Frost in the night and day Moderate weather at S.E. – Early sowed out all of the Thornylee with Grass seeds, & Covered the same with the Harrows A.M. - afternoon Water furred out all of same & partly Rolled some of said field in fine Condition – three Ploughs at Robs Hillock for Bear in fine Tilth, Horses on the Grass for first time Evening Blacklands.

Tuesdy 17th Cold Bleak weather from S.E. & no vegetation – all Ploughs Robs Hillock A.M.- P.M. sowed 4 Bolls of Bear on West side of Robs Hillock and partly Harrowed same very tender & in fine Order, Rolling Grass seeds Thornylee – The Bestials of all kinds are everywhere in great strait for Grass & Many for Fodder there is no demand for any kind of Grain or Meal at present – Bear seed much to do Generally –

Wedy 18th Th^r 52° Clear weather & Drought a S.E. but milder – Ploughs & Harrows at Robs Hillock – Sowed about Two Bolls more Bear – vegetation some better – Some people Scrimped for Fodder as well as Grass.

Thursdy 19th Th^r 53° Strong drought from S.E. & Bold Sun Shine – all Ploughs & Harrows two P.M. Sowed about two Bolls more Bear & partly Harrowed it – rolling nearly out all Thornylee - lead some Broom &c for Kitchen fire &c.

Fridy 20th Th^r 53° very strong Sun Shine & warm P.M.- wind East – all Ploughs at Robs Hillock & Harrows P.M. did out whole of BEAR SEED in fine order be the Crop what it may, 36 Acres – Mercury falling.

Saturdy 21st very fine pleasant dry weather from E^t – 7 Harrows at Robs Hillock all day pretty fine and one Cart to Johnshaven for Great Coals at 8/ – the road over Garvock very good at present –

Mondy 23d Yesterday dark weather & Coldish at N.E.- this day sun shine & moderate E.N.E. – three Carts to Johnshaven for Scotch Coal, and three Harrows at Robs Hillock on least Harrowed places &c Jobbs.

Tuesdy 24th Th^r 57° Fine pleasant dry weather from S.W. – all Harrows & P.M. 4 ploughs at 2^d furr Golds Hillock-Turnips deep-it is tender – putt all young Beasts in the Straw Yard from the Byres.

Wedy 25th Early at 4 A.M. & till 8 a fine shower of rain which however was dried up before noon – 4 ploughs at Golds Hillock – Chaffing Oats N B –

Thursdy 26th Th^r 56° Blowing Stormy weather from W.N.W.- and Excessive drought – 4 ploughs at Turnip Ground as before growing very Stiff – Others at Dighting of Oats N Barn all day to loft & at Sundries.

Fridy 27th Therm^r 54° weather as yesterday, some Slight flying showers of Hail & rain a N.W. – Flited a Cottar Man from Pittengardner – and all Ploughs & Harrows at Golds Hillock now very stiff & Dry – Vegetation very poor and Backward – Caterpillars in Bushes & Trees –

Saturdy 28th Dry weather & Cold till P.M. some showers of rain from the Eastward – all Ploughs and Harrows at Golds Hillock & Cow folds as before too Hard & Lumpy the South End of it for Harrowing – at Sundries –

Sundy 29th This day a good deal of rain particularly in the afternoon from the S.W. – High Wind –

Mondy 30th Therm^r 51° This morning about 6 Came on Heavy rain at S.E. which Continued at Intervals – all ploughs finished at 2^d fur for Turnips and P.M. Harrowing now too wett again, but got the better of the Lumps –

Tuesdy 31st Fair weather mostly but some light skiffs from Westw^d Harrows and Ploughs at 3^d furr for Turnips Inching the Rib before Cowter – it is now tender but foul with Quicken & Knott Grass – Quantity of rain all in the Beginning & End of the Month is Two Inches, and two tenths of an Inch & the Ground is now wet Enough & want only warmth – the fruit Trees are mostly Blasted leaves & flowers and the small fruits Gooseberries Excessive rich but begun to be Injured by Caterpillars – Hear of Oats, Lint, Bear & Maslin is Injured by Frosts & Sluggs as to need being reploughed again – prices of Cattle Immense from 6 to 8/ per stone – Beef from 6^{ds}to 8^{ds} per lb and not to be got – Horses suppered well straw & pasture – lint very Indifferent – Lilac flowering - Pease garden offering to Bloom – Onions & Carrots good.

Diary for the Month of June 1796

Wedy 1st Therm^{r} 54° High wind & strong drought a S.W.-all Ploughs & Harrows at the Turnip Ground still pretty foul – Cattle &c at a Roup Polburn yesterday as high as 8/ per stone &c –

Thursdy 2d Therm^{r} 53° White Frost morning, day fair & strong drought from Southward – all Ploughs at Turnip Ground – and Braking Clods on D^{o} – Hear Cattle got some Check at Glammiss Fair yesterday – particularly large wrought ones.

Fridy 3d Th^{r} 52° Cold Bleak weather from the N.E. & dry – 4 ploughs all day at Turnip Ground – Miday at Barns and Sundries – The Hay Grass thin, shott & very light Indeed.

Saturdy 4th perfect storm of wind & Bleak weather from the N.E.- Ploughs finished 3^{d} furr for Turnips, some of it will need fourth D^{o} – and not be ever Clean or fine –

Sundy 5th Moderate weather from S.E. & Great drought – Checking of vegetation very much Indeed & Everything Backward.

Mondy 6th Th^{r} 55° Clear Sunshine & drought a S.S.E.- all Harrows the whole day for Turnips, not yet fine Enough & the Stone Roller on the Cloddiest of it – Sundries – The Season as Backward as I remember for years.

Tuesdy 7th Therm^{r} 52° Fine Moderate day but Coldish towards noon from S.E.- about 3 P.M. Came on a pretty good shower for upwards of two Hours with thick weather – Ploughs & Harrows at Turnips Ground & Gathering weeds, but Dry enough to Burn – so must be Carted off tomorrow –

Wedy 8th Fair weather & strong drought at S.W. – 4 Carts all day at Carting off weeds, stones &c from Turnip Ground to be ready for Dunging tomorrow – at Cammock Muir fair a Good Many Cows & Small Stotts &c – many Buyers & very high prices – Strawberries setting & Guins D^{o} –

Thursdy 9th Th^{r} 52° last night & this morning some showers of rain and very Blowing weather Cold from the North – Yoked 4 carts to Drive Out Dung for the Turnips almost forty Cartload per Acre & women spreading Ditto after Brakefast, but very little vegetation being so stormy.

Fridy 10th Strong White FROST lay till 6 A.M. & Blasted the Potatoe Stems in many places – 4 Carts at driving out dung all day to Turnips – Hear that Trinity fair went well & Excessive High prices Some Fatt Cattle from 30 to 36 S^{tr} each – Strong drought a South.

Saturdy 11th Therm^{r} 55° Moderate weather from S.W – Early Begun Drilling & Sowing of TURNIPS – a Plough all day & Carts driving Dung out & others spreading it – afternoon Came on rain E^{ly} which stopped ye Rolling Turnips – now appearance of rain & Mild Growing weather –

Sundy 12th Strong drought all day from N.W. & Cold weather – The White Clover just beginning to Flower, and the Ashes little Blown yet – no Cutting Grass – very Backward.

Mondy 13th Blowing dry weather from S.W. – One Plough drilling Turnips all day & rolled all sown P.M.- two Carts driving Dung from Back Yards to D° – set up the Cattle Flakes at Waterlair for young Cattle &c Jobs.

Tuesdy 14th Therm^{r} 55° last night a moderate shower till morning and this day pretty strong drought a S.W. & SunShine – Carts driving Dung & P.M. Earth to Dung Court – One Plough Drilling, Sowed & Rolled all Drilled the Distance of the Rows from 28 to 30 Inches only – Sent all young Cattle 24 to Waterlair to Eat down Rye Grass on a three year Old Field of it very good Indeed &c Jobbs –

Wedy 15th Thunder P.M. Cold Stormy weather from Northw^{d} till 3 P.M. some showers of Hail & rain very Cold – Plough Drilling & P.M. Carts at Driving Dung to Golds Hillock E side – Houghing Potatoes – Observe the Crop all round very Indifferent.

Thursdy 16th Th^{r} 54° Moderate weather from S.W. but not warm – P.M. several showers of rain light – a Plough Drilling all day and sowed what was drilled – three Carts driving of Dung all day to D° – The Bear looking very thin & stinted – & the Oats & Pease full of weeds &c – and the Hay is universally a very Bad Crop in this Corner – the pasture is not Sufficient for Feeding Cattle as yet –

Fridy 17th Th^{r} 57° Fine Growing weather from S.W. some light rain P.M.- Carts driving Dung – & a Plough Drilling – Sowed & rolled -Set up all the Flakes Waterlair – Hear that the Dunsmuir Markett very high –

Saturdy 18th very strong drought from N.W.- Drilling as before Carts driving dung – and P.M. driving in Earth into Dung Court -about Nine Acres Sown & Rolled in Good Condition – Neighbours not yet Begun –

Sundy 19th About 11 A.M. Came on light showers of rain from the South and appearance of more – Observed some of the Artichokes sett & first of Toads in ye Season, few or no Swallows & no Butterflies – Ashes not leaved.

Mondy 20th last night some rain and this day Considerable showers & stormy Cold weather from N.N.E. – finished out all Golds Hillock Turnips about Ten Acres in Order – Observed the Crops about Stonehaven & Aberdeen on Hot Sandy Grounds much better than here –

Tuesdy 21st Cold stormy weather from the Northward Blasting Vegetation all force at ploughing long Burns & part of the Back Yards about three Acres for Turnips very Hard & Lumpy with four Horse Ploughs – at Sundries –

Wedy 22d Moderate weather from the Southw^{d} – Scarcely begun to sow Turnips about Aberdeen – Rode Some Miles on the New formed TURNPIKE road benorth Stonehaven which will be pleasant Instead of the Dreary present One. Ploughs &c as yesterday – Cleaned pidgeon House &c.

Thursdy 23d Frost last night & this morning day fair & Mild from the Southward – Ploughs as yesterday – Turning & Mixing of Dung in the Court – Observed the Grass & Corns much better here, since Sunday last.

Fridy 24th Th^{r} 59° very strong drought from the N.W.- Setting everything all

Harrows & two stone rollers for ye Turnips which by its Hardness will Brake the Tools & Create Immense labour – at Houghing Potatoes &c.

Saturdy 25th Th^{r} 60° very strong drought from S.W. & air warmer than hitherto – all Harrows & two stone rollers – it is very rough and ill to Make – flited the Cattle Flakes at Waterlair Middle of the day &c Jobbs.

Mondy 27th Yesterday a Moderate shower about 11 A.M.- at night Cold & stormy a N.N.W. – this day fair till past noon when Came on Closs & pretty heavy rain for two Hours – all Ploughs at the Turnip Ground & Gathering off weeds of it – begun Cutting Grass for Horses.

Tuesdy 28th Th^{r} 58° This day fair & Strong drought a South – all the Ploughs & Harrows at the Turnip Ground very lumpy & rough – People at Sundries – Do not Observe a Good Crop of anything but Gooseberries.

Wedy 29th Th^{r} 60° very warm weather for the first time this Season from the S.E. – all the Horses Employed on the land as yesterday with two stone rollers on it & Carting off weeds, Clods &c P.M. at Cammock fair few Beastials chiefly Cows & but little Sale of same.

Thursdy 30th Therm^{r} 63° Artichokes size of Goose Eggs. very pleasant & warm weather for the first time this Season and the Cattle Startling – Drought severe – all Harrows & Carting off weeds of Turnip Ground and P.M. four Carts driving out dung to D^{o} pretty thick & well rotten – Quantity of rain this month is Just One Inch – The Hay not Half a Crop and the Oats midling good – but the Bear universally thin & not stocked the latest the Ground red as yett the Pease pretty good in Free land but quite stinted on stiff lands – the Pasture Excessive Scanty & not Half its usual Quantity & yet Cattle are not Purchaseable nor Sheep either, Beef & Mutton about 6^{ds}per lb – few white fish or Salmon the Fruit Trees Blasted, but other Garden Articles good no demand for Barley, Oat Meal from 23/6 to 25/ per Boll Roses just offering to Blow no Pease or strawberries –

Diary for the Month of July 1796

Fridy 1st Th^{r} 60° Hazy Growing weather from S.E.- 4 Carts at putting out dung as before – Others spreading of it-very thick the Crop Mending a little, but in greatest strain of rain.

Saturday 2d This day about $\frac{1}{2}$ past 11 A.M. Came on a Closs & smart rain till $\frac{1}{2}$ past 2 p.M. when faired – drought a N.W.

Mondy 4th Th^{r} 55° Yesterday a severe storm of Cold weather flying skiffs of Rain & Hail at North, this moderate from S.S.E. – Carts Driving dung & off weeds-to Dung Court, a Plough Drilling Rolled all sown P.M. – at Sundries. Jobbs –

Tuesdy 5th This morning about 6 Came on pretty brisk rain which lasted Moderately till 2 fair, and Came on again at 6 – finished sowing of whole Turnips about 13 Acres Twelve drills of D^{o} Yellow sown this day – the Crops and Grass looking fresher – at Sundries – Wind Easterly.

Wedy 6th Th^r 50° a Strong Gale & Stormy with much rain a N.N.W.- the Ploughs at Back Yards for Wheat – Middle of the day Flited the Cattle Flakes Waterlair – planted Curleys for Spring use – Observe the Bear Mending –

Thursdy 7th Moderate weather from N.W.to S.S.E.- about 2 P.M. Came on Closs showers of rain more or less Heavy all the afternoon, it being Fast day all People at y Church.

Fridy 8th Th^r 55° Pleasant fair weather from SSE – 4 ploughs at Backyards Intended for Wheat nearly 2^d furr done – Middle of day Dighted & Sacked about 14 Bolls of Oats for the Mill of Kair Sowed the Spring Plants rather Early, but Ground in Season.

Saturdy 9th Moderate weather – Ploughs did out Backyards – after at the Church – about 3 P.M. Came on very Heavy rain which lasted for about three Hours – dighted & sacked out Oats for the Mill 20B 3f – & some Bear for Poultry –

Sundy 10th Very Heavy rain with Hail on Higher Grounds from 11 to ½ past 1 P.M. & very Cold weather Northward – Have few Strawberries & Pease – everything very Backward –

Mondy 11th Th^r 51° as Severe Stormy & Cold weather from North but dry – sent the Oats to the Mill, and at Church- P.M. leading into Dung Court Earth, and at sundry petty Jobbs – Shore the Sheep –

Tuesdy 12th Moderate weather from Southw^d – Carts to Johnshaven for Coals to the Smith & for family English at 7/6 per Cart – at Paldy fair Sheep Markett very few Sheep and many Buyers about double last years price those who would Feed to 6lb per Quarter about 7/ each – Bought none

Wedy 13th Moderate & Growing weather a S.E. – 2 Carts for English Coals to Johnshaven & two D° Brought over all the Cattle Flakes from Waterlair to Floors & set them up there for the Young Cattle – at Sundries &c –

Thursdy 14th Light rain first part of the day – remainder fair & drought a S.W. – at Sundries – at Paldy fair not near so Many Cattle as usual, but many purchasers and prices beyond whatever Could be Expected in this Country – Cattle sold from Six to Nine Shillings per Stone to what they Could be fed to, even with that of winter added to the Grass – Cattle as high as Twenty pounds & upwards – all owing to the Demand for England on account of the War &c – Garden Pease ready –

Fridy 15th Th^r 61° very pleasant warm & dry weather a South – Cattle startling for about the first time – People Cutting Thistles E^t End of the Haugh all day – they are very abundant where the Crop is shortish – and at Sundries – the Grass rising fast at least the Pasture –

Saturdy 16th Early this morning Came on rain & thick Hazy Growing weather from S.E.- 6 Harrows Braking in the Back Yards- Others Scouring the Ditch on the Head of the Bank for a Midding Layer to put Lime & Dung in Intended for Wheat- Hear the Horse Markett yesterday at Paldy fair very thin, little sale & low prices – The Earliest Bear & Oats only offering to shoot & some of the Pease & Lint

Blooming – The Pasture here now is very fine every where – Fair P.M. – Hay Mending but light –

Mondy 18th Thr 60° Yesterday Blowing stormy weather & Evening rain from S.W.- this day strong drought from D° – Some Harrows at the Backyards & others Thistling the Haugh Oats – Observe both the Hay & pasture Mending & Corns –

Tuesdy 19th Thr 64° last night about 9 Came on light rain from Southward which Continued most of this morning – about $\frac{1}{2}$ past 2 P.M. very Heavy showers all the rest of the day – Thatched an old Bear stack & P.M. Flited the Cattle Flakes – Cattle startling Middle of the day & very warm from South –

Wedy 20th Thr 63° Fair all this day & pleasant warm weather at S.W. some Harrows all day at Backyards & others Thistling – Some people at Mill of Kair with 20B 3f Oats – Meal 18B 3pks Sundries –

Thursdy 21st a Good deal of rain in the night – day some showers Ploughs at Backyards- Sent for the Meal and at Sundries a Considerable Growth –

Fridy 22d Moderate & Growing weather from S.W.- 4 Ploughs all day at Backyards – at Scouring of the East side of Long Burns Turnips & Sundries – Observed the Bear Shooting & Mended Waterlair –

Saturdy 23d Bleak weather from N.W. and about 6 P.M. Excessive Heavy rain for about an Hour – Ploughs at finishing 3d fur Backyards & P.M. at Carting in Earth Dung Court &c- at Bervie & Gourdoun I Observed this day the Crop very various – On a patch on the Sea Confines a little Bear nearly ripe & on the High ground Immediately above it Scarcely Covering the Clod – at a Roup of victual at Bervie, Oat Meal sold from 23/6 to Twenty four shillings & two pence Str –

Mondy 25th Thr 58° Yesterday Growing Grey weather from S.W.- this day the same – Sent three Carts after Midnight to Mathers for Lime Shells to Backyards – and driving of Earth to Dung Court, and P.M. – at Sundries –

Tuesdy 26th Weather as yesterday, flited the Cattle flakes morning after Brakefast four own Men begun Cutting of the Hay long Haugh but very Moderate Crop – at Garvock Sheep Markett a Great Number of Sheep but so High saw little or no Sale of them – Sundries.

Wedy 27th Thr 57° Pleasant warm weather & drought a Westward – 4 Men at Hay Cutting & three people begun Thinning of the First Sown Turnips – in some places Ill Sown but otherwise very promising – at Sundries –

Thursdy 28th Moderate weather from the Westwd – Cutting Hay till Brakefast after all Persons Male & Female at the ACCURSED Markett of Garvock for Hiring Servants which should be abolished – Observed the Crop Indifferent in many places –

Fridy 29th Thr 61° very pleasant warm & Growing weather various – 4 people Cutting Hay & some amongst the Turnips – after 4 P.M. pretty Closs showers & from 5 to 6 D° Obliged to stop work – Sent $13\frac{1}{2}$ B Oat Meal to Mill of Conveth at 22/6 – The Turnips promising well –

Saturdy 30th Th^r^ 55° Frost last night & this morning Slight-day fair various 4 people Cutting Hay & two dressing of Turnips growing apace- It was warm in the morning but Cool afternoon &c

Sundy 31st Dark weather & warm from S.E. but P.M. strong drought so as the Hay could have been Coled – the Bear & Oats partly Shot & Pease well Bloomed – the Crop on the whole but Midling & rather late, the Hay not Half a Crop – the Turnips on light dry Ground Promiseth better & Surer than any other Crop – Quantity of Rain 4 Inches & One Half D° – The Price of Oat Meal from 22/ to 24/ & little or none of it except of Foreign Corn in the Country – Bear Meal 9 to 10^{ds} & now rising No demand for Bear at present & price unsettled from 19 to 21/ – price of Hay not settled Old Hay the Dragoons Montrose are paying 5^{ds} per Stone and it is said from 10^{ds} to one shilling from the Cole

Diary for the Month of August 1796

Mondy 1st Th^{r} Noon 58° Early this morning a smart shower of rain, but day rather windy from S.W. but warm – 4 Men Cutting of Hay & others Hand & Horse Houghing of Turnips – the Pease Podding & promise very well – at Sundries –

Tuesdy 2d Th^{r} 59° very pleasant warm dry weather from S.W.- after Brakefast begun Gathering & Tramp Coling of first Cut Hay into 3 above 100 stones Each – with Cottar wifes afternoon.

Wedy 3d Th^{r} 58° very fine drought from S.W.- Flited the Cattle Flakes morning & at Turnips – after 9 Hands at Gathering & Tramp Coling of Cottar Bank Hay in fine order – a little Frosty last night & a little rain about 3 P.M.

Thursdy 4th fine drought from S.W.- all day at Hay Coling and one plough taking away a furr from Turnips – Too few People & Hired Hands Cannot be got –

Fridy 5th Stormy Blowing weather & some showers of rain and after 7 P.M. Heavy for an hour – after drought – Own People & 4 Hired people at dressing Turnips and one Plough preceeding them – Sundries –

Saturdy 6th Th^{r} 63° Strong drought a S.W.- 10 People as yesterday at the Turnips of which did more than double the Quantity and as well – unaccountable how they Evade at times & not apparent either at the time, all needful now done, too high winds for Medling with the Hay – Sundries.

Mondy 8th Yesterday Bleak dry weather a S.W.- this morning Early Came on Showers of rain from the S.E. more or less for the rest of the day – 4 people Cutting Hay and Others at Horse & Hand Houghing of Turnips – and Sundries –

Tuesdy 9th Thick Hazy weather good for Growing but Bad for the Hay 5 Men Cutting Hay & others dressing Turnips – The Cattle had a Stop in their Sale lately but yarns getting as high it is said as before – Sundries –

Wedy 10th frosty dew morning – day fair till noon when Came on a very Heavy shower with Thunder in N.E.-and more or less rain all afternoon – people Cutt all

Hay on this farm turned over what was Cut a week ago – P.M. flited the Cattle Flakes- & others Horse & Hand Hoeing of Turnips.

Thursdy 11th Slight frost in the morning – day fair & Drought S.W. People at Turnips till Dinner time, after Turned over all Hay Cut – at Sundries –

Fridy 12th Therm^{r} 60° very Heavy white frosty dew – day fair & drought a Southw^{d} – all people & seven Hired Horse & Hand Houghing of Burns Turnips till at Brakefast – Gathered & put into a large stack near 2 Acres of Hay & turned over all that is Cut P.M., but am afraid of rain, Barometer falling –

Saturdy 13th Th^{r} Noon Shade 61° Dew in the night, but no frost – all people at the Turnips morning, after Brakefast begun Gathering & Hand Coling of Cottar Bank Hay in fine order of which did from three to 4 Acres – Very warm & sultry in the forenoon, Cattle Startling – Cooler P.M. a East –

Mondy 15th Th^{r} 60° Yesterday strong drought & warm weather at Westward, This day dry & various from different Quarters – In the morning at the Turnips finished taking away first furr from all of them & nearly first thinning of D^{o} – after own & Several Hired Hands Coled into large Hand ones all Hay on this farm –

Tuesdy 16th Th^{r} 63° Dark Cloudy & very warm weather – 7 Men Cut up the new park of Hay at Waterlair of 5 & $\frac{3}{4}$ Acres a very Moderate Crop – Intend to Thresh it being pretty rich of Rye Grass – Others at Turnips – Middle of day Excessive warm more so than Hitherto during ths Season – Sundries.

Wedy 17th Th^{r} 62° very warm weather & strong drought from SSE – 6 Men Cutting Hay at Waterlair – finished the new Grass park about noon, and begun Cutting the other strong rough Grass & unequal weight – One at the Harrowing of Back yards fallow – The Turnip promising well on free land, but more Backward on Strong – Some people have pulled their Flax, but is mostly a Bad Crop – mine not ready as yet.

Thursdy 18th Th^{r} Noon 69° remarkably warm weather – 7 people Cutting of Hay at Waterlair nearly finished – Others at the Turnips – The Heat of this day as much so as I have felt in this Country – Hear Lammass Muir Markett very thin of Cattle & but few Buyers –

Fridy 19th Th^{r} 67° very warm dry weather from Southw^{d}- Cut up all the Hay before Dinner, after flited the Flakes – Others at Turnips first Thinning finished – towards Glenbervie, Observed some Good & some very Bad Crops – at Lammass Muir Markett at Brechin Cattle not very Brisk but High prices –

Saturdy 20th Th^{r} 66° very warm & sultry weather from the S.E.- people at the Turnips 2^{d} dressing, and three Men Cast up a Canal 38 feet long-7 Broad, & about three to steep the Lint in – filled it with water from Leppie Burn evening-Sundries.

Mondy 22d Yesterday very thick Fogg & moist Air all day – this day warm & drought from S.W.- morning at the Turnips – after a Dozen Hands begun Threshing Rye Grass from the Bent Waterlair Seeming rich – got from 8 to 10 Bolls of it – good & lofted it.

Tuesdy 23d Very strong drought a S.W.- all the day a Dozen of People Threshing

& Coling Hay at Waterlair – at the Sheep Markett Laurence fair very few Sheep & more Men in the Markett than Sheep – they Sold or were rather rouped about one shilling S^{r} per Quarter of what they Could be fed to at farthest double of last year – Bought 9 Each @ 7/6.

Wedy 24th Thick Hazy weather from the N.E.- However Threshed all the Park Waterlair Hay Out & Coled it – Have got in whole from 20 to 24 Bolls fine Seed & lofted it Distillery loft, then & often Turned –

Thursdy 25th Droughty weather from S.W.- Own people pulling of the Lint Robs Hillock very fine Crop – Sold two Stotts three years old at 26£ 2/6 – there were few Beasts in the Markett & many Buyers but prices not so high as at Paldy fair – my father sold a three year old Stott at 18£ Str The Highest price of the Age –

Fridy 26th Th^{r} 62° Some light showers in the night & this day – pulled out & put in water 7 Cartload Green lint & Covered the same – Could not have Coled Hay – at Turnips &c The Oats & Bear filling very well – Sundries –

Saturdy 27th Th^{r} 55° Cool weather & rather rough from N.W. with light showers – flited the Cattle Flakes morning – after own & three Hired people Gathered & Coled about $3\frac{1}{2}$ Acres Hay on uppermost park Waterlair in tolerable order.

Mondy 29th Yesterday Closs weather but little drought – this day some little rain morning, but Broke out drought & Sun Shine from S.E. – after 12 o'Clock own and 5 Hired Gathered & Coled all Hay Waterlair into Hand Coles – rather over won – Sundries –

Tuesdy 30th Th^{r} 59° very pleasant dry weather & warm a S.W. – all the people and one Hired Man at 2^{d} dressing of the Turnips- Ground very Hard & dry – & a plough laying to a fur to the Turnips, very advanced in some places, Observe some people Carting Hay to diff^{t} places.

Wedy 31st Th^{r} 59° Strong drought from S.W. & very fine filling weather – all people dressing turnips too far advanced & laying to a fur to them they are fine Crop – Quantity of rain Just One Inch- being first part of the Month Chiefly – Observe some small Bits of Bear Cut down – the Oats & pease & Bear here will take about 2 weeks as yet – the prices of Grain unsettled the Meal has fallen about $1\frac{1}{2}$ per peck – Hay from 6^{ds} to 7^{ds} per stone but no Markett for it – Pasture very good –

Diary for the Month of September 1796

Thursdy 1st very pleasant and dry weather from Southw^{d} – people at the Turnips Hand & Horse Houghing the whole day – Observe severals leading their Hay – Ground very dry-

Fridy 2d Moderate Breeze from the Northw^{d} & strong drought – all people & Hired at the 2^{d} Dressing of Turnips and a plough laying to a Furrow to them leading at the Turfs all Home 8 Cartload large – Built D^{o} – &c.

Saturdy 3d Cloudy weather & some drops of rain-turned out drought- People

Hand & Horse Houghing Turnips much Stinted by the drought on the thin Grounds – P.M. Flited the Cattle Flakes to Calf Wards & Sundries.

Mondy 5th Yesterday Dark weather & Coldish a N.W.- this day the same – till Brakefast at Turnips, after at leading Cottar Bank Hay with three Carts & few Hands at Horse tramping of D° – the Sowe is about 50 feet long & 15 Broad – Hay in fine Order.

Tuesdy 6th Th^r 60° Coldish weather & some showers of rain morning – day windy & very Bad for the Hay leading – 4 Carts and 16 Hands all day – nearly led Home all the Hay but about 4 Cartful & nearly finished the Sowe –

Wedy 7th very Blowing weather which Increased into a Gale & Stopped finishing out the Sowe, Yoked two Carts & took out & Carted all the Lint well watered 12 nights & spread it on South side Cottar Bank it covers nearly from double to triple the Ground it growed on which nearly twice as much as any I ever had before –

Thursdy 8th Headed out the Hay Sowe morning & after 14 People & three Carts lead about $\frac{2}{3}^{d}$ of all the Hay Waterlair into a Rick 48 feet by 12 & Horse Tramped it in fine order – Moderate S.W. –

Fridy 9th Th^r 69° finished out the Sowe WATERLAIR, very high & neat & will Contain fifteen or 16 H^d of dried Hay on 13 Acres This day amongst the warmest of the Season – Moderate Breeze at South – Severals Have lead home some Bear, & some few Cutting & Gayting of Early Oats for Meal –

Saturdy 10th Thick Misty weather whole day & Everything very wett dressed and tidied Ricks of Hay – and Cutting at Sprotts & Rushes for Thatching of D° – Sent two Carts to Johnshaven for 2 Bolls Scotch Coal &c.

Mondy 12th Yesterday Broke out drought P.M.- this day the same all day a South – Cutting rushes, making of Ropes and Thatched the Hay rick Waterlair -- Some people leading their first Cut Bear – and most people Obliged to stop Shearing for want of ripe victual – Sundries –

Tuesdy 13th Th^r Noon perfect shade 66° Dark weather till noon when Blowing a fresh Gale a SSW Own people Hand & Horse Houghing of Turnips – nearly finished, they are very much Stinted by the late & present severe drought – all Ditches & wet places quite dry – yet the Crop is not Ripening over fast –

Wedy 14th very fine pleasant dry weather a SSW – at Turnips nearly whole finished – much stinted by the drought – at Cutting some Water side Hay which we did not need for Horse Suppers – a fine Crop – most farmers obliged to Ly for want of Ripe victuals, although Many are Cutting up that is not near ripe, I saw some Early Oats leading home this day and then several little parcels of Early Bear already Milled.

Thursdy 15th very warm & dry weather from S.W.- after Dinner mine begun Harvest by Cutting up part of the Waterlair Bear but Middling Crop & ripe – Towards Aberdeen from Stonehaven Some of the Crop more advanced & some of it much latter & very unequal in Quality –

Fridy 16th Closs Drisling rain & dark weather – People Cutting Sprots & rashes for the Hay rick here being to wett to Bind – the price of Oat Meal at Aberdeen 16^{ds}per peck & Bear D° 9^{ds}but no price fixed for Bear – Hay 7^{ds} – Flesh $4^{ds}\frac{1}{2}$ to 5^{ds}per lb.

Saturdy 17th Dark weather till near Evening but people Cutting up Bear all day at Waterlair nearly done and at Giving 4th furr to the Fallow –

Mondy 19th Yesterday fair weather & strong drought from S.W. Returned from Aberdeen – this day fair & drought a Westward people finished the W Bear by 10 A.M. – Threaves 150 on 7 Acres – after at the Auld toun park Pease a good Crop – Sundries – Quey C^{d}.

Tuesdy 20th Bleak weather from the Eastward about 9 Came on rain from the East which Increased & stopped Shearing of Pease by 10 – after dinner begun again & stopped before 5 at Barns Threshing Old Bear & Making of Ropes for a Hay rick – dighting Rye Grass –

Wedy 21st a Considerable deal of rain last night, and more or less so all this day from the N.E.- with stormy winds – Flited the Cattle Flakes Calf wards A.M. lead Home 2 Cartful Rushes &c for Hay Sowe & at Barns Threshing Old Bear, rope Making &c

Thursdy 22d Cold stormy weather from NNE but dry – all at Pease finished about 4 P.M. – after Begun Cutting up the Bear West side Thornylee new Ground & rich Crop rather ripe- at Sundries in Garden &c –

Fridy 23d Cold dry weather from N.E.- all day at Thornylee Bear East side fully ripe – The late Gale & rain Have Broke down much all the rankest Victual –

Saturdy 24th Fine weather till 11 A.M. Cloudy & drisling rain which Increased & stopped the Shearing – after at Shearing of Rushes &c – Thatched the Hay Sowe here, and lead Home the Hay North side river into a stack &c Jobbs – The rain Increased much in the Evening & stormy at S.E.

Mondy 26th Th^{r} Noon 52° Yesterday, pleasant dry weather at E^{t}- this morning very strong white frost which lay till 9 A.M. – at 7 begun shearing Thornylee Bear in Band till Brakefast after stooked – at sundries &c – Observed severals leading pease & Bear P.M. –

Tuesdy 27th Frosty dew in the morning but the Corns dry wind at N.E.- Cut out all Thornylee Bear large Sheaves $249Th^{s}$ and after noon, begun Cutting Oats E^{t} side Kavels – very fine Crops rich & fully ripe – and at Sundries.

Wedy 28th very strong Frost morning, day very warm a N.E. all people at shearing Oats Early in Gayte, after Brakefast stooked D° – no drought Gayte not dry Evening yet observed numbers leading Surely not fitt.

Thursdy 29th Dry shearing & Moderate Breeze from N.N.W. Cut out all the ripest upper part Kavels Oats a very rich Crop, after Brakefast at part Haugh D° – and finish – turned over the Pease North parks – after Dinner all Hands E^{t} side of Robs Hillock Bear, now rather too ripe &c Jobbs

Fridy 30th $Therm^{r}$ Noon shade 51 degrees Frosty dew in the morning but dry

Shearing of Robs Hillock Bear all the day – by far too ripe particularly the Infield of it & the best Crop. Observed most people leading, but surely too soon – Some Have it Heated already – & some spread out in the field Cast it on suppose wet shorn – Wind N.W.- Quantity of rain two Inches – Ground as dry as for seven years at least the wet Grounds & durable dry- price of Oat Meal 14 to 15^{ds} per peck & Bear $7\frac{1}{2}$ a stupid difference – no price or sale of Bear either old or new – The Sale of Cattle rising as high as in July last & Great demand – Manufactures rather dull – it is said the Harvest although not Half here is not so much so to the Westward.

Diary for the Month of October 1796

Saturday 1st very fine Drought from the N.W.- unhooded the stooks whole day at Robs Hillock Bear, nearly done – a fine Crop – Set up the stooks again & at Sundries.

Mondy 3d yesterday strong drought & rough weather from the Westward-this day same – part of the people shearing out the whole of Bear Crop Robs Hillock Threaves 300 nearly-after Brakefast begun leading of Waterlair Bear which did by Dinner with four Carts 12 Cart load into two stacks part rest from West side Thornylee – and Gathering Pease into winraes for leading tomorrow –

Tuesdy 4th last night from 9 to 10 a very Strong Gale at West which Hath Shaken some of the Oats & drove some of the Pease over the road Dyke – finished two stacks of Bear morning, Gathering & lead Home Six Stacks of the Pease, not Hollow in Good order, Guess about one Half of them.

Wedy 5th Early this morning a very Strong Gale from S.W.- day pleasant- lead Home all the pease but a triffle into 10 pretty large solid stacks – and at 4 p.M. begun leading E^t side of Robs Hillock Bear – good order.

Thursdy 6th Th^r 52° from 4 to $\frac{1}{2}$ past 5 A.M. a Heavy rain – fair till $\frac{1}{2}$ past 3 P.M. another Heavy shower of rain & Hail which stopped leading of R H Bear – Thatched 9 Pease stacks – Cuting rushes &c Jobbs – Wind S.W.

Fridy 7th Showers last night with much Lightning in the N.E.- Shearing dry Haugh Oats till noon, when stopped with rain at E^t – after Dinner at Kavels Oats Gayte and Sett up all Down stooks on the farm –

Saturday 8th Frosty Dew in the night & morning – Shore wett Kavels Oats in Gayte till 10 – then took up the Lint in Gayte, after lead Home all the remainder of Bear Robs H. into two stacks & remainder – women shearing of Oats P.M.- dry – at Sundries –

Mondy 10th Yesterday morning some rain 4 p.M.Increased very Heavy till 8 when Came on Strong Wind a N.W.- this day fair & Droughty weather a WNW – Shearing H Oats till Noon when Begun to lead Thornylee Bear into 2 large Stacks Stathels – all Gathered & lead remainder of Pease – & also Bound up & put into a stack all the Lint – good order &c.

Tuesdy 11th White frost morning – fair all day a West N.W.- lead Out all Thornylee Bear morning – after begun with 4 Carts & lead nearly 4 large stacks of

Oats, first Cut part of Kavels – did not finish them – at night Women Bound Gayte Kavels & then shearing of D°.

Wedy 12th Fair & pleasant dry Cool weather from N.W.- lead Home almost whole of Cut down Oats-six stacks – Women & others shearing at Kavels nearly finished –

Thursdy 13th Th^{r} 46° Strong frost white till 9 A.M.- begun shearing but to little purpose till then – at Haugh Oats shearing – at Drumlithie Markett a Considerable deal of Cattle & prices full as high as at Paldy fair or in this Season- the Harvest mostly done on the Sea Coast, but in several places hereabout much to do –

Fridy 14th Hard frost in the night & behind the sun all day – people shearing Oats Dickies Haugh all day – nearly finished all North side river – at the Horse Markett Drumlithie many Horses & few Buyers – Could not sell an Old One of mine.

Saturdy 15th Strong drought & very Cold weather from the North – Early flited the Flakes – all day Shearing of the Oats Haugh very ripe but not losing in Shearing not finished but nearly so on Back Waters.

Mondy 17th more than usual Hard Frost Black & White and Ice thick on the Dubs till noon – at Corn Yard till 9 A.M.- after finished Cutting all the Haugh Oats by noon, after at Red Hall park over ripe & rough work in Cutting but not many Coming out – Meal Oat about 14^{ds} and Bear 7$^{ds}\frac{1}{2}$ – no demand for Grain of any kind – Cutting 2^{d} Crop Clover for Horses

Tuesdy 18th about 4 A.M. Came on Heavy rain from S.W. till 9 when faired – at 11 a.m. after Gayting of Red Hall park Oats whole day being wett –

Wedy 19th Fine Blowing weather from N.W.- Early at Oats Cutting and finished that of whole Crop by One P.M.- after Dinner Cutting rushes &c Bog planting & set up any sheaves or Gayte yet to lead for Cows.

Thursdy 20th Fine dry weather from S.W.- Early yoked 4 Carts and lead Home all Haugh Oats under 4 stacks, women at Binding of Haugh & R H park Gayte & set the latter up –

Fridy 21st Fine Blowing weather from the Westw^{d} till past noon when it shifted to the Northward with pretty Closs rain all the rest of the day – Early yoked 4 Carts and lead Home remainder of whole Crop by 2 P.M. into 3 stacks – Oats 13 – Peas 10 – Bear 8 finished.

Saturdy 22d Th^{r} 55° faired last night about 10 P.M.- all this day Strong Gale from Westward – to stop Thatching part of it – at Barns making ropes and did 20 stacks – Observe some to Cut down at Polburn & some out at sundry places –

Mondy 24th The Feast Yesterday strong drought at Westw^{d} & strong Gales this day Strong Frost & drought a N.W.- Thatched out all the remaining stacks – Women Cutting Rushes for it- Observed severals leading all day –

Tuesdy 25th Frost in the night day fair & very pleasant weather from N.W.-Sent all the Flakes to Waterlair for 17 Young Cattle sent there this day to Eat up very fine Hay Foggage yet untouched – Sundries –

Wedy 26th very strong Frost all day but warm middle of it – Sent three Carts to Johnshaven for Great Coal, and One Plough at Fallow Backyards – Cutting Grass for the Horses suppers Gosk Cattle Waterlair – Sundries –

Thursdy 27th Keen Frost as yesterday white till near 10 A.M.- Sent three Carts to Johnshaven for Coals as yesterday, one Plough at Backyards – P.M. Cutting of Grass For Horse Suppers Waterlair – Cleaning Thorn Hedges.

Fridy 28th Moderate & Mild Dark weather dry – Sent three Carts to Johnshaven for G Coals now 13 of 16 being all to be taken – it takes $7\frac{1}{2}$ Hours Carriage – Gave Horses some Corn & fine Grass after 3 P.M. – One plough at the fallow Backyards – a Hired Man at Threshing Brocks.

Saturdy 29th very pleasant dry weather with drought a Northward – Sent two Cartload of Lint to Drumtochty Mill and two ploughs at Backyards fallow – and at sundries – Hear Crop Scarcely in yet in Garvock –

Sundy 30th This day fine dry weather from the Westward, and the Grounds & roads very dry – The Crop not yet all got in Garvock & Buchan –

Mondy 31st Therm^{r} Noon 40° Strong Blowing weather from N.W. & very Cold drought Sent three Carts to Stonehaven for English Coal lead in there some time ago a 2/4 per Boll 24 stones – One plough at Backyards, Supper the Horses with the Meadow Hay & Gave them Corn – no fixed prices for Grain as yet Bear about 20/ – Oat Meal 16/ per Boll – Fatt Cattle from 7/ to 8 per stone and lean Ones not to be got at any price – Hay from 7 to 8^{ds} to Dragoons Montrose &c Quantity of rain One Inch, Seven Tenths of Inch but as drought Quickly succeeded the rains it Hath been an uncommon fine Month for the Harvest in General – and the Crop is good for the most part & well got in unless it may be some of the first too Hurried –

Diary for the Month of November 1796

Tuesdy 1st Moderate weather till about 11 A.M. Came on pretty Heavy Snow from the Westward which Continued on the Grounds most of the day- three Carts to Stonehaven for English Coal-and One plough Back Y^{ds} Fallow – Barns &c

Wedy 2d Th^{r} 48° pleasant fair weather from S.W. – 4 ploughs at the Gibsons park ribbing & Backyards finished – There is but very little yet ploughed. There are Many people turning stacks towards Stonehaven –

Thursdy 3d Dry weather from S.W. till towards night rain pretty strong – 3 Carts to Stonehaven for Coals – One plough- Sold 2 three year old Stotts at Laurencekirk Markett at 30£ S^{r} & all old Bear 50 to 60 Bolls at 20/ per Boll – the new Bear STRANGE worth more.

Fridy 4th Early this morning Came on a Considerable fall of Snow from N.W. which lay till 11 A.M.- then many showers through the day – all Hands Dighting & Sacking Bear for Montrose whole day.

Saturdy 5th last night and this morning very Heavy rain & severe Gale from N.W.

but Early morning dry & some frost- the Roads Good, & fields pretty dry – 3 Carts Brought home all the Coals of Season & a fine Fanners from Stonehaven now at 3£ 10 S^r new & best Construction – Sundries.

Mondy 7th Th^r 43° Yesterday dry weather from Westw^d - this day same from D^o – 4 ploughs at Gibsons park ribbing done, and at taking in a stack of Oats red H park for Meal &c.

Tuesdy 8th Moderate weather with flying showers slight – Westy – three Carts to Montrose with 15B Old Bear went off before 5 Returned by 8 P.M.- one plough at pease land – begun Threshing Oats by lott for Meal – Bled all the Calfs for fear Quarter Ill & Cut two young ones.

Wedy 9th Flying light Showers from Northw^d but good ploughing weather – 4 ploughs ribbing pease land & after at upper End of Robs Hillock across Intended for Turnips – afternoon took in the last stack, Bear Crop 1795 S Barn

Thursdy 10th very strong frost this day but did not Hurt the Ribbing of N End of Robs Hillock – after the Horses at Grass – setting up pease stooks &c Jobs –

Fridy 11th Th^r 36° very strong frost till 11 A.M. – 4 ploughs at the Robs Hillock from 8 to One – lead some Turnips Put all the young Cattle Byres which fills them Compleatly – Sacking Bear for Montrose & Sundries.

Saturdy 12th Strong frost whole day & Clear pleasant Calm weather – Sent three Carts to Montrose with 15B Bear & Brought out sundries – in return by 8 p.M.- One Plough at Robs Hillock, & lead Home 2 Cartful of Turnips from the Burns – Sundries –

Mondy 14th Yesterday frost & fresh alternately, this day fresh and a shower about 4 p.M. from the Southward- three ploughs at Robs Hillock – and One Cart to the Mill of Kair with 6 Bolls of Oats for family Meal, and afterwards lead in two cartful of Turnips from Burns fine Crop – Brought Home 4B 1f of Malt from Bridge Leppie.

Tuesdy 15th Fresh weather from S.W.- 4 ploughs at the Robs Hillock nearly finished – at Scouring Ditch Head of Bank for Midding Layer Back Yards and Drawing Thatch for the Cottar Houses of Old Bear Straw Threshing-and Sundries.

Wedy 16th Frost in the morning till after sunrise, which quickly vanished, the wind Got into the SE – Mercury sunk and the Heavy rain & storm begun about 11 & lasted till past 4 P.M. which brought down rivers &c – at ploughs and lead Home two Cartload of Turnips Burns – at the Mill of Kair 6 Bolls Oats – got above $15\frac{1}{2}$ pecks Meal per Boll which was more than Expected as the Oats were soft being the last part of the Crop got in & when raining – the prices of Grain very high & in demand – Bear from 20/ to 21/ Str Oats 16 & 17/ – Oat Meal 13^ds per peck & Fatt Cattle 8 to 9/ per Stone, lean ones nearly as much –

Thursdy 17th Frost all this day but Clear & pleasant dry weather and the roads better than before – three ploughs at the ribbing across North Side of Kavels for Bear – water furred Waterlair-and lead some Turnips – Sundries Barns

Fridy 18th Strong Gale from Northw^d with Heavy rain in Early morning and the same in Stormy Showers at Intervals through the day – three ploughs ribbing at Kavels – Others at Dighting of Bear for Montrose, and at sundries –

Saturdy 19th very Stormy weather & Gusts of Snow & Sleet from the North all day- 4 ploughs at ribbing of Kavels for Bear, at sundries – fields no wetter.

Mondy 21st Yesterday Frost & dry weather from the Northw[d] – this day the same, Sent three Carts to Montrose with 13 Bolls Old Bear and one pair Horses at Ribbing and leading of Turnips Goulds Hillock Field.

Tuesdy 22d Hard Frost except in the sun about Middle of day – One plough Ribbing Kavels – Dighting & Sacking of Oats N B for Meal – in a stack Oats N B and lead in two Cartful of Turnips after Dinner.

Wedy 23d Closs rain till Noon but not Heavy – not Cold – Yoked 4 ploughs Kavels – at Barns dighting of remainder of old Bear, & sacking of D° – Sundries and at variety of small Matters &c

Thursdy 24th more or less rain whole day a N.E. – 4 ploughs at Ribbing of Kavels dry Enough – P.M. took in the first stack of Bear of Robs Hillock S Barn.

Fridy 25th Fair & dry weather from the North but no drought – 4 ploughs nearly finished all ribbing of Season Kavels and leading of Turnips – Thatched a Hutt of Cattle Flakes to Hold Turnips from rain, Snow &c –

Saturdy 26th very pleasant & dry weather from N.W. more so than several weeks by past – One Plough finished all the Ribbing of Season – One Cart to Montrose with remainder of Old Bear – and two driving Turnips & Earth – P.M. spread it on y[e] Ley –

Mondy 28th Yesterday Cold stormy weather from North with some showers of rain, all last night & this day very strong drought a N.W. – Carts driving Earth on ley & into a Layer Head of Bank to spread lime upon for the Back Yards Fallow & into Dung Court – This was rich Earth from Scourings of a lint Steeping Hole.

Tuesdy 29th Cold & dry weather from N.W.- Begun by three ploughs the seed fur on East End Haugh for seed fur in very good Condition, made it deep & small and at spreading of lime on Dunghill Layer for Fallow – and at leading Turnips &c Jobbs –

Wedy 30th Therm[r] Noon 28° bitter Cold Evening last night Came on strong Frost & a skimering of Snow from the N.W. which Increased to a very Strong Gale and this morning the River much Frozen Over & Dubs Carrying Horses & Carts with Heavy Gusts of Snow at Intervals – being a Thanksgiving day for the Good Harvest no work without Doors – Excepting sending 7 Bolls of Oats to the Mill – Dighting more Oats N Barn also for the Mill – Quantity of rain 4 Inches – The Demand for Cattle some Slacked, but Fatt ones & Sheep very high about 8/ per stone each if good – Hitherto the Demand for Bear great & price about 21/- per Boll – now likewise stopped for a time – Oat Meal 17/ per peck 13[ds] – There is no judging as yet of the Crop – the Oats are rich & Good but the Bear I suspect the Contrary at least by the Spring – Fodder will be plenty – I do not think the best Turnips above Half a good Crop.

Diary for the Month of December 1796

Thursdy 1st Strong Frost & Cold from N.W.- Came on much more Snow Evening but less frost – 4 Carts driving of Dung to Dunghill for Backyards, Barns &c.-

Fridy 2d Th^r 30° last night a good deal more Snow – this day mild & wind shifted a S.W. – Carts driving Dung & Oats to Mill, at dighting more for the Mill P.M. – at the Crofts at Auchenblae not a score of Cattle no Sale.

Saturdy 3d Fresh weather & very Stormy from the Southw^d- about Noon the wind Shifted to the N.W. with Snow & frost & Strong Gale – 2 Carts leading Dung – one driving Corn to the Mill & P.M. lead some Turnips – took in Oats N B morning –

Mondy 5th Therm^r Noon 30° Yesterday strong Frost & very Cold weather a N.W.- this day the same till a little after noon the wind shifted all at once from the West to the North with a FURIOUS Gale of Snow & drift which Stopped all work without doors & also prevented watering the Young Cattle, but Gave them Turnips – driving Dung & leading of Turnips previously –

Tuesdy 6th The Storm Abated this morning, but the Country Covered with Snow & Blown into the roads, and about the Houses Two feet deep, little or no Traveling on the roads – 4 Carts driving Dung E^t side Robs Hillock for Turnips – the Rivers frozen up & strong Frost Continues &c –

Wedy 7th Th^r 34° Frost Moderate rather Fresh a West with some snow – 4 Carts driving Dung to R Hillock & leading of Turnips Burns, and at sundries Barns &c – Sold first stack of Bear at 20/ per Boll –

Thursdy 8th very Stormy Blowing weather & Excessive frost a N.W.- 3 Carts driving Dung all day to Robs Hillock – others at the Mill 20B-2f – Meal 20:2-2-0 – Digging of Turnips Burns afternoon & Smithy –

Fridy 9th Moderate & Appearance of Fresh or more Snow – Sent three Carts for the Meal & one leading of Turnips from the Burns, very difficult to Digg – they are not Frozen to the Heart – Slapping Bear –

Saturdy 10th Th^r 39° rather Fresh weather from N.W. & Evening a Good deal of rain – people at Barns all day Chaffing and Dighting & Sacking Bear sold – lead Home by Digging 2 Cart load Turnips Burns – the Snow Somewhat vanished, but moderate Thaw –

Mondy 12th Th^r 39° Yesterday pleasant but Moderate Thaw – this day the wind North Westerely & Moderate thaw – Early took in a stack of R.H.Bear S Barn larger $\frac{1}{3}$ than last – and Chaffing Oats N B to loft for family Meal &c Jobbs.

Tuesdy 13th Fresh weather Moderate a N.W. – People at Dighting & lofting of Oats for the Mill & after took in a stack Haugh Oats N B for same purpose – leading of Turnips One Cart from Burns fine Crop – Gardner dressing Fruit Trees – the roads becoming dirty & loose & rivers not Broke.

Wedy 14th Fresh weather from N.W. & Snow mostly Gone but the Frost not so – 4 Carts digging & driving on Earth on Robs Hillock Dunghill from the sides of East

Ditch D° field and P.M. pulling of Turnips into Heaps on field the Ice not Broke on the Rivers; Give all the Young Cattle on the Farm Turnips; fair Feeders –

Thursdy 15th very pleasant Thaw – the Rivers Broke without any Flood or Grounds wett – Yoked 4 ploughs on East End of Haugh Award a fine season for it – few going in Neighbourhood – the Snow is not gone in Hollows & oyr places –

Fridy 16th Th^r 42° Noon Mild weather from the Northward – three ploughs at Haugh Award – One Cart leading Turnips from the Golds Hillock – P.M. at Sundries about Midding Layers – and Sundries – The Cattle Horses Comsume above five Bolls rank Fodder daily & all get Turnips – few ploughs going There is no great demand for Bear at present but it is about 20/ per Boll – Oat Meal scarce about 17/ per Boll – Flesh Meat not to be Had almost & about 5$^{ds}\frac{1}{2}$ to 6^{ds} per lb – prices of Spinning about 16^{ds} per spindle – good Flax about 18/ to 20/ per stone – everything advanced and on the Increase, Irons wood & every thing for the Farm use are dearer than Grain or Cattle proportianly.

Saturdy 17th Moderate & fine dry weather from the Northward – three ploughs at Haugh and one Cart leading of Turnips Golds Hillock and at sundries –

Mondy 19th Yesterday a slight skimering of Snow from the Northward this day same but slight frost at Northward – Yoked 4 ploughs at Haugh all day – at Middings P.M.

Tuesdy 20th Th^r 31° Some Snow & Blowing weather from the Northward with Frost – Yoked 4 ploughs Haugh good work – P.M. lead 2 Cartful of Turnips – at Barns &c Jobs – The Rivers freezing again – few Ploughs Going &c –

Wedy 21st Th^r 28° about 6 A.M. Came on a Strong wind & some drifts from the North till 9 A.M. with strong Frost which Stopped Ploughing when Tried – after driving of Dung & Home some Turnips – at pulling D° P.M.

Thursdy 22d Th^r 29° very strong Frost, wind Wesly – 4 Carts driving of Dung to Robs H. for Turnips – East side of it. Digged up Turnips Burns P.M. & Sundries –

Fridy 23d Th^r Noon 27° very severe Frost & Clear Winter weather from the N.W. One Cart leading of Turnips from the Burns – and other Hands Chaffing Oats N Barn – after Dinner Digging Turnips now very Frozen in – Sundries – the rivers frozen over, roads good, not much Snow –

Saturdy 24th Th^r 26° Uncommon Strong Frost & Chilling wind Clear from the N.W.- three Men Holing Turnips all day and one Cart leading of D°-some days store – at Barns & other Jobbs little of river Open – The Turnips Frozen to the Heart & difficult to Eat.

Mondy 26th Th^r 31° Yesterday Excessive bitter frost & Clear weather from North- this day the same a N.E. with some Gusts of Snow People at Barns dighting Oats for the Mill and at Sundries – Gardner Pruning Forrest Trees –

Tuesdy 27th very bitter Cold weather-with Gusts of Snow & Hail from the N.E. – Sacked & sent to the Mill of Kair 15 Bolls Oats for family Meal – difficult access to it through the river – and lead Home two Cartful Turnips &c Jobs.

Wedy 28th This whole day a Constant Storm of Snow & Sleet from the E.N.E.- but severer afternoon fresh no work without doors, Chaffing Bear S Barn and Could not water the Cattle but gave Turnips in Sufficiency – difficulty of getting Fodder at present.

Thursdy 29th Constant storm of wind & more or less rain the whole day from S.E.- with dark lowring weather and the rivers very much Swelled & Ice broke much Dammage is Apprehended along the Coast, as the Sea is very rough – at Barns & lead Home 4 Cartful of Turnips before noon –

Fridy 30th very Strong Thaw all day from the S.E. and 2 P.M. Came on Constant rain, but moderate weather – Sent other 15 Bolls Oats to the Mill,and one Cart leading of Turnips -took in 5th stack of Oats N Barn before dinner – at Sundries

Saturdy 31st a very Considerable rain & Stormy weather from the S.E. last night & part this day – the Ice Shoals mostly vanished and the Frost quite out of the Ground – little dammage done the Banks of the River – Quantity of Rain this month is Two Inches & Nine Tenths of an Inch – Total of the year is 28 Inches – prices of Grain not steady or settled – Bear about 20/ & Oat Meal 16 to 17/ – Beef & Mutton about 6^{ds}per lb – Good Flax about 18 to 20/ per Stone 22lb Ams^{d} & Yarn & Cloths proportional – the Fodder is Spending Apace – not above 2$\frac{1}{2}$ Acres Turnips used of Twelve and one fifth – Cattle Thriving fast by them.

Diary for the Month of January 1797

Sundy 1st Therm^{r} Noon 46° Moderate weather from Northw^{d} with some light rain at Intervals – the Frost quite Out of the Ground – the Pasture & Stems of Turnips, quite Spoiled by the late severe weather – Scarcely any Snow in Sight –

Mondy 2d very Closs Hazy weather & more or less rain from N.E. all the day, three Ploughs at Haugh Award, a tolerable season for it – and One Cart leading Turnips – Golds Hillock dry ground – Thatched a Cottar House P.M.&c Jobbs

Tuesdy 3d Bleak but dry weather from S.E.- three Ploughs at the Haugh – One Cart leading of Turnips as before, and at Cottar House thatching a second one – Sundries –

Wedy 4th Blowing Stormy weather from S.E.- a good deal of RAIN all the day- no Ploughing – at S Barn dighting of Bear, and P.M. lead Home some Turnips &c –

Thursdy 5th Th^{r} 35° Thick Hazy weather all day & drizling rain being Christmas *Old Style*, no work of the farm Going on – Dirty roads – at Johnston &c Observe a Considerable deal of red land &c –

Fridy 6th Dark Mild weather whole day – three Ploughs at the Haugh as before dryer – One Cart lead Turnips P.M.- people at the Mill with 30B Oats = 31:3-1:2 Meal – Observe of late Several of the Fruit Trees in the Garden Barked with Hares & much dammaged –

Saturdy 7th Closs Thick Foggy weather – Carts leading Home the Meal seeds &c & Turnips Golds Hillock – One Plough at the Haugh now still dryer & at Sundries. Evening Girnelled the Meal &c –

Mondy 9th Yesterday thick weather & drisling rain from Southward- this morning pretty strong Frost & Moderate weather a N.W.- 4 ploughs at Haugh Award and at Barns lofted Bear, Intended for a short time & Sundries –

Tuesdy 10th Thick frosty white Rhime & land pretty Hard but not to stop Ploughing at Haugh three ploughs – and one Cart leading of Turnips Burns – took in a stack Oats for Seed supposed morning – and P.M. Pulled & threw into Heaps Turnips for fear Snow & Frost – Sowed some Garden Pease

Wedy 11th rather fresh weather from SSE & some showers of Snow – four Ploughs at Haugh – P.M. at pulling of Turnips for fear of more Snow, not Improbable –

Thursdy 12th Th^r^ 31° very strong Frosty weather a S.W.- 4 ploughs at Haugh and tolerable good work – at Barn Chaffing of Oats N B and at Garden & Sundries &c – The Cattle Eat a great deal of Fodder &c –

Fridy 13th last night an uncommon Severe Frost till near morning Fresh wind at S to S.E.- about Noon it began to Blow & was Increased into a storm of wind & sleet but which abated by five p.M.- 4 Carts driving of Dung to R Hillock & lead some Turnips Burns &c –

Saturdy 14th Strong wind Westerly & the frost mostly Out of the Ground – 4 ploughs at the Haugh good order P.M. Dighting Oats N Barn and at Sundries. This Hath been a very favourable Thaw and a short delay of work – so much the better –

Mondy 16th Yesterday frosty weather wind Northerly – this day frost & moderate drought a N.W.- 3 ploughs at Back Waters – & one Cart leading of Turnips from the Burns – Over Garvock a sheet of Ice & frozen Snow, Muir & Gutturs almost Impassable for Travelers to Muirton.

Tuesdy 17th Th^r^ 32° Black frost but moderate – 2 ploughs at Back Waters but Indifferent work & not much of it, lead some Turnips & pulled more of them Burns- Horses at Smithy – the prices of Grain Considerably fallen Bear to 18/ & Oat Meal to 15/6 per Boll – Fatt Cattle few to sell about 7 to 8/ per stone – the roads are now at least the publick is good & the rivers not Frozen – at Sundries –

Wedy 18th Th^r^ 37° Fine pleasant weather betwixt Fresh & Frost a S.W.- Sent three Carts to Montrose with 16 Bolls of Bear formerly sold deliverable at Gurdon and one Cart leading of Turnips Burns &c –

Thursdy 19th Some rain in the night & this morning, but not much a S.W., yoked 4 ploughs at Back waters in good Condition, and at Barns & Sundries –

Fridy 20th very pleasant fresh weather from S.W. – three ploughs at Back Waters, One Cart leading Turnips from Golds Hillock, and at Barns P.M. – Gardiner Planting a Row of Larches at Foot of the March Dyke on the North side of the River opposite the House –

Saturdy 21st Closs Foggy weather & Mild a S.W.if any wind – 4 Ploughs nearly finished all the Back Waters – and P.M. lead Home 2 Cart load of Turnips & at S Barn Chaffing of Oats & Sundries –

Mondy 23d Yesterday dark weather wind fresh a S.W.- this morning a strengthened Westerly & some showers of rain – three ploughs at Haugh nearly finished, and one Cart leading of Turnips Golds Hillock – Gardner Planting Larches alongst the March Dyke betwixt Sheperds Haugh – above New Bridge.

Tuesdy 24th remarkably pleasant Sun Shine weather from S.W. – three ploughs at Floors Ley and one Cart leading of Turnips from Golds Hillock – at Barns & Gardner Planting in different places –

Wedy 25th Slight frost in the night day stormy weather from S.W.- three ploughs at Floors ley Exceeding rich & deep and one D° Steering North side of Golds Hillock after the Turnips for Bear – at Barns dighting Oats.

Thursdy 26th almost Constant rain all this day from the SSE and dark Gloomy weather – Ploughs as yesterday – at Barns & Gardner planting –

Fridy 27th Blowing Stormy weather a S.S.E. and about 2 P.M. Came on Heavy rain Increased wind – three ploughs at Floors Ley and one Cart leading Turnips- morning took 3d Stack Bear S Barn – Sundries –

Saturdy 28th Slight Frost in the morning – day pleasant Sun Shine – three ploughs at Floors Ley and one Cart leading Turnips Golds Hillock – at Barns & Sundries – Gardner planting Larches.

Mondy 30th Yesterday towards night a severe Gale of Wind & rain from S.W.- this day very Stormy but little rain a S.W. three ploughs at the Floors Ley – and one Cart leading Turnips – Planting Larch amongst the firrs Thornylee –

Tuesdy 31st much rain & stormy weather from the Westward – about 8 A.M. Came on a Closs Fall of Snow which did not last till 10 & melted away in the low Country before 1 P.M. but Lay on the Hills – People dighting & lofting of Oats Intended for seed & took in first Pease N N Barn – Quantity of rain this month is just Two & one Half Inches – The prices of Grain Have Fallen very Considerably of late & very dull sale of all kinds at present owing to much Importation from Ireland & other Places – Bear 17 to 18/ – Oat Meal about 15/ – Butcher Meat scarce about $5^{ds}\frac{1}{2}$ to 6^{ds}per lb – no Farms yet sold – Manufactures very Brisk & Spinning High 16 to 1/6 per Spindle – Turnips not Half used here. Some Spring Flowers appearing.

Diary for the Month of February 1797

Wedy 1st Blowing weather & drought a West – all the People at Taunton fair, a good Crowd but less Demand from Angus than usual and Consequently little rise on their Wages &c – first Stack of Pease.

Thursdy 2d Strong Blowing weather from S.W.-with some skiffs rain- 2 ploughs at Floors Ley – One Cart leading of Turnips & One D° driving Oats to the Mill of Kair.

Fridy 3d Strong Blowing weather from S.W.- three ploughs at the Floors Ley – One Cart leading of Turnips Golds Hillock beginning to Sprout in the Stems. Gardner Planting Larch amongst 7 year old Firrs – at Barns Chaffing Bear P.M. – & Sundries –

Saturdy 4th Th^{r} 45° Moderate & very pleasant Spring like weather from S.W.- three Ploughs Ley, One steering of the Golds Hillock after Turnips – Gardner root Dressing some Espalier Plumb Trees – all now Productive Labour – Some Growth in the Ground.

Mondy 6th Yesterday dry Spring like weather a S.W.- this day the same – 2 ploughs at Ley – One Cart at leading of Turnips & one Driving Oats to the Mill – a Highland Sholt Excessive Ill with a Tinge or Colick for 14 hours but Recovered – Snowdrops.

Tuesdy 7th Strong drought a S.W.-no frost – three Ploughs at Ley – One Cart leading Turnips – P.M. Men at leveling down an Old Dyke Calf Wards – the Gardner Planting Elms &c Bog Atry planting & Sundries.

Wedy 8th Remarkably fine dry weather without Frost a S.W.- three ploughs as yesterday & one Cart leading of Turnips Golds Hillock – Gardner planted or made up Early Cabbage, Curleys &c and furred them up and at planting Forrest wood as before –

Thursdy 9th Fine weather as yesterday – 4 ploughs at different places, 2 at Old Toun park seed fur for Oats & Grass seeds – P.M. took in 2d stack Pease N B – Observe some Pears & Several other Trees Budding.

Fridy 10th Th^{r} 49° Uncommon pleasant & fair weather a S.W. – all the Ploughs as yesterday – at Sundries – Saw about four score of Good Cattle pass by for England from Aberdeen Shire & this Neighbourhood, Many of them Fatt & said to be very High priced –

Saturdy 11th Weather as yesterday but Cooler a S.W.- three ploughs at Old Town park- and one last leading of Turnips-Backyards - at leveling down an old Dyke West side Floors to Make it possible one Continued plain of whole Haughs.

Sundy 12th This day about 11 o'Clock small rain and the wind suddenly shifted into the North with Sleet & Snow till Midnight – but not frost or deep –

Mondy 13th Fresh Hazy Mild weather a S.W. & Snow mostly gone – three ploughs at Old Toun park & one Cart leading Turnips -Sundries – at a Roup of Corns with Fodder at Polburn the whole Sold Offered from 16 to near 19 shillings per Boll –

Tuesdy 14th Storm pleasant weather till about 11 A.M. Came on a strong Gale from N.W. with Gusts of Snow & Evening Frost, 4 ploughs as before – P.M. Pulled some Turnips into Heaps

Wedy 15th last night it Blowed a perfect Hurricane with a Clear Hard atmosphere from N.W. which did not abate till about four this morning with so Hard a frost as to stop ploughing in many places, two ploughs & one Cart leading of Turnips Backyards – at the Mill of Kair Oats B33.3-2 Meal-36.0.2 – Several of our Tiled Houses damaged and a Rick of Hay on Waterlair tirred on the North Gavel & some scattered – many Houses in the Neighbourhood Damaged, rivers near frozen over –

Thursdy 16th Moderate weather from the S.W.and little Frost – Two Carts

Brought Home the Meal &c and other two leading of Turnips Backyards – P.M. repairing of a Hay Ruck ruffled at Waterlair.

Fridy 17th pretty strong Frost morning wind S.W.- yoked 4 Carts to drive Dung West side Robs Hillock till 11, then at Barns, Emptying Meal Girnells the rest of the day – Gardner planting B Atry –

Saturdy 18th Frosty whole day Exceeding fine pleasant weather a SSW- One Cart to Arbuthnot for 2 Ash Trees Bought there, and one leading of Turnips & all others Chaffing Bear S Barn – In 3^{d} Stack Pease.

Mondy 20th Yesterday fine pleasant but Blowing drought a S.W.-this day same – 2 ploughs at Old Town park- one Cart leading Turnips – and all others Dighting & lofting of Bear &c

Tuesdy 21st White Frost in the morning & day very pleasant sunshine wind a S.W.- three ploughs at Old Town Park – one Cart leading of Turnips for Transplanting into Garden for seed-picked Ones morning took in a stack Oats – P.M. turning Dunghill – Mizerign flowred.

Wedy 22d Strong Frost quite White till 10 A.M.- Ploughs however at work – & leading Turnips – a Cart to Laurencekirk with a pair of wheels to Ring with very thick Iron a full Inch ring – Sundries.

Thursdy 23d Fresh dry weather from S.W. – Yoked three ploughs on S.E. park 3 year old Ley, very fine Sward Waterlair, very rich – One plough steering of the Golds Hillock after Turnips for Bear &c Jobs.

FRidy 24th Uncommon fine weather & warm Sun – three ploughs as yesterday – One Cart leading Turnips – and P.M. at Sundries – Gardner Dunging & Digging part Nursery Ground for Early Potatoes –

Saturdy 25th Pleasant warm weather free of Frost night & day – three ploughs at Waterlair, and one Cart leading of Turnips – Planted Early Potatoes P.M. Turning Dunghill with Lime for Back yards –

Mondy 27th Yesterday & this a Continuance of very steady fine weather – notwithstanding the Change of the Moon yesterday evening-which often Alters the weather – 3 ploughs at Waterlair and one Cart leading of Turnips - pulled all the Yellow Ones & put picked ones into the House after Cutting off the Stems – for Culinary uses –

Tuesdy 28th Th^{r} 48° some Crocuses flowred Slight Frost in the morning – day very fine weather a S.W.- three ploughs at Waterlair Ley and One Cart leading of Turnips being all of the Backyards – Quantity of rain is only three Tenths of an Inch. Observe some Oats sown this day at Polburn on ley – roads very fine – Prices of Grain of all kinds quite fallen and little demand Oat Meal 14/6 – Bear about from 15 to 16/ – Pease not Saleable – Cattle of all kinds beyond all former prices – about Ten shillings per stone either fit to Kill or what will feed to it with the Summers Grass, Sheep, Swine &c same Exorbitant rate which Cannot Last – Hay to Dragoons Montrose 10^{ds}per stone – Manufacturing said to be stopped much Idle dread of French Invasion.

Diary for the Month of March 1797

Wedy 1st Closs Foggy weather from the Southward – three Ploughs at Waterlair – and one at Steering the Golds Hillock after Turnips for Bear – at Sundries –

Thursdy 2d Cold Bleak dry weather from the S.E.- three ploughs at Waterlair & one here as yesterday – P.M. at Turning Dunghills – some Sowing Pease –

Fridy 3d very Bleak dry weather from S.E.- ploughs as yesterday at Midding layer Robs Hillock – and Gardner Planting Ash BogAtry – and at Sundries –

Saturdy 4th very Cold Chilly dry weather from S.E. – Ploughs as yesterday – One Cart leading Turnips – offering to Shoot & form the Flower – at R Hillock Dunghill turning - Gardner Planting Bog Atry –

Mondy 6th last night a slight fall of Snow from N.E. this day Milder with light rain – filled the Barns forenoon – after at Waterlair for Pease and leading of Turnips – Gardner planting –

Tuesdy 7th Variable weather & drought Chiefly a S.W.- ploughs as yesterday – at Bervie some Oats sown & Bear & ploughed down – Observed a good deal of Turnips between this & that place & several stacks of Hay, but the Barn Yards not near so full as last year at this time –

Wedy 8th Th^r 46° White frost in the night day warm wind Northy – three ploughs at Waterlair – One Cart leading of Turnips – at Sundries P.M. – little vegetation – it is reported Cattle have fallen much of late.

Thursdy 9th Slight frost in the morning which Makes the Grounds raw till Afternoon – Wind Easterly – being a General FAST no work done this day.

Fridy 10th Cold Drought a E.by S.- three ploughs finished out all the Tillage at Waterlair for this Crop fine Season by Dinner time- One Cart leading Turnips – Observe several Neighbours begun Oat seed this day.

Saturdy 11th last night some showers of Hail from the N.E.- this day drought & Grounds dry – Sowed Six Bolls of Pease or 7 D^o Waterlair & Harrowed same – One Cart leading of Turnips all the day – Yoked twice for the first time this season – Sundries –

Mondy 13th Th^r Noon shade 45° Yesterday fore part Frost, day fair a N.E.- the last night so Hard Frost as to stop Harrowing in the morning Yoking – Sowed the Waterlair Ley park about 17 pecks per Acre or $7\frac{1}{2}$ Bolls & Partly Harrowed the same very stiff – Sundries –

Tuesdy 14th Uncommonly pleasant dry weather from S.E.- 9 Harrows at Waterlair ley stiff all day – One Plough old Water Tracts Floors – and P.M. leading of Turnips – Women at Barn Chaffing & Dighting of Kavels Oats for seed – Sundries.

Wedy 15th very Strong frost last night & this morning to prevent Harrowing morning yoking – at ploughing two old Water Tracts Floors – and P.M. 8 Harrows Cross Harrowing Waterlair Ley by a double Tine Across-very Ill to Make.

Thursdy 16th Frosty in the morning – day dark Coldish weather from N.N.E.-

finished out Waterlair ley A.M. and P.M. Sowed 8 Bolls Oats on the Floors Ley and partly Harrowed it- at Barns &c dighting & sacking seed Oats – no kind of demand or sale for any kind of Grain, nor for Hay or Fatt Bestials – Imaginary deficiency of Money, and of French Invasion of this Country absorbs & Employs the Whole Thoughts of the Idle & Speculative defenders of the Realm.

Fridy 17th Dark fresh weather till 1 P.M. sun Broke out – Wind N - 9 Harrows all day at Floors – Sowed on it 11B and one Half or some better than a Boll per Acre Includes Water Tracts & most of it very rich – leading of Turnips – Still advancing to Flowring – Garden &c.

Saturdy 18th Th^r^ 45° very Hard Frost morning to stop Harrowing till P.M.- at E^t^ End Haugh with Ploughs – P.M. 9 Harrows at Floors, new Broke up Ground very severe on them – dighting & sacking seed Oats & leading of Turnips &c.

Mondy 20th Yesterday fine dry weather a S.E.- this day same only a light shower from 12 to 2 p.M.- Sowed the Backwaters 4B & mostly Harrowed it $2\frac{1}{2}$ double Tine and two Bolls on Dickies Haugh partly Harrowed Barn – Men & women after Brakefast dighting Oats –

Tuesdy 21st Dark Hazy dry weather – wind Easty but very light – Early Begun sowing remainder of Haugh Oats – Sowed 16 Bolls & partly Harrowed same – Cottars & others dighting seed Oats & to loft till dinner time – Sundries.

Wedy 22d Dark dry weather and Blowing a S.W.- all Harrows at Haugh whole day – lofted some Oats (12 Bolls) – took in a stack of D° S B – leading of Turnips and at sundry Jobbs-

Thursdy 23d Blowing weather from S.W.-all the Harrows A.M.at Haugh- P.M. three ploughs on the Heading D° & three Harrows – P.M. Ploughed also the Gallownere for Lint Intended.

Fridy 24th Th^r^ 50° Slight frost morning & wind Moderate Westerly-finished all the Oats seed & Harrowing Haugh $23\frac{1}{2}$ B A.M.– P.M. Sowed the Old Town park about 7 Acres with 8 Bolls Oats which Finished the whole 47 Acres with $50\frac{1}{2}$B Oats about 17 pecks per Acre overhead.

Saturdy 25th last night Came on rain which Continued all the night from N.E. and till 2 p.M. leading of Turnips and Ploughs at Golds Hillock Steering for Bear. Others Dighting & Sacking Rye Grass for Aberdeen.

Mondy 27th Yesterday fair weather frost morning – this day same & some drought P.M.- three Ploughs Steering after Turnips Golds Hillock – One Cart to Aberdeen with 48 Bushels Rye Grass &c.

Tuesdy 28th Frost last night & this morning – day dark Cold weather from NNE – A.M. Harrows at Old Town park for Grass Seeds and P.M. Broke in the Fallow Backyards for Bear – the Cart from Aberdeen returned before 8 P.M.- with every articles wanted – roads were very bad &c.

Wedy 29th Fine pleasant weather a E^t^ with Moderate drought - Driving Turnips – Out some Dung & P.M. Harrowing the Backyards fallow foul – preparing Grass Seeds for Old Town park – Sowed Onions, Carrotts &c Garden.

Thursdy 30th Fine dry weather variable from NNW to NE- Sowed part of the Old Park with Grass seeds & Harrowed it but stopped by the wind – all Harrows Braking Land for Bear & for Turnips- people Gathering Weeds &c Backyards to Burn &c.

Fridy 31st Th^r^ Noon 53° Fair weather variable till 2 p.M. Came on a Heavy rain which did not last above an Hour – Harrows Braking in Turnips land Robs Hillock A.M.- P.M. ploughs steering of Gibsons Park for Bear – Quantity of rain during this month is One Inch & two Tenths – The Oats & Pease seed mostly over, but yet a good deal to do in some places – no kind of vegetation of Grass and the Turnips nearly Flowring – no kind of sale of Grain and of Oats, little for Fatt or Lean Cattle – no sale of Hay – the late Derangement of the London & other Banks seems to have stagnated much all kinds of Business & occasioned many Bankruptcies &c –

Diary for the Month of April 1797

Saturdy 1st Fine drought all day a S.W., but Cold – the Ploughs as yesterday & leading of Turnips – Dighting & Lofting of 16½ Bolls Pease &c Jobbs –

Mondy 3d Yesterday Cold & Bleak weather from the Eastw^d^- this morning very Hard & White Frost, fair but a E^t^- Ploughs at North park – Sowed out Old Town park with Grass & Seeds & Harrowed it P.M. – about 3 P.M. Came on rain & stormy weather a N.E.

Tuesdy 4th much rain during the night & stormy at E.N.E.- whole day no work without doors – people at Dighting & lofting Oats from S B – the river somewhat swelled & fields wett – not very cold.

Wedy 5th Cold Bleak weather from N.E. with light showers of Hail & Snow on the Hills – all Ploughs at 2^d^ furr Robs Hillock for Turnips already Clean & fine – and leading Turnips – In a Stack Oats S Barn Miday – all kinds of vegetation Checkqued & Grounds raw –

Thursdy 6th Cold dry weather from N.E. with skiffs of Hail – 4 ploughs at Robs Hillock all day – at Sundries Dung Court – Gardner dressing Borders &c –

Fridy 7th Cold Stormy weather from N.E. with much rain the whole and no Horse work – people at Turning Dung in the Dung Court &c – the rivers Considerably out and fields very wett – it will Injure the Crop Sown –

Saturdy 8th Constant rain almost till 1 P.M. faired, wind N.E. lead Home 6 Cartful of Turnips, & turning Dunghill in the Court rest of the day – fields very wett –

Mondy 10th Th^r^ 42° Yesterday Cold Bleak weather – drissling rain a N.E.- this day same, all Ploughs at Robs Hillock & leading of Turnips Some of them Flowred – very little growth in the Ground & fields Excessive wett –

Tuesdy 11th Cold Bleak weather but dry till three P.M. when Came on much Heavy rain a N.E. which Stopped the Ploughs – at Sundries – N N Barn &c –

Wedy 12th Closs foggy weather and Sometimes Drizling rain various people at Barns dighting & sent 16B Oats to the Mill of Kair, and lead some Turnips – lands very wett.

Thursdy 13th Th^r 43° Pleasant Sunny day – W^{d} East^{y} – 4 ploughs at the Robs Hillock all day – Midday took in a stack of Waterlair Bear for seed, Malt &c – Sundries –

Fridy 14th White Frost in the morning till 7 A.M.- day very pleasant drought Easterly – all Ploughs at Robs Hillock & West side Kavels, Fouled by Wheat 4 years ago & will need a steering Fur & Dunging to be Equal to what of same park was Turnips & Barley.

Saturdy 15th Variable weather a Considerable shower of rain from Noon till One P.M. after several more of same – all Ploughs at steering as yesterday – Observed some people Harrowing their Oats nearly Breered – for fear of Caking -at Sundries –

Sundy 16th Cold Bleak weather from N.E. – afternoon some showers of Hail very Heavy P.M. – Country White on ye Hills &c and Even on the plains –

Mondy 17th Cold Bleak weather but mostly dry a N.W.- the Ploughs as before – P.M.Sent two Carts to the wood of Pittarrow for Firr Hagg, weedings. The Oat seed not near finished in this Corner as yet – and the Earliest Sown Oats & pease but few of them above ground, little growth of Grass, or vegetation of any kind – a total stagnation of late of Grassland also of Cattle of all kinds, Owing to a stagnation of the publick & private Credit, an Artificial & false Alarm.

Tuesdy 18th Dry weather from N.W.but Hail shower at a Distance - Ploughs at Kavels – Carted off all weeds from the Backyards not dry enough to Burn – lead Home some Cartful of Turnips. Dighting Bear for Malt – 1st Ewe Lamb^{d}.

Wedy 19th White Frost night & morning – day drought at N.W. Ploughs & Harrows at Golds Hillock for Bear – lead Home the whole remainder of Turnips &c.

Thursdy 20th Frosty in the morning day strong drought a S.E. – Ploughs & Harrows at steering Backyards for Bear very fine – Grounds now in fine order –

Fridy 21st Th^{r} 44° Cold Bleak but dry weather from the East – all the Ploughs at Backyards steering – at Mill of Kair with 15B-3f Oats Meal – Observe some people begun to sow Chester Bear – the March violet & other field flowers not appearing.

Saturdy 22d very Cold stormy weather from E.N.E.but fair till near 3 P.M. a shower of sleet during the rest of day all ploughs &c Backyards – Brought 17 Bolls Meal, Seeds, Dust &c on One Cart – Sundries Th^{r} 43°

Mondy 24th Yesterday a Considerable Growth some showers P.M.- this day same, and D° – ploughs finished all the steering Backyards of season for Bear, and of preparing further the Ground Gallownere for Lint.

Tuesdy 25th Fine moderate weather from S.W.and a Continued vegetation – Begun Ploughing Seed fur West side of Golds Hillock for Bear – and giving 3^{d} fur to Potatoes grounds Robs Hillock – P.M.Sowed $4\frac{1}{2}$ pecks Dutch Lintseed in very fair Condition for it – Gardner at Sundries.

Wedy 26th Slight frost night day pleasant & Moderate weather from S.W. – 4 ploughs all day at seed furr and at dighting Bear for seed – Sundries –

Thursdy 27th Moderate drought S.W., all ploughs A.M. – P.M.Sowed 4B- 6p

Bear West side Golds Hillock and partly Harrowed the same – Dighting Seed Bear – at Laurencekirk fair a Considerable number of Cattle fatt & Lean – few Buyers - fatt about 17/6 per Stone and Lean about 6/ to 6/6 – no Sale for Grain &c – rain at night.

Fridy 28th More or less rain all this day but afternoon Constant stormy from S.E. no work without P.M. – dighting & lofting Pease, Malt Barn Cutting Potatoes for seed, and Sundries – Horses on Hay.

Saturdy 29th Moderate weather from S.E.- but threatened rain – all Harrows whole day at Braking in Kavels from Rib for Bear with the top being softer after the rain – at Barns dighting seed Bear – and in Hay N N Barn for Horses, & other Cattle.

Sundy 30th Ther^{m} Noon 49° Fair weather from N.W. varied to S.E. till 5 p.M. Came on Constant & pretty Heavy rain all the Evening. The Quantity of rain is three Inches – The Breer of Oats look well & thick but it is Excessive Bad for the Barley seed – the Grass very Scanty & little saving of Fodder – no demand however for Hay, or for any farm produce whatever & the prices Constantly falling – Pork Hath fallen nearly One Half & it is owing to Many Imaginary Conjectures about Political Matters, not properly understood, & deranged Mercantile Speculations of kinds – no growth of field pease – Hawthorn not foliated – Imperial Crown just flowred.

Diary for the Month of May 1797

Mondy 1st Variable weather from diff^{t} points – ploughs & Horse Carts driving dung to the Potatoes Ground – towards Stonehaven some Begun Bear seed, but little done –

Tuesdy 2d More or less rain all day, but about 5 p.M.became Heavy & stopped the ploughs E^{t} side Kavels before six – and the gardner planting Potatoes by the Dibble, considerable rain all the Evening –

Wedy 3d various weather from the Westw^{d} but with Flying Showers - 4 ploughs all day at Kavels – good season but too wett to sow or Harrow, and the first sown Bear is advancing too fast before the sowing of the Grass seeds but Cannot Help it – finished planting about $\frac{1}{2}$ Boll Potatoes by Dibble –

Thursdy 4th Fine dry weather till 1 P.M. Came Heavy & Closs rain various – Sowed out Golds Hillock with three Bolls more Bear 7B 6f total & p.M. sowed the whole 9 Acres 4B 1f rye Grass & 60 lb Am^{d} red Clover – gave a rake to the first sown Bear before sowing the Grass.

Fridy 5th Cold dry weather from N.W. and Grounds fit to Harrow in the Grass seeds in the afternoon tolerable order – Ploughs at Kavels Seed fur for Bear – in Hay.

Saturdy 6th Fair weather but Cold & Stormy from the N by E till 2 p.M. some Excessive Heavy Hail Showers which Whitened all the Country & lay on Back places – three ploughs at Kavels morning – P.M. all Harrows Broke in Backyards pretty dry & fine – at sundries

Mondy 8th Yesterday morning a very Hard Frost, Ice on dubs the day Cold &

stormy at various – this morning Frost day fair & strong drought a S.W.- ploughs at Kavels and the Long Burns – and three Harrows Backyards –

Tuesdy 9th Frost in the Morning, day Cold & Stormy a N.E. but dry Ploughs at long Burns for Bear & at Kavels Braking in for Bear a little rough – at Barns dighting & sacking Bear for seed –

Wedy 10th very Cold & Stormy dry weather from the NNE – all Harrows at Kavels & part Backyards sown with 6 to 7 Bolls – and partly Harrowed the same, at sundries, Vegetation quite Checked –

Thursdy 11th Th^r^ 43° Uncommonly severe Cold & Stormy weather a NNE – Ploughs Gibsons park A.M. – P.M. begun driving of dung to West side of Kavels for Bear, an appearance of rain; a shower when late but little.

Fridy 12th The weather more Moderate a S.E.- Carts driving of Dung all day – some spare Harrows P.M.- In Hay middle of day - Others spreading Dung &c Pasture worse than Ten days ago – Sundries –

Saturdy 13th a Considerable shower morning from N.E.- 3 Carts Dunged out West side Kavels part from Cottar town and P.M. 4 ploughs at ploughing down same – fair P.M. and drought a S.W. – at sundries –

Mondy 15th Yesterday pleasant dry weather from the Southward – this day same, and for the first time the appearance of Summer weather – Sowed the Long Burns with Grass Seeds & Harrowed it – Ploughs & Harrows at West side Kavels mostly sown, but not Harrowed – at Barns &c Jobbs –

Tuesdy 16th Pleasant dry weather from the East – Sowed out all the Kavels total 10 Bolls Bear & mostly Harrowed it out, very ill to do though tender, by the open Winter Grew in the Rib – Water furring the long Burns, and Rolling Golds Hillock very fine & partly Breered – In Hay Midday – The prices of Grain still falling, also Cattle of all kinds, Evening appearance of rain from ENE – Grass Scarce a full bite, and the young Cattle eat straw at night – Sowed some Early Pease Garden –

Wedy 17th last night about 11 Came on Heavy rain a E^t^ which Continued till 7 A.M.- day fair & Growing weather – all Carts dunging the Backyards & others spreading D° – Sundries &c.

Thursdy 18th fine dry weather & strong drought from S.W.- drove out all the Dung to Backy^ds^ & spread it all – P.M. yoked three ploughs on it, and three Harrows at West side of Kavels rough – at Bervie Markett Great many Cattle of kinds and Considerable Sale by no Means Cheap but lower than before.

Fridy 19th Fine Pleasant drought at S.W. & Growing weather – 4 ploughs at Backyards fine Tilth – dighting of seed Bear, and of Rye Grass – Sundries –

Saturdy 20th from One morning till 4 Heavy rain a S.W. and some showers at Intervals in the day – ploughed out all Backyards A.M. and P.M sowed out whole of it & mostly Harrowed it out –

Mondy 22d Yesterday some smart showers of rain from Westward – this day Drought & Growing weather from S.E.- begun Dunging of Gibsons park from the Town for Bear, and at Sundries at Lkirk road Making.

Tuesday 23d Early this morning strong White frost, and the Appearance of rain from S.E. all day – three Carts whole day at Dunging of Gibsons Park & Others spreading of D^{o} – Sundries –

Wedy 24th Fine dry weather from S.W. & Moderately mild – Dunged out all for Bear seed of season and begun ploughing Gibsons park, at Barns and Sundries – Grass rising apace.

Thursdy 25th very pleasant & Summerlike weather from S.E.- 4 Ploughs at G Park for Bear & others spreading of Dung on D^{o} all day nearly finished – at the Barns miday – the earliest Bear looking very well.

Fridy 26th very fine Grow weather from South W with some light showers – finished all Barley seed 36 Acres seed B- f- in very fine Condition & should be a good Crop if the Season is but tolerably favourable, & begun to Brake E^{t} side Robs Hillock for Turnips P.M.- Sundries – Sowed Turnips Garden for Harvest use &c Jobs.

Saturdy 27th Blowing stormy kind of drought a West with very slight showers – Braking in the Robs Hillock for Turnips, and Two Rollers going over the Golds Hillock & long Burns Bear to keep in the Sap &c Jobs –

Mondy 29th Yesterday showers of rain Heavy towards Evening this day very pleasant till 1 P.M. Came on rain & stormy weather from N.E. – Harrows at Robs Hillock in the morning Sowed Gibsons park with Grass seeds & Instantly Harrowed it – P.M. Water furred same.

Tuesdy 30th very Cold stormy weather from N.E. but dry – Carts at Flitting a Cottar a wright from Redmyre, and others Harrowing & rolling P.M. – at Sundries –

Wedy 31st Frosty early morning, day showery & stormy P.M.- all Ploughs giving 3^{d} furr East side Robs Hillock for Turnips across – In Hay Horses Half Corn – The Breer of Kinds looking very well,except the Bear on Stiff Land - Hay & Pasture good – the Caterpillar ravaging the Gooseberries much, much earlier than we observed before - The Quantity of rain One Inch & Nine Tenths and the Grounds very dry, where usual wett – The Markett for Grain dull & little sale Bear for small Shells 14 to 14/6 & Oat Meal 13/4 to 14/ Hear of no sale of Farms as yet – no Sale of Pease or Hay; the Bear seed not generally finished.

Diary for the Month of June 1797

Thursdy 1st very strong drought a S.W. windy – 4 ploughs at Robs Hillock as before – Chaffing of Bear S Barn – Sowed the last Pease of Season Garden Earliest kind.

Fridy 2d Weather as yesterday – & farm force likewise Employed as before, at Barns Miday – the Growth Checked by drought & frosty nights &c –

Saturdy 3d Strong frost in the morning, day very strong drought from various – Ploughs did out E^{t} side Robs Hillock 3 furr & Harrows P.M. &c Sent 40 Stones Hay to Dr Fordyce at Lkirk – at sundries –

Mondy 5th Planted Collyflowers Yesterday very strong drought & Cold weather from North this day more Moderate wind shifted to the S.E. and severe drought – driving Ashes of fuel for the Turnip dunghill E[t]side R Hillock – and P.M. Put up the Cattle Flakes Waterlair & Young Cattle into them –

Tuesdy 6th about 5 A.M. Blowing weather & a shower a S.W.- P.M. same – Begun Dunging E[t] Half of R Hillock thick for Turnips, miday at Barns dighting Bear &c –

Wedy 7th Th[r] 60° uncommon Strong drought a S.W. all day – three Carts at driving out Dung whole day – and dighting & sacking Bear S Barn – at Cummock Fair hear there were many Bestials, but no great Sale – sundries –

Thursdy 8th Moderate weather a E[t] some skiffs rain, but apparently Heavier on the Hills to the Westward – all Carts at Dung driving as before, In stack Oats S Barn – The Caterpillar & other Insects destroying the Gooseberries and the Fruit Trees – Cutting Thistles amongst Lint –

Fridy 9th very Strong drought a NNE & SunShine – the Carts at Dunging as yesterday, and owing to its smallness exceeding tardy in putting out – at sundries – Observe all vegetation much Stinted Indeed –

Saturdy 10th Drought & weather as yesterday – Still driving out Dung to E[t] side R Hillock Rather thick laid on – in Hay Miday for Horse Suppers – Cows get Cut grass on the planting behind the House – Sundries –

Mondy 12th Yesterday moderate weather & strong drought at Eastward about midnight Came on rain & stormy Cold weather from the N.N.E. which lasted till near noon, Carts driving Dung to Robs Hillock all day E[t] side not quite finished It is Heavily Dunged above 40 Cart load per Acre - Gardner sweeping off Caterpillars Which Have destroyed a very Considerable part of the Gooseberries –

Tuesdy 13th Some very slight showers first part of the day – Driving out Dung & in Earth to Dung Court P.M.- begun Drilling of TURNIPS, fine mould – others spreading of Dung &c Jobbs

Wedy 14th Thick Hazy weather from N.E. & Growing – begun Sowing of Turnips & rolled some P.M. – Others driving in Earth Dung Court- and at sundries, Jobbs – Observed some Sheep from Trinity Muir, very dear – &c.

Thursdy 15th Cold Bleak dry weather a N.E.- showers on the Hills - Drilling & Sowing & rolling Turnips and others spreading of Dung, much distressed for want of Hands both Men & Women to carry on sufficient work – Heard Trinity Muir went well.

Fridy 16th & 17th Saturdy Bleak dry weather from S.E. and towards Evening Gathering thick like rain, Grounds growing very dry, and running wells drying up, the Pasture & Hay very much Checked, as is all kinds of Grain Crop & much of the latest Bear not Breered – The Turnip land on Stiff ground not Redemable at all- Continued Sowing of them in very fine Tilth – P.M. flited the Cattle flakes at Wester Waterlair – Houghing Potatoes &c Jobbs - Caterpillars on Gooseberries Checked.

Sundy 18th Grey weather & drought a S.W.- about 6 P.M. a shower of rain but of no Continuance – Our Fruit Trees Garden & Gooseberries quite destroyed by Caterpillars & other Insects, Observe the Hay in the Neighbourhood a very light Crop Indeed –

Mondy 19th Th^{r} 52° Early this morning Came Closs & pretty Heavy rain a N.E. which lasted till about Noon & went pretty deep - People Gathering Dung at the Court A.M. – after all Ploughs at Robs Hillock.

Tuesdy 20th last night & this morning excessive strong White & Black frost – wind at N.E.- fine sunshine – the Potatoe Stems quite destroyed – Ash leaves – Ploughing & Harrowing West side Robs Hillock –

Wedy 21st Strong frost this morning – Observe on my way to Aberdeen yesterday, every part of the Crop looking worse than here but the Potatoes not touched within three Miles of the Sea, People at turning Dung Court –

Thursdy 22d Strong drought a S.E. and Coldish weather – Observed the Crop North side Don very Indifferent – Visited the Camp of about 1500 Men beeast Don Mouth, Cold Quarters for Men & Horses, some latter dead – people at sundries

Fridy 23d Coldish dry weather from N.E.- rode on the New Turnpike road from Muchalls to Stonehaven & two Miles on this side of it-fine road if finished – people Turnip Ground

Saturdy 24th Cold stormy weather from North, Carts driving Dung all day, & women weeding out lint Haugh tolerable, the Hay & pasture bettered this week for all the drought – Prices of Cattle rising –

Mondy 26th Moderate Growing weather from S.W. with some light showers, people driving dung all day, and Others Spreading D^{o}- P.M.- one plough P.M. Drilling West division of Turnip field – the first sown Have been Breered for several days past –

Tuesdy 27th Frosty morning day some light showers & Growing weather various – all Carts driving dung whole day and one plough drilling & sowing – P.M. rolled what was sown in fine order – Women spreading Dung, In Hay as yet for Horse Suppers – and at Sundries – Pease Garden Podded.

Wedy 28th Th^{r} 57° Fine Growing weather and Showers at Intervals through the day – one Plough Drilling & Sowing Turnips, the first sown finely breered – Brought over the Cattle Flakes from Waterlair & set them up on N.W. Corner of Black lands-removed the young Cattle here the Pasture exceeding rich on Bridge Shade &c Sundries.

Thursdy 29th Th^{r} 52° about 8 A.M. Came on pretty heavy showers of rain from the East which lasted till 1 P.M. and wet the grounds more than hitherto – driving in Earth to Dung Court to Manufacture with Pigeon, Hens, Swines & other Stronger Dungs for the last of the Turnips West side Robs Hillock – The river out both yesterday & this afternoon but suddenly Subsided – Heared that the Markett at Megrie Ury was very dear & rather dull but great Number Came West from it &c –

Fridy 30th Frosty every morning, day fair & Growing from S.E. – People

Collecting Dung for the last of the Turnips, Pigeons &c – Getting in some In Holding fuel – Sowed & rolled all the Turnips Drilled, about 2 Acres yet to do Observe many Have Scarcely begun Sowing as yet – the Ground is now best for them in order – The Quantity of rain this month One Inch & two Tenths – no kind of demand for any kind of Grain at present – nor do I hear of any of the Farms being Sold as yet, the Earliest Bear here offering to Set to shooting – the Hay Crop in general very thin & light & many places not worth Cutting – the Oats & Pease tolerable, but the Bear in stiff Cold land Scarcely Breered as yet & very thin Pasture short & scanty, yet an Immense Demand for Cattle to the Southward.

Diary for the Month of July 1797

Saturdy 1st Thick Hazy weather and several showers during the day from S.E.- people Collecting & moving of Dung for the last of the Turnips W side Robs Hillock, a Cart to Montrose for Coals, and sundries &c &c.

Mondy 3d Yesterday morning a Heavy shower – this day a very Considerable fall of rain from 9 A.M. to 4 P.M. and towards 6 Arose a strong Gale & stormy a N.W. – Carts driving dung morning – Rest of the day Gathering & Spreading of D° – begun Cutting Grass for Horses.

Tuesdy 4th very Stormy with rain all Night, day fair & more Moderate a N.W.- Carts dunged out all the Turnip for season, Plough drilling at D° – Others Spreading of Dung, Observed the Crop towards Jacks Bank tolerably good, but too much rain –

Wedy 5th Early this morning Came on rain from the Eastw[d] which Continued most of the day – Drilled out the whole of the Turnips from 13 to 14 Acres & sowed same in fine order – flited the Cattle Flakes P.M. – &c.

Thursdy 6th Bleak Stormy weather, and some skiffs of rain a North – being Fast day no work on the farm, in a Ride over Thornylee, Arthur House, & Sooty Wells, Observed the Crop good on the two latter –

Fridy 7th Dark Gloomy weather but growing wind a S.E.- sent Ten Bolls Wlair Bear to Jacks Bank Distillery – and a Cart for stones to Church yard dykes Fordun from Laureston, rolled out Turnips W side – Sundries –

Saturdy 8th Th[r] 62° very pleasant warm Summer weather for first time a S.E. Sent 15B Oats to Mill of Kair, and some stones to Fordun, and Carting in Earth to Dung Court, Women Thistling Oats N park & sundries – Evening very thick Fogg – Great vegetation –

Mondy 10th Shore the Sheep Yesterday warm weather and drought from Southward – this day same – people at Cottar Yards for Curleys and Thistling at Floors Oats very good – also at the Church - a very great Growth in the ground – Cattle Bridge Shade.

Tuesdy 11th Fine warm growing weather from SSE – people at Cutting Thistles Floors Oats, Houghing Potatoes, and Planting Curleys Cottar Yard spare, put Cattle on the Bridge Shade very rich pasture indeed yesterday – the Cattle startling this two or three days past, Bear beginning to shoot, driving in some Earth to Dung Court – some little rain & appearance of a shower evening – no fire for some days.

Wedy 12th last night, and at Intervals a Considerable deal of rain in showers a S.W.- Sent a Cart to Johnshaven for Small Coals, Others Thistling, and P.M. flited the Cattle Flakes, some of the lint & Bear laid – fine weather –

Thursdy 13th fine growing weather from SSW & fair, Sent three Carts for Fir wood to Garrol Wood for farm use, and one D^{o} with Oats to Mill of Kair, Thistling Haugh and Cut up the Turnip seed Garden for seed, Midling Crop. The One Half of the Potatoes are destroyed by Crows &c

Fridy 14th Th^{r} 70° Noon remarkably Pleasant & warm weather a South – Whole day three Carts to Johnshaven for Coals – and one D^{o} to Mill with 12 Bolls at a draught-180 Stones – the Cattle Could not get Settled to Pasture for Heat & flies – after 8 P.M.- everything growing rapidly & women Thistling Haugh Oats – Sundries –

Saturdy 15th very pleasant Summer weather a South – 4 Carts to Johnshaven for Coals, returned before 2 P.M. being 49 Barrells P.M. Housed them & Cutting Grass for Horses Suppers – Cattle startling all day – not near so warm as yesterday – Gardner Houghing Potatoes devoured by Crows – &c Jobs.

Sundy 16th Fine pleasant dry weather from Southw^{d} – rode through the Fields in Neighbourhood, some of them a good Crop & others very Midling – little Sale of Cattle and no demand for any kind of grain – Oat Meal & Bear about 13/4 per Boll – the first sown Bear shooting, also Oats, & pease & lint Blooming – the Pasture much Mended, but not the Hay. All the Wheat Neighbourhood Extreamly Luxuriant more so than that begun Shooting only – no Garden Pease or Strawberries yet.

Mondy 17th early this morning Foggy and about 8 A.M. Came on Smart Showers which lasted till 1 P.M.-then fair. Begun Cross Ploughing Brae Water lair with a very strong Plough & Six Horses two drivers, Others Sundries –

Tuesdy 18th Th^{r} 60° Fine Growing weather from S.W.- about $\frac{1}{2}$ past 1 P.M. Heavy rain for a short time, about 5 P.M. another heavy shower – the Plough at Waterlair, and a Cart leading Earth for the Dung Court, learn the Sheep Markett Paldy, prices higher than ever –

Wedy 19th Early this morning & at Intervals all day heavy showers from S.W.- the six Horse plough Waterlair – P.M. Flited the Cattle Flakes, and at Sundries –

Thursdy 20th Frequent showers at Intervals, at Waterlair as before – at Paldy Fair many Cattle and prices high, the Conse quence not near so many sold as last year in it Good Cattle & stotts about 7/ per stone, but numbers unsold.

Fridy 21st Fine pleasant Growing weather from S.W. – Plough Waterlair and a Cart carrying of stones off D^{o} – Sowed the seeds of different kinds of Spring Plants &c Jobbs.

Saturdy 22d very pleasant & warm weather from Southward – about midnight sent 4 Carts to Mathers for 16 Bolls Lime shell for Brae Waterlair – and Cutting Grass afternoon for Horses – Gardner Earthing up Potatoes –

Mondy 24th Th^r 64° Yesterday pleasant growing weather S.W.- afternoon a good deal of rain & thick weather – this day drought, Some of the Bear Swayed – 4 Carts to Mathers for 16B Lime for Waterlair – and at Sundries.

Tuesdy 25th very warm weather & strong drought from S.W. – Plough at Waterlair 2^d furr finished – begun taking away first furr from first sown Turnips and people begun Thinning them, very tender and Easily done if the Hands were good which they are not – We Have Had Pease & Strawberries for several days.

Wedy 26th remarkable pleasant summer weather & very warm – all the People amongst the Turnips – P.M. Flited the Cattle Flakes Black lands, and at sundries –

Thursdy 27th Early this morning a thick Fogg a E^t- about 9 A.M. a Heavy shower of short duration, but stormy rainy till late afternoon, Hand & Horse Houghing Turnips till dinner time – after at sundries repairing roofs of stacks &c Jobbs.

Fridy 28th In the morning till 9 A.M. a good deal of rain ENE day fair – people Horse and Hand Houghing Turnips all the day – Bear much lodged in many places &c.

Saturdy 29th Thick Hazy growing weather a E.N.E. till 9 A.M. when Closs & pretty heavy rain the rest of the day – 4 Carts to Mathers for 16B lime Waterlair, at Turnips till Brakefast – after little or nothing done without doors.

Sundy 30th Early this morning Came on very Closs rain & sometimes heavy through the whole day a N.E. & stormy – the rivers a good deal raised, and the best Bear & Lint quite layed – also the Wheat, but few of the Oats – roads very Bad &c.

Mondy 31st Much rain all this day at Intervals – rivers, fields and roads full of water – driving in Earth to the Dung Court -and P.M. Flited the Cattle Flakes – 4 people hired begun Cutting of Hay West side of Thornylea, very moderate Crop – The Quantity of rain in the Gage is four Inches & one Half-same as July 1796. The Cattle Marketts said to be declining, and no sale for any kind of Grain or farm produce and Hear of no Farm rent bestial yet sold which is very uncommon so late in the Year Meal Oat & Bear about 13/4 what is sold – Butcher Meat very high from 5^{ds} to 6^{ds} per lb – prices of spinning high 16 to 18^{ds} per spindle – the lint layed everywhere, some Bear Oats & Wheat. Some of the Earliest Bear filling & turning – field pease Podded &c.

Diary for the Month of August 1797

Tuesdy 1st very pleasant dry weather a SSW – two Carts at laying on Ditch Scourings on the N.W. part of Blacklands after Flaking that is very poor Soil & spreading it – people at the Mill of Kair with 30 Bolls Oats – Meal 30B – 3f 1 peck. Heared the Sheep Markett Garvock excessive dear

Wedy 2d Cloudy troubled Air, but dry till near noon, when showers all round, Increased soon after, and stopped the most of the work without doors Thunder in the Northw^d from One to 2 p.M.- by Six the Water of Bervie raised & Increased to 9 P.M., I suppose with Dammage.

Thursdy 3d Almost Constant rain & warm weather whole day and the Country perfectly deluged, at Garvock Fair very dull and little sale at reduced prices – Servants wages also on the fall of which very much need – Engaged none – Sold two Stotts –

Fridy 4th Several light showers variable, hot & Closs – Carts driving Earth B yards – flited the Cattle Flakes, Cutting up of Rushes for Covering two old Bear Stacks, Turnips &c Jobs – One Inch, 6 tenths of rain fallen three days; Immense

Saturdy 5th Frosty Rhime morning, day showers at Intervals – Carts driving Earth on B lands, new Thatched 2 Barley Stacks & repaired one of Oats-thinned some Turnips – Sundries.

Sundy 6th Almost Constant rain after Noon particularly toward night a N.E. The Mercury Barometer 29 Inches & rising.

Mondy 7th Th^{r} 61° Constant rain till 9 A.M. when faired & some drought from S.E.- driving Earth & P.M. thinning Turnips – Sold all the Bear threshed at 13/ to Mr Gillies & pease 11/ the Bear is not good – & some old smelld &c –

Tuesdy 8th About 5 A.M.Came on a Heavy storm of rain & wind from the S.E. which lasted till dinner time & Brought down the rivers as large as for some years by 5 P.M. – Flited the Cattle Flakes Blacklands, at sacking Pease, reding up Barns & various Jobbs –

Wedy 9th Blowing stormy weather with rain from 5 A.M. till 9 – After some drought a SSW – driving of Earth as before, Others at thinning of Turnips – 2^{d} division of Sowing, good, & at Sundries –

Thursdy 10th Blowing weather from S.W. with some light skiffs rain - People at dressing of Turnips all day – Observe some Neighbours spreading out their Hay Coles near two thirds Spoilt & useless of them – mine in Bout little Injured –

Fridy 11th very strong White Frost in the Morning – day fair but several showers from one to three P.M. – dressing Turnips, and P.M. sacked out the Pease for Montrose 21 Bolls – at Sundries – the Hay Cut first mostly lost.

Saturdy 12th Fine dry weather all day – three Carts to Montrose with 21B-and Brought out Coals in return – Others dressing Turnips – Observed a good deal of Hay Coled in the Neighbourhood this day –

Mondy 14th Yesterday fair weather, showers on Braes, this day same but Grass wett till 11 A.M.- at Turnips till Brakefast- after begun to Gather into win rows West side of Thornylee Hay (into Tramped Coles) – in good order & very little damaged at all – Owing to precariousness of the weather no people but our own.

Tuesdy 15th Some light showers Early morning, day D^{o} but did not prevent Coling Hay Thornylee after Brakefast in good order large Tramp Coles – at Turnips before Brakefast – at sundries &c &c.

Wedy 16th Th^{r} 63° Fair weather from S.E.& Moderate drought – Early begun Gathering Hay Thornylee & Coleing before noon in good order – 8 hands after Brakefast, Cannot get People to work it as fast as it ought to be &c Jobs Whether the lint be filling or not it is not rotting.

Thursdy 17th very pleasant dry weather all day from S.W.- Own & some hired too at Gathering Hay Thornylee all day- 5 Cutters begun Cutting Hay red Hall Park- a pretty Heavy Crop & in fine Condition – Observe many Coling this day all around – Hear there is a demand for Bear up to 15/ & Learn that at Lammass Muir the prices of Cattle rather fallen and duller of late – the Beastials much in demand two year old stotts – the price of Bear rather on the Increase.

Fridy 18th Foggy weather Early morning, day fair &c till about betwixt three and four Came on Closs & pretty heavy rain from North N.E. & stormy wind which stopped both Hay Cutting & Colers before 5 P.M. –

Saturdy 19th about 9 A.M. Came on heavy rain from the N.E. which lasted till 2 p.M. then lightly afterwards – People at Sundries & turnips at D° – the Crop Compleatly layed almost good or Bad of it particularly betwixt this & Glenbervie &c –

Mondy 21st Yesterday chiefly fair, but all along the Braes showers of rain, this day same but rain here Evening. One plough taking away fur last sown Turnips West side of Robs Hillock, Others thinning D° – and four Horses Harrowing Waterlair Brae with the Drag. Brashed Cow Calved.

Tuesdy 22d fair weather from N.W. and a little drought, at Horse & Hand weeding Turnips – Harrowing Wlair and finished what was left wett Thlee Hay – at Sheep Markett L fair it is said excessive dear.

Wedy 23d fine dry weather but Coldish from N.W.- Fannerd & Sacked 20B Bear & sent it to Gurdon Shipped there – Others at Turnips thinning all day – 5 Men Cutting $5\frac{1}{2}$ Acre park Waterlair for seed, Evening sacked the remainder Bear loft for Gurdon $19\frac{1}{2}$ B –

Thursdy 24th Fair weather but little drought a South – Early sent 4 Carts with the Bear to Gurdon & brought back 12 Barrell English Coals – a year old stott got its near hind Knee Joint Dislocated-reduced it believes – was Informed Lfair Cattle sold very high & much Sale 3 year olds at 21 Guineas – my Fathers at Johnston sold for———

Fridy 25th a Considerable deal of rain from S.E.in the morning day some heavy showers about 2 p.M.- 2 Carts to Johnshaven for Coals to the Smith and others amongst the Turnips all the day, at Sundries &c Jobs –

Saturdy 26th Th^r 58° Fair weather this day, and Some drought a S.W.- all people amongst the Turnips in great need of dressing – P.M. Flitted the Cattle Flakes, the best of the Bear and Oats are still lying & will do so – at sundries –

Mondy 28th Yesterday fair weather a S.W.- this day the same – all People at the Turnips whole day, the ground now pretty dry, and at sundries –

Tuesdy 29th Thick Hazy weather from S.E.-till 3 P.M. Came on Heavy rain which stopped work without doors – before at Turnips Horse & Hand Houghing – at Sundries. Some begun to Inn Hay which will be Stopped Short.

Wedy 30th Fair & dry weather till about noon Came on heavy rain which dispersed workers without doors, till 3 P.M. – after Carting on Earth on Smithy park ley – at Barns Chaffing Oats – the river Came down very large at 4 p.M. but fell in again towards night –

Thursdy 31st Th^r 53° Fair weather from the S.W.-all day – people at the Turnips, and after Brakefast Turned over the Red Hall park Hay-then again at Turnips – Observe a great deal of the Oats & Bear about Fordun quite lodged & spoiling & pease Fliting Fast & not filling of late – Quantity of rain in the Gage is five Inches & One Half Inch – enough to drown the Crop & there is a Second growth Beginning-Prices of Cattle said to be on the Increase, as is all kinds of Grain – Meal & Bear about 14/ to 15/ - Hay $3^{ds}\frac{1}{2}$ to 4^{ds}per stone but most of it quite Dammaged – Beef & Mutton from $5^{ds}$to 6^{ds}per lb and little of it good & Scarce. Manufactures Brisk & spinning high at 16 to 18^{ds} per spindle-Eight Ounce Yarn – Pasture now decreasing – the lint not judged half a Crop &c. Gooseberries finished nearly.

Diary for the Month of September 1797

Septr 1st Fridy Frosty Rhime in the night, about 8 A.M. Came on rain which Increased whole day, people at the Turnips – & flited the Cattle flakes after dinner –

Saturdy 2d the night very stormy from N.W. with rain and the Rivers much out this morning, day fair & drought N.W. all people at Turnips, finished first Houghing, Hirers in a very Slovenly Manner most others worse –

Mondy 4th Yesterday mostly fair – this day D° & Drought a S.W.- after Brakefast began Harvest on Gold's Hillock, full ripe & Good Crop, at sundries –

Tuesdy 5th Strong Frosty dew, day appearance of rain & stormy weather afternoon so as to prevent the Hay Coling red Hall park own & Hired being 15 people – Tramp Coled about 5 Acres into 6 Coles, in good Condition & sound – Stormy weather and appearance of rain & very Cold Increased towards night –

Wedy 6th the rain Came on Heavy & Strong Gale a S.W. from 7 to past midnight or Early this morning, and Brought down the Rivers – fair & drought all the day, before dinner pulled all the FLAX & brought it up to the Water Floors, and P.M. Coling Hay red Hall park – Severals begun Shearing.

Thursdy 7th Fair & dry weather from S.W. till 3 P.M. when Came on rain Severe towards night & stopped people begun to Thresh rye Grass Waterlair – Got finished the red Hall Park Hay into large Coles Snug – & put some of the lint into the Water – truly disagreeable & Bad weather –

Fridy 8th about 9 last night Came Suddenly a Violent Gale from West by North & drought which Blowed down of the Hay Coles put said day – many Stooks, & other things, but abated early this morning – Covering lint – and at Coleing Hay Waterlair in good order never turned into Tramp Coles, but Stopped by 5 P.M. with Heavy rain all night – put flax into water.

Saturdy 9th Th^r 50° about 6 A.M. Came on a furious Gale of wind & rain from N.N.W. which lasted till 10 – then stormy Cold weather all day – flited the Cattle

Flakes, and driving Earth on Blacklands – Country deluged, & Crop laid & Broke &c

Mondy 11th Yesterday fair weather, and drought – this day Grey and threatening rain whole day, but kept off – Own people Gathering & taking of Waterlair Hay, good order, whole day till very late, Observed Some people leading at Kair only a few days Cutt, Harvest will now be General

Tuesdy 12th Th^r 48° about 9 last night Came on Closs & Constant rain most of the night & morning & wet the grounds & roads very much – Morning put up again the two Coles red Hall park - last night finished all the Hay of leaving 4 stacks there in the field good order – this day begun again shearing Bear but stopped by rain at noon, after dinner at Waterlair Pease, rich enough but very little Fodder, being withered.

Wedy 13th Stormy weather & some Showers all day from SSW, but after Noon,frequent & heavy rain,so as to stop all work without doors, finished Cutting all the Pease, and at the Barns dighting Oats to loft-the Turnips are growing worse daily, no Hay Foggage rising – truly bad weather.

Thursdy 14th Fair & dry weather from S.W. till near Six P.M. Came on a shower which stopped the shearers of Bear, it is much Broke with the wind & rain & much lost in the Cutting down Accordingly – Cut some rushes Early – Pulled & tied up the Onions on pease sticks Garden.

Fridy 15th very strong white frost – Early shearing rushes after 9 begun shearing remainder Golds Hillock Bear, and the ripest of Backyards D°- it is very free, Twined and Makes very Foul work in Cutting up; but fine Weighty Crop & fine Grain – There is about three Inches of RAIN fallen already in this month –

Saturdy 16th White Frost in the night, & drought till 3 p.M. when Came on rain from Southw^d though not heavy – stopped shearing till near 4, begun again but in Half an (Hour) Obliged to give it up-by uncertain & stormy a S.E. – Observed some people Leading very late shorn Bear – The prices of Meal to 16/ and very scarce at Montrose & other places; the Turnips every where failed in this Country more than I ever Saw before no Hay Saved hardly to Montrose, but mostly in petty Coles & Consequently in bad order, little Hay Foggage –

Mondy 18th Yesterday morning Heavy rain till 6 – then fair and drought rest of the day – this day Showrey till near two p.M.- faired till near Six, at Rushes till dinner & about from two to three hours at the Back Yards Bear –

Tuesdy 19th Heavy frosty dew till past nine – at Barns, after at finishing Backyards Bear Th^r– then at Head of the Kavels D° very ripe – Appearance of rain from the Eastw^d – the first Cut not fitt to lead –

Wedy 20th a Great deal of rain last night and this morning and little fair whole day – at Barns Threshing Oats, making ropes, took out some of the Lint from the Water not properly done, P.M. Flited the Cattle Flakes the rivers much out, and the Earth Deluged –

Thursdy 21st Heavy Frosty dew till after Brakefast, & not dry till 10 A.M. and obliged to Dissmiss for a time till Called by fresh rain & appearance of Broken weather – Employed our own People in taking out the Lint out of the water 11-12 & 13 nights not Equall done, too much rain over it & spreading of it.

Fridy 22d Frosty dew in the morning, day fair & some drought from N.W. – Spread all the lint before dinner, it Covers a good deal Ground, P.M. amongst the first sown Turnips R Hall with Hand Houghs – lost Crop Generally.

Saturdy 23d a Severe & Constant storm of rain & wind begun at 6 A.M. & Continued whole day & night a S.E. which still more deluged the Country & must ruin the Crop – People at Barns threshing & at Sundries –

Mondy 25th Yesterday fair weather & drought night Heavy frosty dew after Brakefast begun Shearing again Kavels Bear N End ripest – appearance of fair weather a S.W. – Sundries.

Tuesdy 26th Strong & frosty dew, Cutting Kavels Bear in Band - till Brakefast then Binding – the first Cut and 2^{d} offering to Grow some of the sheaves – The Stooks fit to lead without picking & opening, yet some are doing so –

Wedy 27th no Frost this morning day Closs warm weather Easterly and no drought – people shearing Kavels Bear in Band till past 10 A.M.- after D^{o} Bound – finished it late Evening – Observed some Neighbours leading Certainly very unfit for it – Others Hutting on Fields –

Thursdy 28th Th^{r} 59° fine dry pleasant weather from S.W.- Shearing all day dry Gibsons park Bear seemingly good Grain & Crop Threaves 105 – which finished whole of BEAR CROP – Observed many leading pease & Bear, but it Cannot be in Condition.

Fridy 29th Early begun shearing of Oats E^{t} End Haugh Gayte after Brakefast begun leading Waterlair Pease 4 Carts into three stacks in fine order for all the rain and thatched two of them – finished by 7 p.M.; many who Carried in too Early are said to be Obliged to Carry out again & to turn over for Heating &c of them – which is the worst of all Husbandry &c Jobbs –

Saturdy 30th Closs foggy weather & very moist atmosphere whole day and appearance of rain from S.E. – People at the Haugh E^{t} End a short foul Crop – Hear much of the laid Wheat & Oats have little or nothing in them almost all the Hay is yet in the Fields & much of it in small Coles & useless rotten, prices of grain Oat Meal & Bear about 16/ per peck 13^{ds} – Beef & Mutton from 5^{ds} to 6^{ds} per lb not good – the Turnips are totally a lost Crop almost every where, 4 or 5 Acres not Equal one Good – the Quantity of rain this month is large Five Inches which is enough to ruin the Crop – the Wheat rising fast in the South & Harvest as late it is said as here- Observe some growing in the stooks of Bear &c – Indeed upon Closs Examination a great deal spoilt Cattle two & three year Olds sell well & fatt Beasts but others very dull & little sale – many Farmers have Rushed in their grain too soon & I learn from all quarters it is heating & rotting & they are turning it over & out again.

Diary for the Month of October 1797

Mondy 2d Yesterday fair weather but little drought, this day some shearing Haugh Oats, after Brakefast begun & picked first shorn Bear about 3 Acres of which found from none to 7 sheaves damaged, from two to three average & lead the Won into a Good stack, rain & stormy weather began about 4 p.M. but did not much hurt the lead.

Tuesdy 3d Fair weather till 3 p.M. Heavy rain for two hours - thatched the Bear stack – shearing Haugh Oats partly Gayte, partly stooked – Observe some Sheaves of the 2^d Cutt Bear Actualy HEATING in the stook –

Wedy 4th very Strong White frost till Brakefast – after shearing Oats in Gayte till past 11 then dry & Bound – this will make long & tedious Harvest & Bad every way.

Thursdy 5th Strong White frost which Spoilt the Potatoes Stems – after Brakefast begun shearing Haugh Oats & Continued all day not dry till near noon – The Bear first Cutt neither fitt for Wling or leading & Grows –

Fridy 6th fair weather at S.E. but little drought – Shearing whole day Oats finished Backwaters a good Crop – after at the Floors Old Ley flaked dry – lead home all the Bear Wylings good order into a Hutt Corn Yard – Sundries –

Saturdy 7th a pretty smart shower of rain from 4 to 5 A.M. day mostly fair a S.E.- at Floors Oats very rich Crop but not very rank. the latter, fine grain but Observe some Coming out in Cutting &c Jobbs –

Sundy 8th Strong drought last night & this morning till 7 A.M. Came on Stormy weather & rain from S.E. which lasted till near 11 – some showers P.M.- but not lasting.

Mondy 9th White frost in the morning & Corns wet most of the day - at Floors Oats – at noon Yoked 4 single Carts and lead in 2 stacks of E^t side of Kavels Bear in Excellent order without picking or Hollow stacks.

Tuesdy 10th Begun leading West side of Kavels Bear by Six o'Clock, finished by 11 – then Brought home the Gibsons park not so well win, after Wyling of the first shorn a Month ago on Golds Hillock from one to Six Sheaves Obliged to Gayte, Built 4 stacks this day – Came on light rain about 4 p.M. but little of it –

Wedy 11th Drought all this day a S.W. moderate, Thatching Shearing and Wyling & picking of Backyards Bear but in very Midding Order – Built One stack & a Half of it.

Thursdy 12th very fine drought & dry shearing morning floors Oats - before begun & lead Home all the Bear 10 stacks – Others Thatching all of them, at Drumleathie fair a good many fine Cattle but no Sale – had 3 Stotts but brought them Home.

Fridy 13th Coldish dry weather till 11 A.M. Came on rain from S.W. which lasted till 3 p.M. but not heavy – took up the lint into stooks it will turn out very Bad, Thatched out all Bear –

Saturdy 14th Oats wett till 11 A.M. then stooking them, finished all the Crop on

Whiteriggs evening-Observe the Gibsons Park Bear offering to Heat on the N.E.side notwithstanding the Cold dry weather a N.W.- am Informed that some Farmers in the Neighbourhood have Turned Every stack in the Yard, Cushny &c- and lead much of it out to the fields again – Bad Work –

Sundy 15th Strong drought all day from the Westward – drew out some of the Bear that was offering to Heat to Make Air Enter into the Stacks – Observed to Johnston much to shear and much out – both Bear & Oats &c –

Mondy 16th Dark lowring weather a S.W. all day – and Came on slight rain before 4 p.M.- Early begun Shearing of the Water lair ley Oats 7 Acres, in Gayte and stooks, but did not finish – Observed People at leading recently Shorn Bear & Oats; Sundries –

Tuesdy 17th Strong Frost which wet everything till near noon - however finished out shearing of all Crop before Brakefast at Waterlair,in Gayte – after turned over two stacks of Bear from Gibsons park, the part of the stack viz the lower Half being Injured, built up the two tops into a stack again Smaller- Others at Cutting of Rushes &c till dinner time then Binding up the Floors Gayte & stooking it – Flited the Cattle Flakes also Blacklands- Hear of many people leading out their Crop again, and others leading Green Bear new Cutt such Extremes, there is no Material Drought as yet – prices of Grain unsettled – Beef &c $5^{ds}\frac{1}{2}$ to 6^{ds} per lb – Only One Inch rain in the Gage till this date.

Wedy 18th Destruction of Dutch Fleet about 3 A.M. Came a furious gale & wind & rain from North and snow on the Hills, Waters out much – faired till 4 P.M.- rain – took in a stack of last years Bear N Barn & trifling sundries – this is really tedious & Bad weather – strong wind again at night a West – Bitter Cold all this day & very severe on the Cattle out –

Thursdy 19th all this day a strong Gale & drought a N.W. moderate in the Evening, putting Gayte up before Brakefast, and after at Binding of D^{o} – Built on Turlies the Turned Bear in tolerable order, and at noon begun leading Oats Floors in fine Condition, Frost again in the Evening as usual –

Fridy 20th Strong Black & White Frost, Ice about an Inch thick on Water and into the Ground more, the trees Ash &c quite Denuded – 4 Carts all day at leading Oats, six large stacks nearly finished – Others Binding of Gayte – & Sundries – the Frost Continued whole day & some & stormy after 4 P.M.

Saturdy 21st last night & all this morning a very severe Gale of Wind from N.N.E. with a deluge of rain which brought down the rivers higher than for fifteen years by past - it was highest about 10 A.M.- it brought down much Corn, timber Bridges &c – believes it hath taken off some of mine but fortunately what was most Exposed was lead Home the two preceeding days, or much loss – the rain Continued more or less – Shifted the Flakes blown down by the wind, it broke the Espalier Frames & other Mischief.

Mondy 23d $8\frac{1}{2}$ Inches in Water Gage Yesterday almost Constant rain a N.E.- this frequent showers till noon when fair but no drought a N.E.- people Gathering some of Haugh Corn Carried down by the Flood – about 15 Threaves on the Haughs above & below Mill of Garvock, which is supposed all lifted, at Barns dighting old Oats &c.

Tuesdy 24th almost Constant rain & the very worst weather for every thing wind little a S.W. – Brought up the Oats & Gayted them on the Haugh & West side of Corn Yard – Bled all the Calves to prevent the Quarter ill, dighted & lofted remainder of last years Crop of Oats; the Haughs will not Carry Carts.

Wedy 25th Blowing stormy weather from N.E.- and almost Constant rain – no work can be done without doors; it is miserable weather – P.M. Flited the Cattle Flakes, it is almost destruction for them to Lay out, and I Cannot get them all Housed –

Thursdy 26th Blowing weather from NNE but dry – Set up some of the Gayte Old Town park – threshing Old Bear own Hands, and at sundries – Observe the stooks wet to y[e] Heart but not growing.

Fridy 27th a sort of drought from North – with flying skiffs of rain or sleet & snow on the Hills – People at Barn Threshing Bear, and some at sundries – Corns quite wett

Saturdy 28th Pretty Hard Black Frost & drought a N.W.- at Barns, and after Brakefast set up all Gayte, and opened & Gayted such sheaves of the stooks as were wet to the Heart, but none of it Fit to lead, Others did so – sent the Cattle to Waterlair fine Grass.

Mondy 30th Potatoes all Secured Yesterday very Hard Frost and little drought – this day more drought a N.W. & frost, all Shearers &c after Brakefast Binding Gayte, Opening stooks, &c Confused & Indifferent work, One of the Oat stacks lead before the Heaviest rain not right Headed wet on the East side to the Easin & Obliged to be taken down & Gayted, the dammage both to stacks & Oats out is very Considerable.

Tuesdy 31st fine Frosty drought a N.W.- very Early Yoked 4 Carts & drove Home all the Old Town Park Oats & P.M. also all the Waterlair D[o]- both good order being about 13 Acres this day – women & others Binding of the Gayte & at Sundries – Built the stacks in kind of Hollow – Quantity of rain this month is Five Inches & One Quarter, chiefly after the Middle of the Month & Hath done Infinite Mess – chief to the Crop Crop more than Commonly Imagined Such Oats & Bear as have been Milled turn out poorly the first from 10 to 13 Pecks & the latter from 16 to 18 pecks – I look upon the Crop as much dammaged before shearing in many places, & Generally as ill got & Bad order as either in 1792 or 1795 years. the prices of Cattle are at least 15 per C[t] lower than last year at this time & less demand – prices of the Flesh Meat from 5[ds] to 6[ds] per lb – Bear Old 18/ – Oat Meal 16/ no Sale of any part of this Crop or demand for it.

Diary for the Month of November 1797

Wedy 1st Fine Moderate drought shifted from N.W. to S.W.- finishd In Gathering of all the Crop, One of the worst Harvests I ever Had & Crop Certainly much Injured beyond Conjecture, Cutting rushes &c for Thatching – Gave the FEAST.

Thursdy 2d Strong drought a West all day – dismissed all women morning – Men thatching, Eddering &c – at L kirk fair not many Cattle & no Buyers, prices fallen much both of Fatt & lean Beasts, Beef fallen $1\frac{1}{2}$ per lb – Sold all old Bear.

Fridy 3d Fine dry weather a West – people at finishing out the Corn Y[d] – Weighed & sent 100 stones Hay to Dr Fordyce L kirk – very fine Hay from a Cole Thornylee – at sundries.

Saturdy 4th very pleasant dry weather free of Frost, Men at Sundry Jobs about Corn Yard & Barns, was Obliged to take down a very large stack of Oats- Having Broke down the wand on the Pillar, it stands by Great Weight & rebuilt it again on same repaired stathill, Thatched it &c &c.

Mondy 6th Yesterday morning Came on rain a East, and in the Evening very stormy & Heavy rain till midnight Waters out – this day fair & pleasant wind a S.W.- Flited the Cattle Flakes, and at sundries, Barns, Corn Yard &c &c.

Tuesdy 7th self Indisposed Closs thick & Hazy weather a S.E. whole day. Carts driving Dung to Haugh Intended for Turnips and to render the Court Accessable, at Barns threshing Bear, Suppered a few Cattle Intended for Turnips – but keep them as long as possible while fresh weather –

Wedny 8th fine SunShiny weather and wind a S.W.- Carts driving Dung to Haugh, and to Clear the Court – at Barns and stacking up Thornylee Hay stacks to stand the Winter not having either time or people lead it home.

Thursdy 9th a White frost last night, but did not ly long – Carts driving dung to Haugh, and at Barns threshing Old Bear for Fodder to Bestials, Bedding the dung Court with river wreck Gardner dressing Hedge –

Fridy 10th Frost in the night, day very pleasant warm sun shine, People at repairing the Thatching of the Old Hay rick Waterlair, & the stacks of new on D° to stand the Winter, People sowing Wheat on ley.

Saturdy 11th pretty strong Frost in the night day very pleasant weather – people at Barns & thatching the Hay Stacks 6 large on red Hall Park – Numbers turning their stacks neighbourhood, rather late & amongst others WHEAT which I never Saw – the Cattle prices much fallen.

Mondy 13th Yesterday strong Frost which lyes always notwithstanding Brisk sun shine till 11 A.M. – the Stacks in the Barn Yards smoking like so many KILNS, it is said many of them lost – this day same kind weather with a little Breeze a South – turned over One of my Oat stacks into two Small Hollow ones – Housed all the Cattle this night –

Tuesdy 14th not quite so strong frost as before, thatched the Turned stack of Oats & thatched them, drove some stones to the new Bridge, and water Furring Waterlair very wett too much so to Plough – Barns &c Jobbs – sent $\frac{1}{2}$ Boll Barley to Kair for Pott Barley &c –

Wedy 15th some little showers it seems in the morning day very fine dry weather & some drought a S.W.- lands Excessive wett, not Ploughable. Brought Home the Cattle Flakes to Town, threshing, and rebuilding a stack of Oats built on unsound wood & Swayed the stack – this Carleness & Bungling of the Grieve in Managing the Corn Yard will Occasion a very Considerable loss of the Oat Crop, being so free in Handling & Sundries – Gardner laying of very rank Curley Greens &c.

Thursdy 16th Fair weather, less Frost & some drought a S.W.- the Ploughs Have Water Furred the Parks of Waterlair and this day the Kavels, Sent to Hill Garvock for a Cartful of Turfs, but very Midling dry but will serve for resting of Fires, at Barns and Sundries – the Young Cattle are sent to Waterlair Grass Every day – quite loose in the Belly – put the Flakes round West side of the Corn Yard & put up a Shade for Turnips – Hear of some Corn Yards on N Waterside quite rotten & only fitt for the Dunghill –

Fridy 17th very fine dry weather from S.W. and Grounds and Roads much better – People at Dighting of Bear for Stonehaven all the day – and at Cleaning of Hedges around the Bog Atry Planting sundries –

Saturdy 18th Strong drought & stormy weather – with some very Heavy Gusts from N.W.- at 7 sent 4 Carts to Stonehaven with 14 Bolls Bear sold to Mr D Greig – no more being threshed and Brought Back 8 Bolls D° 24 stones each of English Coals, returned by 5 P.M.- Housed all Cattle by 3 P.M.- Snow &c

Mondy 20th Yesterday very strong Frost with skimming of Snow laying on the ground whole day – this day fresh weather from the Southw[d] & snow vanished – Cattle at Waterlair - 4 ploughs at Back water ribbing across for pease, and Sundries.

Tuesdy 21st Fresh weather and Squally at Intervals, 4 ploughs at Haugh ribbing – at Barns at sundries – the Gardner hath dressed all the Hedges & most of his work is done for the present. He removes &c.

Wedy 22d a Prodigious Strong Gale with Heavy rain at N so as to stop all work without doors whole day and kept Cattle within also – People at Barns reding up and threshing Old Bear near finished &c.

Thursdy 23d Strong Frost from N & skimmering snow but the Country as White & appearance of standing winter – took in the last stacks of Old Bear S B – at Barns &c Jobbs – no work without doors –

Fridy 24th The degree of Cold very strong & people Complaining of it very much – people Threshing. it being term time, few Hands & many of them Fliting –

Saturdy 25th Closs winter Continues all day – and Waters & roads frozen – People at Barns, and sundries. about 2 Sunday morning a strong wind arose and the weather became fresh & the Snow left the low Country.

APPENDIX 1

1788 Acco^{t} of the Rental of the Lands of Airthurshouse, Thornyly &c

	£	Sh	d
Decr 20th 1788			
Received & Granted Receipt to John Milne on Stamped Paper for the Sum of Seven Pounds fourteen shillings & Nine pence Halpenny S^{tr} as rent for the first Half years payment of part of Thornylie possessed by him for Crop 1788. JB	7	14	$9\frac{1}{2}$ p^{d}
Decr 20th 1788			
Paid to the order and Discharge of the Rev^{d} Mr Took Fettercairn for the Ministers Stipend at Garvock for Crop 1788 the sum of £2-6-$3\frac{3}{4}$	JB		
Decr 31st 1788			
Received from David Brand at Sutty Wells for John Scott the sum of Twelve pounds Ten shillings & Ten pence S^{tr} as first half years rent of his possession of said Farm, & part of Thornylee and that for Crop 1788 – JB	12 JB	10 –	10 p^{d}
Jany 27th 1789			
Received from Alex^{r} Christie at Airthurs the sum of Seventeen Pounds, Ten shillings and three pence S^{tr} as first Half years rent of his Possession of said Farm & part of Thornylee and that for Crop 1788. JB	17 JB	10	3 p^{d}
June 25th 1789			
Received from David Burn Tennant in Thornylee for his possession there the sum of Three pounds and ten shs S^{tr} as in full of his rent for Crop 1788 & Granted Discharge accordingly –	3	10 JB	
July 6th 1789			
Received from John Milne at Mill of Garvock the sum of Seven pounds forteen shillings, and nine pence half penny in full of his rent for his possession of part of Thornylee & Discharged same	7	14 JB	$9\frac{1}{2}$ pd
July 17th 1789			
Received from Alex^{r} Christie at Arthurs the sum of Seventeen pounds ten shillings and three pence S^{tr} & Discharged him for Crop 1788	17	10 JB	3 pd
July 20th			
Received from David Brand at Sooty Wells the sum of £12-10-11 S^{tr}& discharged him in full for Crop 1788 –	12	10 JB	11 pd
Novr 30th 1789			
Received from John Milne at Mill of Garvock the sum of £7-15 S^{tr}as the first Half years rent of part Thornylee.	7	15 JB	0 pd
Jany 8th 1790			
Received from John Scott now Tennant at Sooty Wells in place of David Brandt, but only an assignation from the latter to the former in terms of said Tack the sum of £12-10-10 S^{tr} for first Half years rent of said Possession, for Crop 1789.	12 pd	10 JB	10

	£	*Sh*	*d*
Jany 26th 1790			
Received from Alexander Christie in Airthurhouse the sum of Seventeen pounds, ten shillings & three pence S^{tr}as the payment of his first half years rent of possession from one for Crop 1789	17	10 pd JB	3
June 11th 1790			
Received from John Milne at Mill of Garvock the sum of Seven L^{ds}forteen shillg^{s} & 7^{ds} in full for Crop 1789	7 pd	14 JB	7
June 25th 1790			
Received from John Scott in Sutty Wells twelve pounds, ten shillings and Eleven pence S^{tr} in full rent. Crop 1789	12 pd	10 JB	11
Octr 12th 1790			
Received from Alex^{r} Christie in Arthurs the sum of Seventeen pounds, ten shillings & three pence S^{tr}being in full of his rent for the Crop 1789 –	17 pd	10 JB	3
Decr 3d 1790			
Received from John Milne at Mill of Garvock £7-15 S^{tr}as first half of his rent of Part of Thornylee for Crop 1790	7 pd	15 JB	0
Decr 22d 1790			
Received from John Scott Sutty Wells the sum of Twelve pounds, ten shillings & ten pence S^{tr}as first Half of his rent for Crop 1790 –	12 pd	10	10
Decr 25th			
Received from Alex^{r} Christie at Arthurs the sum of £17-10-3 S^{tr}as first Half years rent of his possession and that for Crop 1790 –	17	10	3 pd
June 7th 1791			
Received from John Scott in Sutty Wells Twelve pounds, Ten shillings & Eleven pence S^{tr} which with former payment is in full of his rent for Crop 1790 – JB	12	10 JB	11
June 13th 1791			
Received from John Mill at Mill of Garvock the sum of Seven pounds, forteen shillings & seven pence S^{tr}in full for Crop 1790 –	7 pd	14 JB	7
June 17th 1791			
Received from Alex^{r} Christie at Arthur House the sum of £17-10-3 which with a like sum Decr 25th 1790 is in full rent for Crop 1790 – JB	17 pd	10 JB	3
Decr 15th 1791			
Received from John Scott at Sutty Wells the sum of Twelve pounds,ten shillings & ten pence S^{tr} for first Half years rent of his possession Crop 1791	12 pd	10 JB	10
Decr 23d 1791			
Received from Alex^{r} Christie at Arthurs the sum of Seventeen pounds ten shillings & three pence S^{tr} as first Half years rent for Crop 1791	17 pd	10 JB	3
Decr 27th 1791			
Received from David Burn Tennant at Thornylee the sum of Seven pounds S^{tr} being his full rent for his Possession for the Crops 1789 & 1790 to Whitsy 1791	7 pd	0 JB	0
Decr 29th 1791			
Received from John Milne at Mill of Garvock for first Half years rent of his possession of Thornylee for Crop 1791	7 pd	15 JB	0

	£	Sh	d
June 5th 1792			
Received from John Milne Mill of Garvock the sum of £7-14-7	7	14	7
S^{tr} being in full of his rent for Crop 1791	pd	JB	
June 19th 1792			
Received from John Scott Sutty Wells £12-10-11 S^{tr} & 7$^{ds\frac{1}{2}}$ as	12	10	11
School Masters Salary being in full Crop 1791	pd	JB	$7\frac{1}{2}$
June 21st 1792			
Received from Alex^{r} Christie at Arthurs the sum of £17-10-3 S^{tr}	17	10	3
which with an Equal sum formerly paid is in full of his rent for			$7\frac{1}{2}$
Crop 1791 – also 7$^{ds\frac{1}{2}}$ Schoolmaster Salary	pd	JB	
Novr 28th			
Received from John Milne of the Mill of Garvock for first Half	7	15	0
Year rent of his Possession of Thornylee for Crop 1792	pd	JB	
Decr 6th			
Received from John Scott Sutty Wells the sum of Twelve			
pounds, ten Sh & ten pence S^{tr} as the first Half years rent of his	12	10	10
Possession & that for Crop 1792 –	pd	JB	
Decr 31st			
Received from Alex^{r} Christie at Arthurs the sum of 17£ and ten	17	10	3
Shgs, three pence S^{tr} as first half of his rent for Crop 1792	pd	JB	
June 5th 1793			
Received from Mr J. Milne at Mill of Garvock the sum of £7-14-	7	14	7
7 S^{tr} being with former payment in full for Crop 1792	JB	pd	
D^{o} 17th 1793			
Received from Robert Jolly Tennant at Waterlair the sum of			
Twelve pounds & one shilling S^{tr} as Half years rent of said	12	1	0
Possession & for Crop 1792	JB		
July 5th 1793			
Received from John Scott at Sooty Wells the sum of 12-11- St	12	11	
which with the former sum is in full of his rent for Crop 1792 –	pd	JB	
Decr 7th 1793			
Received from Robert Jolly at Waterlair the sum of Twelve			
pounds & one Shg St as first Half years rent of His Possession	12	1	0
therepd JB	pd	JB	

Account of the Rental of the Lands of Thornylee, Arthurs, Sooty Wells & Waterlairs

Augst 24th 1793			
Received from James Andrew at Bridge of Leppie the sum of five			
Guineas as the last Half of his rent for his possession of Part of			
the lands of Waterlair & Discharged Him Accordingly for	5	5	0
Croppd JB 1792 –	pd	JB	
Decr 18th 1793			
Received from James Andrew the sum of Three Guineas as first			
Half of his rent for his Possession of part of lands Waterlair &	5	5	0
that for Crop 1793	pd	JB	
Decr 21st			
Received from John Scott Sutty Wells Twelve pounds Ten			
shillings & ten pence S^{tr} as first Half years rent and that for Crop	12	10	10
1793	pd	JB	

	£	Sh	d
Jany 2d 1794			
Received From John Milne at Mill of Garvock the sum of Seven pounds fifteen Sh' first Half years rent of his Possession of Thornylee	7	15	pd JB
June 3d 1794			
Received from John Milne at Mill of Garvock the sum of Seven pounds forteen sh & seven pence S^{tr} – which with the sum of £7-15 S^{tr} is in full of his rent for his Possession of Thornylee Crop 1793	7	14 JB	7
June 13th 1794			
Received from Robert Jolly at Waterlair the sum of 12£ 1 Sh S^{tr} which with like sum before as full Crop 1793	12	1 JB	pd
D^{o}			
Received from John Scott Sooty Wells the sum of Twelve pounds,Ten Shg & Eleven pence S^{tr} & $7^{ds\,1}{-}_{2}$ Schoolmasters salary which with former payments is in full Crop 1793	12 Salary pd	10 JB	11 $7\frac{1}{2}$
March 15th 1794			
Received from Alex^{r} Christie at Arthurs He Having fallen late of the Customary term but allowed Interest,the Half Years rent due at Whits^{y} 1793 – and that due at Martinmass 1793 – & Security for that at Whitsunday 1794 part in Bills	35 Schoolmaster paid	1 JB	$1\frac{1}{2}$ $7\frac{1}{2}$
July 10th			
Received from James Andrew for the last Half of his rent of Waterlair & that for pd Crop 1793 £5-5 S^{tr}	5	5	pd JB
Novr 25th 1794			
Received from John Milne at Mill of Garvock for his first Half years rent of Possessions of Arthur House & Thornylee for Crop 1794 the sum of	25 paid	5 JB	3
Decr 2d			
Received from Robert Jolly at Waterlair the sum of Twelve pounds & one shilling S^{tr} as first Half years rent for Crop 1794	12 pd	1 JB	1
Decr 13th			
Received from John Scott at Sooty Wells the sum of £12-10-10 S^{tr} as first Half years payment of his rent of said Possession and that for Crop 1794 – which is received by JB	12	10	10 pd
April 14th 1795			
Received from James Andrew Bridge of Leppie the sum of £5-5 S^{tr} as the first Half years rent of his Possession of Water Lair for Crop 1794 –	5 pd	5 JB	
June 11th 1795			
Received from John Scott Sooty Wells the sum of Twelve pounds Eleven shg & $6^{ds}\frac{1}{2}$ which with former payment is in full of Crop for 1794	12	11 pd	$6\frac{1}{2}$ JB
June 13th			
Received from Robt Jolly the sum of Twenty pounds,one Shillg & $6\frac{1}{2}$ which with £12-1 St formerly paid is in full for Crop 1794 –	12	1	$6\frac{1}{2}$
June 15			
Received from John Milne Mill of Garvock the sum of £25-6-$\frac{1}{2}$ which with the former payments & schoolmasters fee of $\frac{1}{2}$ is in full for Crop 1794 of Both Possessions –	25	6	$\frac{1}{2}$

	£	Sh	d
July 9th 1795			
Received from James Andrew at Bridge of Leppie the sum of £5-5-S^{tr} which with a like sum is in full of rent Waterlair Crop 1794	5	5 JB	pd
Nov 24th			
Received from Rober Golly at Waterlair as first Half years rent of Possession for Crop 1795– & £2-6-0 S^{tr} for Pasture let him at Wester Waterlair	12	1 pd	0 JB
Decr 8th			
Received from John Scott at Sooty Wells as first Half years payment of his rent for said Possession the sum	12	10 JB	10 pd
Decr 10th			
JB pd Received from Mr Jo Milne at Mill of Garvock the sum of £25-5-3 S^{tr} as first Half years rent of his Possession of Arthurs & Thornylee & that for Crop 1795	25 paid	5 JB	3
May 14th 1796			
Received from James Andrew at Bridge of Leppie the sum of five guineas – first Half rent for Crop 1795	5 JB	5	0 pd
May 28th 1796			
Received from John Scott Suttie Wells the sum of Twelve pounds Eleven Sh & 6p S^{t} in full of rent Crop with precedings paid is in full 1795	12	11 JB	$6\frac{1}{2}$ pd
D^{o} 28th			
Received from Robt Jolly at Waterlair Twelve pounds One shilling & $6\frac{1}{2}$ which with preceding paid is in full of all rent for Crop 1795	12	1	$6\frac{1}{2}$
June 1st			
Received from Mr Jo Milne at Mill of Garvock the sum of Twenty five pds Six shilling & $6^{d}S^{tr}$ which with former payment is in full for Crop of Thornylee & Arthur House Crop 1795	25 paid	6	$6\frac{1}{2}$ JB
Oct 29th 1796			
Received from James Andrew for his Possession at Waterlair in full of Crop 1795 – to Whitsunday 1796	5 pd	5 JB	0
Nov 24th 1796			
Received from Robt Jolly for Waterlair the sum of Twelve pounds & One Shg S^{t} first Half rent Crop 1796	12 JB	1 pd	0
Nov 28th 1796			
Received from John Scott at Sooty Wells the sum of Twelve pounds, Ten Shg & ten pence S^{tr} as first Half years rent Crop 1796	12	10	10 2^{ds}
Decr 1st 1796			
Received from John Milne Mill of Garvock the sum of 25£ five Shg & three pence as first Half years rent of Arthurs & Thornylee Crops 1796	25	5 JB	3
May 30th 1797			
Received from Rob Jolly at Waterlair the sum of Twelve p^{ds} One Sh & 6^{ds} which with the Martinmass like sum for full Crop 1796	12	1	$6\frac{1}{2}$ JB
June 3d			
Received from James Andrew five Guineas which with former payment is up to Martinmass 1796 – and Two pounds S^{t} of first Half Crop 1796	5	5 JB	0 pd

June 9th

Received from John Scott Sooty Wells the sum of Twelve pounds, Eleven shillings & Six pence S^{r} which is in full for Crop 1796 — 12 11 6

Rec^{d} from John Scott at Sutie Wells the sum of 12 : 11 : 4 being the full amount of rent for 1797 –
Rec^{d} from John Milne of Milne of Garvock the sum of in full for Crop 1797

Account of the rental of Whiteriges

Rec^{d} from John Scott at Waterlair		
Rec^{d} from John Scott at Sooty Wells the sum of 12 : 11 : as his Martinmas rent	12	11
Rec^{d} from John Milne of Mill of Garvock Decr 13th his Martinmass rent	25	5
Rec^{d} from Robt Jolly in Waterlair the sum of as Martinmas rent	12	1

APPENDIX 2

Received from Mr James Badenoch Esqr of
Whiteriges the Sum of one Shilling & Sixpence
Ster money – Salryde to the Schoolm^{r} of Garvock
Crop 1788 & is Discharged by Willm Warder

On reverse of the above receipt is the following:-

1789
£0 – 1 – 6
paid
Schoolmasters
Salary –
of Garvock

12/2$^{ds}\frac{1}{2}$ due to J M –

March the 4

On the 25th of February 1793
Corn sent to the Mill 31 Bolls
Bear lofted 6 Bolls 2 firlots 2 pecks
Bear for the horses – 3 firlots –
Bear to Fr^{s} Walker – 2 firlots
Bear sent to Ja^{s} Andrew for Malt 4 Bolls
Pease Lofted 2 Bolls – 1 peck

On the reverse of the above note is the following:

B	f	p	l
0	2	–	–
	3	–	–
4	4	–	–
6	2	2	–
11	3	2	

Being total of last threshed stack of Barley
Dighted March the 4th 1793 Infamously Bad
JB
John Lys left of Threshing March 4th – Entered Nov 24th

APPENDIX 3

1793

Consumption of Oat Meal further of Crop 1792 –

D° 23 th		D° Baking	0	2	0
D° 26 th		D° family	0	2	0
March	2 d	D° family	0	2	0
"	4 th	"	0	2	0
"	6 th	D° Baking	0	2	0
"	8 th	D° family	0	2	0
"	12 th	D° D°	1	0	0
"	15 th	Jas Welsh	0	2	0
"	"	John Tailor	0	2	0
"	18 th	Family	1	0	0
		B 1	4	3	p

Expenditure of Meal Crop 1792

"	17 th	Family use	1	0
"	23 d	" "	1	0
"	28 th	" "	1	0
"	30 th	John Tailor	0	2
"	"	Will Gall	0	2
June	3 d	Family use	1	0
"	7 th	" "	1	0
"	11 th	" "	0	2
"	14 th	" "	0	2

D° Continued 1793

			B	f
June	19 th	Family use	1	0
"	24 th	For D°	1	0
"	"	Will Gall	0	2
"	"	John Tailor	0	2
"	28 th	Family use	0	2
July	3 d	" "	0	2
"	9 th	" "	1	0
"	13 th	" "	0	2
"	15 th	" "	0	2
"	18 th	" "	1	0
"	22 d	John Tailor	0	2
"	23 d	Will Gall	0	2
"	22 d	Baking family	0	2
"	27 th	Family Baking	0	2
"	29 th	" "	1	0
Aug[st]	3 d	" "	0	2
"	6 th	" "	0	2
"	9 th	" "	0	2
"	12 th	Family use	1	0
"	15 th	" "	0	2
"	19 th	" "	1	0
"	"	Will Gall	0	2
"	"	John Tailor	0	2
"	22 d	Family use	1	0

1793 Meal Accn[t] Carr[d] forward

Sept[r]		B	f
16th	Family use	1	0
"	John Tailor	0	2
"	Will Gall	0	2
19th	Family use	0	2
23d	" "	0	2
26th	" "	1	0
30th	" "	1	0
Oct[r] 2d	" "	0	2
7th	" "	1	0
8th	To Pitscallie for his Mill inc Crop 1791	2	0
9th	For Family use	1	0
11th	" " "	1	0
14th	" " "	1	0
"	Will Gall	0	2
16th	Family use	1	0
18th	John Tailor	0	2
"	Family use Baking	0	2
21st	" " "	0	2
24th	" " "	0	2
28th	" "	0	2
Nov 1st	" "	0	2
" 4th	" "	0	2
" 7th	" "	0	2

"	26th	"	"	1	0
"	30th	"	"	0	2
Sept^{r}	2d	"	"	1	0
"	5th	"	"	1	0
"	9th	"	"	1	0
"	12th	"	"	0	2

Meal Crop 1793

			B	f
Nov^{r}	25th	John Tailor to make his full year of winter Meal One Half Boll advanced	0	2
"	"	Will Gall	0	2
"	29th	Family use	0	2
Dec^{r}	2d	" "	0	2
"	6th	" "	0	2
"	11th	" "	0	2
"	16th	" "	0	2
"	21st	" "	0	2
"	23d	Will Gall	0	2
"	"	John Tailor but to be deducted	0	2
"	26th	Family	0	2
"	31st	"	0	2
Jan^{y}	3d	"	1	0
"	8th	"	1	0
"	14th	"	1	0
"	18th	"	1	0
"	20th	Will Gall	0	2
"	"	John Tailor reserve due	0	2
"	24th	Family use	1	0
"	31st	" "	1	0
Feby	5th	" "	1	0
"	11th	" "	1	0
"	15th	" "	1	0
"	21st	Family use	0	2

Expenditure Meal Crop 1793

1794			B	f
May	2d	Family use	0	2
"	5th	" "	0	2
"	8th	" "	1	0
"	12th	" "	0	2
"	"	Will Gall	0	2
"	"	Jo Tailor	0	1
"	15th	Family use	1	0
"	21st	" "	1	0
"	26th	" "	1	0
"	31st	" "	1	0
June	5th	" "	1	0
"	9th	" "	1	0

	11th	" "	0	2
"	16th	New Meal	0	2
"	20th	" "	0	2
"	23d	" "	0	2

Expenditure Meal Crop 1793

1794			B	f
Feby	17th	Will Gall	0	2
"	"	John Tailor 0 1 1p 3 pecks being deducted		
"	21st	Family use	1	0
"	25th	" "	1	0
March	3d	" "	1	0
"	10th	" "	1	0
"	15th	" "	1	0
"	17th	Will Gall	0	2
"	"	John Tailor	0	2
"	20th	Family	1	0
"	24th	"	1	0
"	31st	"	1	0
April	4th	"	1	0
"	9th	"	1	0
"	14th	John Tailor	0	2
"	"	Will Gall	0	2
"	"	Family use	1	0
"	17th	D^{o} to Bake	0	2
"	21st	Family use	1	0
"	25th	" "	1	0
"	29th	" "	1	0

Expenditure Meal Crop 1793

July	11th	House	1	0
"	16th	House use	1	0
"	24th	" "	1	0
"	26th	" "	1	0
"	30th	" "	0	2
Aug^{t}	1st	" "	1	0
"	"	Will Gall	0	2
"	5th	Family use	1	0
"	9th	" "	1	0
"	11th	" "	0	2
"	12th	" "	1	0
"	19th	" "	1	0
"	21st	" "	1	0
"	26th	" "	1	0
"	29th	" "	1	0
Sept^{r}	1st	W Gall	0	2

"	10th	Will Gall	0	2
"	16th	Family use	1	0
"	19th	" "	0	2
"	21st	" "	0	2
"	23d	" "	0	2
"	26th	" "	1	0
"	30th	" "	1	0
July	1st	" "	1	0
"	3d	" "	1	0
"	7th	Will Gall	0	2

"	2d	Family use	1	0
"	5th	" "	1	0
"	8th	" "	1	0
"	11th	" "	1	0
"	13th	" "	1	0
"	17th	" "	1	0
"	19th	" "	1	0
"	22d	" "	1	0

Meal Crop 1793 – Expended

			B	f	p
Sep^{r}	25th	Family Meal	1	0	0
"	27th	" "	1	0	0
"	30th	" "	1	0	0
Oct^{r}	1st	Will Gall	0	2	0
"	3d	Family Meal	1	0	0
"	6th	" "	1	0	0
"	9th	" "	1	0	0
"	13th	" "	1	0	0
"	14th	Supper Meals	2	0	1
"	16th	Family "	0	2	0
"	20th	" "	1	0	0
"	24th	" "	0	2	0
"	27th	Will Gall	0	2	0
"	"	Family use	1	0	0
"	30th	" "	1	0	0
Nov^{r}	3d	" "	0	2	0
"	6th	" "	1	0	0
"	10th	" "	1	0	0
"	12th	Will Lyel 4p Meal at 1/ per peck	0	1	0
"	14th	Family use	1	0	0
"	17th	" "	1	0	0
"	22d	" "	1	0	0
"	25th	Will Gall	o	2	0

Expenditure Meal Crop 1793

			B	f	p
Nov^{r}	26th	Family use	1	0	0
Dec^{r}	1st	" "	1	0	0
"	6th	" "	1	0	0
"	12th	" "	1	0	0
"	15th	" "	1	0	0
"	20th	" "	1	0	0
"	22d	Will Gall	0	2	0
"	25th	Family use	1	0	0
"	29th	" "	1	0	0
Jany	2d	being last Crop 1793			

1795

Jany 7th of new Crop 94 family 1 0 0

Transferred to the Account on a separate paper of Crop 1795

JB

Expenditure Meal Crop 1795

			B	f	P
		Continued			
Sepr 1796	20	To John Tailor	0	2	0
"	22d	Family use	1	0	0
"	24	" "	1	0	0
"	26	" "	1	0	0
"	"	Will Gall	0	2	0
"	30	Family use	1	0	0
Oct^{r}	3d	" "	1	0	0
"	6th	" "	1	0	0
"	10th	" "	1	0	0
"	13th	" "	1	0	0

			B	f	P	L
Nov^{r}	17th	For Family of new Meal Crop 96	1	0	3	2
"	22d	" "	1	0	0	0
"	23d	Will Gall	0	2	0	0
"	25th	Family use	1	0	0	0
Dec^{r}	1st	" "	1	0	0	0
"	5th	" "	1	0	0	0
"	"	To John Tailor	0	1	3	0

deducting 5 pecks for his Harvest Meal at his House & to advance his first half Boll

"	17	" "	1	0	0
"	18	" "	1	0	0
"	21st	" "	1	0	0
D°		John Tailor	0	2	0
"	22d	Family use	1	0	0
"	25	Will Gall	0	2	0
"	28	Family use	1	0	0
Nov^r	2d	" "	1	0	0
"	7	" "	1	0	0
"	11th	" "	0	2	1½
"	12	From Kair Meal	1	Repaid	

repaid by new Meal.One Boll Due by John Mitchell JB

B 83 2 1½

for Winters Meal from Martinss

"	15th	Family use	1	0	0	0
"	19th	" "	1	0	0	0
"	"	John Dorwood Supper Meal after deduction	0	0	3	2
"	"	Will Gall	0	2	0	0
"	20th	John Tailor	0	2	0	0
"	24th	Family use	1	0	0	0
"	28th	" "	1	0	0	0
"	31st	" "	1	0	0	0
Jany	5th	97 " "	1	0	0	0
"	9th	" "	1	0	0	0
"	13th	" "	1	0	0	0
"	19th	" "	1	0	0	0
Jany	19th	Will Gall	0	2	0	0
"	"	John Tailor	0	2	0	0
"	23d	Family use	1	0	0	0
"	27th	" "	1	0	0	0
"	31st	" "	1	0	0	0

Nov 1796 22d paid to D Lyell for 2 Women shearers

	B	f	p	L
Supper Meal a peck each	0	2	1	½

To John Dorwood – –

Octr 8th 1796 Lent a Boll Oat Meal to John Mitchell to be repaid in a month JB
Repaid Novr 22d 1796 JB

Expenditure of Meal
Crop 1796 Continued

1797			B	f	p
Feby	6	Family use	1	0	0
"	10	" "	1	0	0
"	14	" "	1	0	0
"	17th	Will Gall	0	2	0
"	"	John Tailor	0	2	0
"	20th	Family use	1	0	0
"	24th	" "	1	0	0
March	3d	" "	1	0	0
"	7th	" "	1	0	0
"	13	" "	1	0	0
"	"	Will Gall	0	2	0
"	14	John Tailor	0	2	0
"	18	Family	1	0	0
"	24th	"	1	0	0
April	1st	"	1	0	0
"	6th	"	1	0	0
"	10	Will Gall	0	2	0
"	11th	Family use	1	0	0
"	14	John Tailor	0	2	0
"	17	Family use	1	0	0
"	21	" "	1	0	0
"	26	" "	1	0	0

Expenditure of Meal Crop 1796

1797			B	f
June	22d	John Tailor	0	2
"	24	Family use	1	0
"	28	" "	1	0
July	3d	" "	1	0
"	7	" "	1	0
"	13th	" "	1	0
"	17	" "	1	0
"	20	" "	1	0
"	"	John Tailor	0	2
"	25	Family use	1	0
"	29	" "	1	0
Aug^t	4	" "	1	0
"	7	" "	1	0
"	12	" "	1	0
"	17	" "	1	0
"	19	To John Tailor	0	2
"	21	Family use	1	0
"	25	" "	1	0
"	30	" "	1	0
Sep^r	4th	" "	1	0
"	7th	" "	1	0
"	10	" "	1	0
"	13	" "	1	0

May	1st	" "	1	0	0
"	6	" "	1	0	0
"	9	Will Gall in full	0	1	0
"	11	John Tailor old term	1	0	
"	"	Family use	1	0	0
"	15	" "	1	0	0
"	"	Sold to Jas Dunbar	1	0	0
		Wright 1 Boll 15/ per Boll			
"	20	Family use	1	0	0
"	24	" "	1	0	0
"	27th	John Tailor	0	2	0
"	29	Family use	1	0	0
June	1st	" "	1	0	0
"	6th	" "	1	0	0
"	10th	" "	1	0	0
"	15th	" "	1	0	0
"	20th	" "	1	0	0

"	16	" "	1	0
"	19th	" "	1	0
"	22	" "	1	0
"	"	John Tailor	0	2
"	27th	Family use	1	0
"	29	" "	1	0
Oct^r	2d	" "	1	0
"	4th	" "	1	0
"	7th	" "	1	0
"	10	" "	1	0
"	13	Farm "	1	0
"	"	John Tailor	0	2
"	14	Family use	1	0
"	17 19 23 27	"	4	0
Nov^r	3d 8 13	"	3	0
"	17	" "	1	0
"	22	" "	1	0

(The following additional Accounts of Expenditue of Meal Crop were found in an earlier diary written by James Badenach.)

Account of Expenditure of Meal
Crop 1794 Commencing Jany 7th
1795 —— per J.B.

1795			B	f
Jany	7th	Family Use	1	–
D°	13th	D° D°	1	–
"	16th	D° D°	1	–
"	19th	Will Gall	0	2
"	21st	Family Use	1	–
"	26	D° D°	1	–
"	31	D° D°	1	–
Feby	6	D° D°	1	–
"	11	D° D°	1	–
"	16	W Gall	0	2
D°	D°	Family D°	1	–
D°	21st	D° D°	1	–
"	27	D° D°	1	–
March	4th	D° D°	1	–
"	9	D° D°	1	–
"	13	D° D°	1	–
"	16	Will Gall	0	2
"	19	Family use	1	–
"	24	D° D°	1	–
"	28	D° D°	1	–
April	2^d	D° D°	1	–
"	3	D° D°	1	–
"	11	D° D°	1	–
"	13	Will Gall	0	2

			B	f
June	15th	Family use	1	–
"	18th	Family use	1	–
"	22^d	D° D°	1	–
"	27	D° D°	1	–
July	2^d	D° D°	1	–
"	6th	D° D°	1	–
D°		Will Gall	0	2
13th		Family use	1	–
17th		D°	1	–
		Poultry	"	"
"	20	James Loggie	0	2
"	21st	Family use	1	–
"	27	D° D°	1	–
"	29	David Clark Cottar	1	–
Aug^st	1	Family use	1	–
"	3^d	Will Gall	0	2
"	6th	Family use	1	–
"	10th	D° D°	1	–
D°		James Loggie	0	2
"	15	Family use	1	–
"	19th	D° D°	1	–
"	24th	D° D°	1	–
		B: 29–2–		

Expenditure Meal Crop 1794

Aug^st			B	f
	26		0	2
	28th	Family use	1	–

"	16	Family D°		1 –
"	20	D° D°		1 –
"	24th	D° D°		1 –
"	28	D° D°		1 –
May	2^{d}	D° D°		1 –
			B	27 0

Expenditure Meal Crop 1794

			B	27-0-0
				B f
May	7th	Family use		1 0
"	11th	Will Gall		0 2
"	12th	Family use		1 –
"	15	D° D°		1 –
"	20	D° D°		1 –
"	25	D° D°		1 –
"	28th	D° D°		1 –
June	2^{d}	D° D°		1 –
"	7	D° D°		1 –
"	8th	Will Gall		0 2
"	11th	Family use		1 –
"	12	J. James Loggie		0 2

Nov^{r}	2^{d}	Family use		1 –
D°		James Logie 3 pecks to himself & 3p 2ll to his wife for Supper Meal to Martinmass	0	1-2-1
"	7	Jean Guthrie & Sister Supper Meal	0	1-2-1
"	9	Family use		1 –
"	12	Will Lyels Supper Meal	0	0-3-1
"	14th	Family use		1 –
		B:25-1-3-3		
"	22^{d}	D° D°	1	0 0
		Expend^{re} Contin^{d} 1794		
"	27	Family use	1	0 0

			B	f	P
1795 Nov^{r}					
	16th	David Clark for his Meal	0	0	3
D°		Ian Graham D°	0	1	0
"	19th	Family use	1	–	–
"	23	Will Gall	0	2	–
"	24	Family use	1	–	–
"	30	D°	0	2	–
Dec^{r}	9th	D°	0	2	–
"	16	D° D°	0	2	–

Sep^{r}	1	D° D°	1	–
"	2^{d}	Will Gall	0	2
"	5	Family use	1	–
"	7th	James Loggie	0	2
"	11	Family use	1	–
"	14	D° D°	1	–
"	16	D° D°	1	–
"	21st	D° D°	1	–
D°		David Clark	0	2
"	25th	Family use	1	–
"	28	Will Gall	0	2
"	30	Family use	1	–
Oct^{r}	3	D° D°	1	–
"	6	James Loggie	0	2
"	8	House use	1	–
"	12	D° D°	1	–
"	15	D° D°	1	–
"	19	D° D°	1	–
D°		David Clark	0	2
"	22	Family use	1	–
"	26	D° D°	1	–
"	27	Will Gall	0	2
"	30	Family use	1	–

			B	f	p
1796					
Jan^{y}	18th	Will Gall		0	2
"	21	Family use	1	0	0
			10	1	1

Expenditure of Meal Crop 1795

			B	f	P
Jam^{y}	25	Ja^{s} Loggie	0	2	0
	28	Family use	1	0	0
Feby	2	D° D°	1	0	0
"	8	David Clark	0	3	0
D°		Family use	1	0	0
"	15	Will Gall	0	2	0
"	16th	Family use	1	0	0
D°		James Loggie	0	2	0
March	4th	D° D°	1	0	0
"	7th	David Clark	0	2	0
"	11th	Family use	1	0	1
"	14	Will Gall	0	2	0
"	16	Family use	1	0	0
"	23	D° D°	1	0	0
D°		James Loggie	0	2	0
"	28	House use	1	0	0
April	5th	David Clark	0	2	0
"	8th	House use	1	0	0
"	11th	Will Gall	0	2	0

			B	f	P
"	28	D° D°	0	2	–
1796					
Jany	4	D° D°	0	0	3
		B	5	0	2
Brought from other pages above			82	0	0
		Total Bolls	87	0	2

Meal Crop 1795

1795			B	f	P
Dec^r	1st	Ja^s Loggie	0	2	
"	3^d	Family use	1	0	
"	9th	D°	0	2	
"	14th	David Clark	0	2	
"	16	Family use	0	2	
"	21	D° D°	0	2	
"	24	D° D°	0	2	
"	26	Will Gall	0	2	
"	28	Family use	0	2	
1796					
Jan^y	4th	D° D°	0	3	1
"	9th	D° D°	1	–	–
"	12	David Clark	1	–	–
"	15	Family use	1	–	–
"	14th	Family use	1	0	0
"	18	James Loggie	0	2	0
"	20th	Family use	1	0	0
"	27th	D° D°	1	0	0
May	2^d	David Clark	0	1	0
D°		Family use	1	0	0
"	6th	D° D°	1	0	0
"	13	D° D°	1	0	0
"	11	Will Gall	0	2	0
"	17	James Loggie	full	0	3
"	19	Family use	1	0	0
		B	25	–	3

Expenditure Meal Crop 1795

			B	f	P
May	26	Family use	1	0	0
"	28	John Tailor Cottar		2	0
June	2^d	Family use	1	0	0
"	6	Will Gall	0	2	0
"	8	Family use	1	0	0
"	10th	D° D°	1	0	0
"	16	D° D°	1	0	0
"	21	D° D°	1	0	0
"	24	D° D°	1	0	0
"	29	D° D°	1	0	0
D°		John Tailor	0	2	0

1796			B	f	P
July	4th	Family use	1	0	0
D°		Will Gall	0	2	0
"	8	Family use	1	0	0
"	13	D° D°	1	0	0
"	19th	Family use	1	0	0
"	24	D° D°	1	0	0
"	29	D° D°	1	0	0
Aug^st	1st	D° D°	1	0	0
D°		John Tailor	0	2	0
D°		Will Gall	0	2	0
"	8th	Family use	1	0	0
"	10th	D° D°	1	0	0
"	15	D° D°	1	0	0
"	20	D° D°	1	0	0

1796			B	f	P
Aug^st	20	John Tailor	0	2	0
"	26	Family	1	0	0
"	29	D°	1	0	0
D°		Will Gall	0	2	0
Sep^r	2^d	Family use	1	0	0
"	6	D° D°	1	0	0
"	10	D° D°	1	0	0
"	14th	D° D°	1	0	0
"	19	D° D°	1	0	0
			30	0	0

APPENDIX 4

Note of Crop, Shearing &c 1794

		Supper night
Mondy 1st Sept^{r}	All People 20 – 13 Men & 7 Wowmen begun shearing up Pease Water Lair which finished Evening –	1
3d	After Brakefast all People at Bear Cutting Black Lands rather ripe & light	2
4th	After Brakefast People at Bear as yesterday	3
5th	D° D° at Barley & Oats afternoon	4
6th	– at Back Yards & after Haugh Oats –	5
8th	after Brakefast all People at H- Oats	6
9th	All day at Haugh & Back Yards Oats	7
10th	at shearing Oats & Turning Pease	8
11th	Leading Black Lands Bear & Water Lair Pease	9
12th Fri	All people all day at G. Hillock Oats	10
13th	all people Cutting Oats red Hall park Thatching Pease, Bear &c stacks	11
15th	After Brakefast near noon at Water Lair Bear	12
16th	After D° at red Hall p. Oats till dinner	–
17th	after 12 noon-at Cutting up red H Park Oats	13
18th	Shearing Gibsons Park Oats from Brakf to Dinner – stopped by rain	–
19th	Cutting rushes morning-after Old Town park Oats	14
20th	after Brakefast at Gibsons park Oats – dismissed	–
24th	all shearers at Water Lair Bear finished	15
25th	all People whole day at Shearing of Thornylee Bear very light Crop & much lost	16
Friday Sept 26th	All shearers whole day at Thornylee Bear – Threaves – 190 – & finished	17
27th Saturdy	All people at Leading Haugh Oats 2 Stacks Black Lands Barley 1 small D° & Thornylee Bear and shearing Back Yards Oats severely shaken	18
29th	All people at leading W Lair Bear & Ley Oats Golds Hillock shearing Oats &c	19
30th	People women shearing all day Burns Oats and 4 Carts leading Oats all day	20
Octr 1st Wedy	All people shearing Out Oats by 4 P.M. Others Lead Home all Bear & Oats ready before Dinner	21
2d	People at Cutting up rushes for Thatch till Brakft-after Raking of Thornylee Bear &c	22
3d	All people at Cutting Rushes till Brakefast after Raking Water Lair Bear-Thatching stacks	23
4th	a few Cottars & near Hands	
6th	The Feast & but little work to the Crop	24
7th	part people at Corn Yard, Cutting rushes and after Bound & Led in long Burns Gate	25
8th	Thatching stacks till near noon when Brought in all the Crop Evening &	

Thatched all of them but one 26

Women
24 Nights

It Consists of Stacks of

Oats	14
Bear	6
Pease	3
Total –	23 –

Oct 27th –	One firlot Bear Pott Barley Kair
D^{o}	$4\frac{1}{2}$ Bolls Oats to Mill for Meal. 4B-6pecks
" 28th	Of Brock,Bear,Gatherings & 5B-lofted Meal Barn
" 29th	Took in 11 to 12 Th^{r} Black lands Bear for Malt and for Horses
Novr 3d	Took 4 Bolls of it for Malt to Family use
" 7th	took rest of said stack NB – finished D^{o} 14th
" 14th	2^{d} stack Black lands Bear SB – Stone stathill finished the 21st D^{o}
" 22nd	Took in NB small stack of Black L^{ds} Barley

Note of Shearing Crop ; 1795

		Supper Meals
Thursdy Sepr 17th	12 Shearers, Half Men & three Bandsters begun Harvest Cutting up long haugh Bear fine Crop – 60 Th^{s} at 11 A.M.	1
" 18th	about 10 A.M.- at above Bear all dry	2
" 19th	Early All Shearers at Bear.Stopped Dinner time	
" 21st	Early at Cutting up Cottar Bear Bank	3
" 22d	all D^{o}-finished Cutting up whole Bear at Whiteriggs-452 Th^{rs} by 4 P.M. -after Waterlair	4
" 23d Wedy	Whole day at Waterlair Bear finished at night 93 Threaves very fine Grain5	
" 24th Thursdy	Begun Cutting Pease Thornylee – but stopped at Brakefast time again after Dinner	6
" 25th	People finished Cutting all Pease before noon – after Cutting Rushes for Thatch	7
" 26th	D^{o} at Cutting Rushes &c and took out the Lint & spread it N Barns	8
Octr 1st Thursdy	after Brakefast begun leading of Cottar Bank Bear 4 Carts in tolerable order – & Women Shearing Oats B Yards	9
" 5th	Early all people Shearing of Oats at B Yards Gayte after stooked-and stopped by Closs rain by Dinner time -	

Note of Shearing, and Leading of Crop 1795

Oct^{r}		*Supper Nights*
Tuesdy 6th	Early all people at Back Yards & Golds Hillock Oats all day dry	10
Wedy 7th	Early at Cutting of Oats G Hillock and after 10 A.M. yoked & drove in all Bear of Crop by 1/2 past 6 P.M. –	11
Thursdy 8th	All day at Cowfolds Oats dry till 10 A.M. & then at West side R Hillock and turned the Pease	12
Saturdy	Early begun shearing R Hillock Oats wet till past noon then stooked	

10th	them	13
" 12th	Early all people at Oats R H & Gibsons park	14
14th	After Brakefast begun shearing of Oats Backy^{ds} in Gayte all day wet	15
16th	At Noon Set up all down Sheaves & shearing Gayte N B Oats	16
17th	All day at Auld Toun park Oats	17
20th	All day at finishing out Gibsons park and after at R Hillock dry p.M.	18
21st	Early all people shearing dry all R Hillock Wedy Oats & then N Burns sheared after turned ye Pease, & at Waterlair Oats	19
22d Thursdy	After Brakefast begun shearing Out the Thursdy remainder of Waterlair Oats being Whole of the Crop of Season	20
Fridy 23d	Morning, people Cutting Rushes &c for Thatch till Early Dinner when Begun leading last	21
Octr Mondy 26th	Cut of R Hillock Oats-with 3 stacks tolerable a little after 5 A.M. yoked 4 Carts & lead Oats Busily till near 8 P.M.	22
" 27	at Leading till Brakefast Wyling Golds Hillock	
" 29	at Leading of Oats, Wyling & Binding Gayte	23
" 30	Leading Home Oats, & after all Pease 14 Cartful into three Stacks – some Hollow	24
" 31st	Lead in all the Crop but a few Cartfuls by Brakefast-Thatching, Cutting rushes &cJobbs	25
Wedy Novr 4th	Gave the In Gathering Feast and Brought in the Picked & worst of the Oat Crop into a stack – The Crop Consists of –	

Bear	6	Stack large & solid	
Oats	16		D^{o}
Pease	3		
Total	25	–	D^{o}

in a Regular Square of five each in a Handsome Manner without a slant or a sway-and whatever way the Crop may turn Out it is at least as well got in as the Malverse weather would permit and much better I believe than any Barn Yard of any Quantity in the Neighbourhood but there is much reason it will not be Productive to its Bulk, or the Quality as good as it should have been –

APPENDIX 5

		B	f	p
Novr 1st	Barley Made for pott	0	1	2
" 8th	Lofted Oats	1		
" 17th	a stack of Wyling,Oats SB- threshen by W Lyel & own serv^ts the first days Wages	9	3	2
" 23d	A Stack Oats N N Barn on Lott			
finished	ye Hollow one only in Yard			
28th	and ye next worst got in			

Acc^t Hay & Rye Grass Sold Crop 1795

To the East Lothian Fencible Cavalry at Montrose delivered in April 1796 1402 Stones of Hay at 8^ds per Stone is 46-14-8 S^tr which received payment of on 7th May 1796

Jas Badenach

Expences in delivering of same about 1-4-8 S^tr JB

Rye Grass Sold at 16/ per Boll – April 30th 1796

Reith Tipperty	$3\frac{1}{2}$ Bolls	£2-16-0
To – Clark	1 Boll	0-16-0
To George Greig	1 "	0-16-0
James Andrew	$\frac{1}{2}$ Boll	7/ paid
Robt Reid Fordun	1 Boll	
	after it was lofted	16/ paid JB

May 30th Lofted
Rye Grass 9 Bolls & one firlot
put into Short loft JB
June 13th Robt Garland Cairnlaw $\frac{1}{2}$ Boll
" 17th Rev^d Mr Spark $\frac{1}{2}$ "
at 14/ per Boll 7/ paid JB

Whiteriggs November 16 1795
Acount of David Clarks harvest meal
56 dyeats 5 pecks and 1 Lib of meal
Acount of supper meal to Jean Graham
25 nights 3 pecks 1 Lib and 1 peck for Sabbath
David Clark received 3
and supper meal to Jean Graham 4
7 *pecks*

On the reverse of above account was the following:-

David Clark Got Prior to Martinmass			
1795 of Meal	B	f	P
	2	2	3
Meal to be deducted for 56 Meals in Harvest or added to his Account	0	1	1 ½ Lippie
Bolls	3	0	0
Due David Clark at Martinmass	0	1	0

Whiteriggs March 16th 1796 Sold to James Christie Butcher at Arbuthnot 40 Fatt Sheep at 11/ 3[ds] to be lifted in parcels & paid for as lifted & the whole to be taken off in a Month hence –

Delivered the first –	2	March	16th	&	received	in Cash	£2–10–0 pd
Delvd the Second	2	"	26th	&	"	"	0– 8–8 pd
	2	"	31st	&	"	"	1– 4–0 pd
	3	April	7th	&	"	"	1–12–6 pd
	3	"	11th	&	"	"	1–14–0 pd
	3	"	14th	&	"	"	1–15–0 pd
	3	"	21st	&	"	"	2– 0–0 pd
	10	"	23d	&	"	"	5–12–0 pd
Total	28						16 6 2

The following statement was on the reverse of the above account:-

Bro[t] forw[d]	28	Bro[t] forw[d]	16 6 2
May 2d Delivered Jas Christie–	3	More Sheep two of them red marked	£ 1–14–0 pd
" 10th " "	4	Wedders received from	2– 8–0 pd
" " " "	3	of which a Blue Ewe marked	1–10–0 pd
Total price of 38 May 13th	38	& Money received this date	£21–18–2

Sheep at 11/3 is 21-7-6 May 16th delivered 2 Wedders 2 1– 7–6
one of which allowed for the Ewe & Lamb from him, £ 0–10–8
and the Other for the Balance of 10/8 with sixpence
more which I Received in full of whole account settled

Jas Badenach

Shearers Engaged per day at 1/ and at 8[ds]
Septr 5th = 5 Men 4 Women
" 6th – 3 " 4 "
" 7th – 3 " 2 "
being 11/ and 6/8

GLOSSARY

(The *Concise Scots Dictionary* published by Aberdeen University Press should be consulted for further information about Scots words used in the Diary.)

alongst *adv* along
ams weight according to the standard of Amsterdam
award *n* of a crop of grain, *esp* oats grown for the second year on the same land
back, bauk *n* ***1*** an unploughed ridge **2** a ridge still apparent in a field after the bauks were tilled
backsett, backset *n* something which hinders or causes a relapse
ban *n* promise, agreement, contract
band, (in) **braid (or broad)** of corn lying unbound on the harvest field
bear *n* bear, barley; *specif* four- or six-row barley, hardier and coarser than ordinary two-row barley
beastial, bestial *n* domestic animals, livestock
bene, bane *n* bean
benorth to the North of
beet *n* a sheaf or bundle of flax
bent *n* coarse grass
bitt, bit, bite *n* mouthful
blate &c, bleat &c *adj* backward in growth
bole *n* a recess in a wall; a small opening in a wall
boll *n* a dry measure of weight or capacity varying according to commodity and locality, *e.g.* a – **of meal** = 140 lbs (approx 63.5 kg)
bout *n* the extent of ground covered as a plough etc moves across a field
(in)**bout** *of corn or hay* lying in rows after being cut
brake *n* ***1*** the breaking up of a storm, frost, ice etc **2** a layer or deposit of snow
brash *vt* break to pieces, smash; *vi* bring up liquid into the mouth
breer, breard &c *n* the first shoots of grain etc
brew *vt* prepare by mixing, mix; *vi* be in a state of preparation
brew looms *n* bowls or tubs used in brewing
brockley, brocklie *adj* uncertain, hazardous
brocks *n* scraps
buffing *vt* threshing (grain) without untying the sheaf
bus, busk *vt* prepare
buss *n* a thicket, clump of trees
byre *n* a cowshed
cauld *adj* cold; *of land* stiff, clayey
causeway, causey *n* ***1*** a paved area, a roadway **2** the paved or hard-beaten area in front of or around a farmhouse **3** the cobbled part of a BYRE or stable
cess *n* a tax ***1*** the king's or land tax; **2** a local tax
chader, chalder *n* a dry measure of capacity *of grain* 16 Bolls
chows *n pl* small coal, nuts
claggy, cleggy *adj* glutinous, sticky

clat, claut *n* a hoe; an implement for scraping dung, dirt etc
closs, close *n* courtyard, farmyard; *adj of the atmosphere or weather,* stagnant; without motion or ventilation; oppressive; *of work* constant, unremitting
clump *n* cluster of trees or shrubs
cole *n* a haycock *vt* put up (hay) in cocks
colewort *n* a particular species of cole, *Brassica oleracea,* or cabbage
commonty *n* land beyond the head-dyke used for grazing, cutting turves, whins etc owned by one or more proprietors
cottar *n* tenant occupying a cottage with or without land attached to it; a married farmworker who has a cottage as part of his contract
coomb, coom *n* the wooden frame on which an arch (*freq* of a bridge) is built; an arch or vault
couple *n* a pair of rafters, forming a V-shaped roof support; one of these, a principal rafter
craft, croft *n* a smallholding
crowdie *n* oatmeal and water mixed and eaten raw
cunnle- (cannel-) coal *n* a coal so called because it burns with a bright flame like a candle.
dag *n* *1* a thin drizzling rain **2** a heavy shower (of rain)
daggy *adj* drizzling, moist, misty
deal, *n* a share; a certain quantity or measure (of coal)
dibble *n* a pointed instrument used in gardening and agriculture to make holes for planting seeds, bulbs &c
dight, dicht *vt* put in good order, dress; sift or winnow (grain); sift (meal)
divering, daver *vi* stagger
drag *n* a large heavy harrow
dragle, draggle *v* *1 vt* draggle, bedraggle (a garment etc) **2** mix (flour, meal etc) with water *3 vi* move slowly or wearily
draught *n* *1* a load; **2** two or more cartloads brought at one time
draw strae *etc : thatching etc* pull straw through the hands so that pieces short of the required length fall to the ground
dub *n* a pool *esp* of muddy or stagnant water
dust *n* particles of meal and husk produced in grinding corn
ear *vti* plough
easin(s) *n* *1* the eaves of a building; **2** the corresponding part of a haystack
edder *n* a straw-rope used in thatching a haystack *vt* rope (a stack) in order to secure the thatch
fanners *n chf in pl* a winnowing machine, grain-sifter
farrow *adj, of a cow* not in calf; having missed a pregnancy, *locally* either still giving milk or not
farry coo a cow that has had a calf, but failed in a subsequent year
faugh *adj* pale; pale brown, yellowish
feal, fail, fale *n* *1* turf as a material for building or roofing **2** a piece of turf, a sod
firlot *n* the fourth part of a Boll
flake *n* *1* a hurdle or framework of crossed slats, *usu* portable and used as a gate, fence etc **2** *in pl* a temporary pen for sheep or cattle made of such a framework
flan, flam *n* *1* a blast, a storm **2** a gust of wind *vi* blow in sudden gusts
flit *vt* remove, transport from one place to another; die
flower, flour *n* wheaten flour
fog *n* *1* grass left in the field in winter **2** moss, lichen *freq* used as thatching material or for packing walls
fother *n* fodder, food for cattle and horses
frably *adj* friable

fund, found *n* *1* a foundation, base **2** a ring of stones and brushwood on which a haystack is built
furr, fur *n* *1* a furrow made by the plough; the strip of earth turned over in the process **2** the deep furrow or trench separating one RIG from another **3** a deep furrow or rut cut by the plough to act as a drain for surface water
garron[1] *n* a small sturdy type of horse, used *esp* for rough hill work
garron[2] *n* a wooden beam
gather *vt* plough so as to throw the soil into (a central ridge)
gavel *n* *1* gable **2** the part corresponding to a gable in a structure other than a building
gayte, gait *n* a single sheaf of grain tied near the top and set up to dry
girnel *n* *1* a storage chest for meal etc **2** a granary, storehouse; *vt* store (food etc)
gosk, gorsk *n* coarse rank grass produced by cattle droppings in a pasture
grain *vi* yield fruit
graith *n* materials or equipment (for a particular purpose, job or trade)
grew, grue *vi* feel horror or terror, shudder, shrink in horror or fear
guin, gean *n* the wild cherry
guller, goller *n and vi* roar, shout, bawl, yell
guttur, guttar *n* *1* gutter **2** *chf in pl* mud, mire, muddy puddles
haar *n* a cold mist or fog
hacke, heck *n* a rack, a slatted wooden or iron framework *chf specif* a rack for fodder in a stable etc or on a portable frame for use in the open
hagg, hag *n* brushwood; felled wood used for fuel
hain *vti* enclose (grassland or a wood) by a hedge or fence; keep unused, preserve from grazing or cutting
hale *vt* haul, drag, pull (up)
harling *adj of weather* rough
hasterns *n* early-ripening oats or peas
haud *&c vti* hold, keep
haugh *n* a piece of level ground, *usu* alluvial, on the banks of a river, river-meadow land
hay foggage *n* the second growth of grass in a hayfield
heave, heve (**hove, hoven**)*v chf of grazing animals* blown up with having eaten too much fresh green fodder
hengh hang
hirst *n* a barren, unproductive piece of ground, *usu* a hillock, knoll or ridge
hoary *adj* mouldy
hodge *v* shake, quiver *vt* - **up** hitch up, heave
hole *vti* *1* hole **2** dig, excavate; dig up, loosen from the ground *esp* potatoes or turf
holme, haulm *n* the stem or stalk of grain of all kinds, or of pease, beans, etc
hood, huid *n* hood *chf* one of a pair of sheaves of corn placed on the top of a STOOK or corn stack as a protection against the weather
hose *n* the sheath enclosing an ear of corn
hough *vti* cut with a hoe, dig, dress
hove *vt* cause to swell, distend *vi* become swollen or distended, swell, expand
hudge *vt* amass, heap up
hull, hill, *n* *l* a hill **2** (1) a common moor where rough grazing rights are shared by the community (2) any piece of rough grazing on a farm
hutt, het, hot *adj* hot
hutt, huit *n* a small stack of corn or hay etc built to protect the crop temporarily from the weather before its removal to the stackyard
in holding (fuel) fuel constantly held in the fire *e.g.* gorse, thorn
Indian Cress a plant *Tropoeolum majus* , a favourite garden flower

kirn *n* a churn
knit *of joiner's or smith's work* secured by a joint, welded
landing *n* the journey of a plough from one side of a field to another and back again, a BOUT
laigh, laich *adj* low
lay *vt* re-steel (a plough-iron etc)
leap *vi, of frost or snow* thaw
ley. lea, lay *n* ***1*** lea, ground left untilled; ground once tilled but now in pasture **2** *specif* second-year or older pasture following hay
lodge *vi* be or lie **on** (the ground etc), **in** (a place)
lot, lott, *n* an allowance of corn paid to the thresher as part of his fee **-man** a corn-thresher
malverse *n* a breach of trust, a piece of grave misconduct
masloch, mashly *n* ***1*** mixed grain **2** bread made from mixed meal
meslin, maslin *n* a mixture of different sorts of grain, as of wheat and rye
midding, midden *n* ***1*** a dunghill, compost heap, refuse heap **2** a domestic ash-pit
milch *adj* giving milk; as a *milch* cow
miln *n* mill
mizerign Mezereon *Daphne mezereum* deciduous bush with deep rosy-purple or white flowers
mould, muild *n* ***1*** mould, earth etc **2** soil broken up in the process of cultivation, loose earth, lumps of earth
offer *vti* try
Old Candling Candlemass Day Old Style *i.e.* 13th February
O.S. Old Style Calendar before the introduction in 1752 of the New Style when Wednesday, 2nd September was followed by Thursday 14th September
Osnaburg *n* a species of coarse linen cloth, originally made at and imported from Osnaburg in Germany
overhead, owerheid *adv* in gross, overall
oyr other
peat-stane *n* a gable stone supporting a coping-stone; a coping-stone; the keystone of an arch
pinn *vti* to stop up a small hole; commonly applied to masonry
plenish *vt* furnish (a house etc)
Quarter Ill or Black-quarter an enzootic apoplectic disease peculiar to cattle, indicated by lameness of the fore-foot and blackness of the flesh
quick *n* a live fence or hedge formed of some growing plant, usually hawthorn; quickset
quicken *n chf in pl* couch grass
rank *adj* ***1*** rank **2** stout, strong
raw *adj* ***1*** raw **2** *of corn-sheaves* damp, not fully dried out
redd, red, rid &c *vt* ***1*** clear (a space, a passage) **2** clear (land) by reaping, ploughing etc **3** clear out (a ditch, channel etc), remove rubbish or silt from
redland *n* land cleared of its crop, bare after cropping or ploughing
rest *vt* cover or damp down (a fire) for the night
rib ploughing plough every alternate furrow, turning the soil over onto the adjacent unploughed strip
ribwort, rib-grass *n* plant of the genus Plantago, the *P. Plantagineae*
rick; reek &c *n* smoke
rig *n* a ridge of land; a strip of land between two furrows
rigging stones *n* stones used as ridge stones of a roof
risk, reesk &c *n* a piece of moor or marshy ground covered with natural grass
riskue, reeskie &c *of ground* having a REESK

roup *n* a sale by public auction
rubbage *n* rubbish
ruck *n* a hay- or corn-stack of a standard shape and size
ruffle *vti* disorder, disarrange
sadden *vt* cause to sink or settle down; beat down; pack hard
scath, skaith *vt* scathe, injure
scrimp *adj* scant *adv* scarcely, almost but not quite
scrimple *vt* shrivel with heat; crumple, crinkle
set *1 vi and in passive, of plants and animals* stop growing, have the growth checked **2** *vtr* let by contract, lease
shade *chf* **sheed** *n* a strip of land, a distinct or separate piece of ground
shade, &c *n* *1* shade **2** a shed, a roofed structure used as a shelter or store
shade *vt* shed, separate
shirp *vi* wither
shirpy thin, shrunken
shel, shell *n* *1* shell **2** a scale of a balance
shock *n* a STOOK
sholt, shelt &c a Shetland pony; also applied to any pony, *usu* a GARRON
shoot or **shot blade** *n* the leaf enclosing the corn-stalk and ear
shot, shott *n* a young pig after weaning
siver, syver *n* a covered drain; a sewer; a gutter
skimmering a light sprinkling (of snow)
slap, slop *vt* *1 building* make a gap or break in (a wall etc) or for (a door, window etc) **2** thin out (seedlings &c) **3** beat, thresh
slawky *adj* slimy
smithy &c smiddy &c *n* smithy **-coom, -culm** a small smokeless type of coal suitable for smith's work
snod *vt* level
sock *n* a ploughshare
sole *n* *1* sole **2** a sill, a supporting or strengthening beam, *esp* in a window or door-case
sottle *vi* **sotter** saturate, soak
sour *vti* *1* sour **2** *esp of water on lime* macerate, soften, slake
sow, sowe, soo *n* *1* sow **2** a ridge-shaped mass, *e.g.* a large oblong stack of hay or straw
sprot *n* a rush, the reedy plant
stale *n* the foundation, made of a layer of stones, brushwood etc, on which a corn- or hay-stack is built
stanners *n pl* shingle
stannery *adj* stony, gravelly
startle *vi* rush about wildly, stampede, *esp* of cattle when stung by flies
stathel *n* *1* the foundation of a stack of grain, built of stone, wood etc, to protect from vermin and damp **2** the main part of a corn-stack; a stack in the process of building or dismantling
steer, steir, stir &c *vti* *1* stir *vt* **2** plough, *specif* replough in spring (land already ploughed in the autumn)
stint *vi of plants* shrivel, droop
stock *vti* *1* stock **2** *of plants* send out shoots, sprout
stook *n* a shock of cut sheaves, *usu* ten or twelve set up to dry in a field
stot, stott *n* a young castrated ox, a bullock, *usu* in its second year or more
strake, straik &c *n* *1* stroke, a blow **2** the motion or marks of a harrow; the ground covered by one journey of a harrow
strick, streek *vt* stretch

streight, straucht, straight &c *vt* make straight, straighten; smooth, set to rights
stuble *n* stubble
sway, swey *vti* ***1*** sway **2** (1) make to sway or swing, move (an object) to one side (2) press down, bend to one side
swingle-tree *n* a cross-bar by which horses are attached to a plough or carriage
tack *n* a lease, tenancy, *esp* the leasehold tenure of a farm etc
tame *n* team
tatch, tash *n* a rope or strap *esp* one for tying bundles of flax
tath, tathe *n* ***1*** the dung of cattle or sheep left for manure on their grazing land **2** coarse rank grass which grows on ground thus manured
teethe *vt* point (a wall etc) with mortar
threave *n* a measure of cut grain, straw, reeds or other thatching material, consisting of two STOOKS, *usu* with twelve sheaves each but varying locally
ting *vt, also vi (of cattle)* stuff (oneself etc) to the point of acute discomfort
tire *n* tier, a row, *specif* of stones or turf on a wall
town, toun *n* a farm with its buildings and the immediately surrounding area
triffling *adj* trifling, trivial, being of small value or importance
turlies, tirlies *n* ***1*** trellis **2** a grid or rack for drying articles
tirr *v* ***1*** *vt* take the top layer off (a piece of ground) **2** *also* **turr** strip or tear off (a covering), *esp* thatch, roofing etc
tramp cole a cock of hay compressed by *tramping*
tumuli *n pl* hillock
turn *vti* ***1*** turn **2** turn (cut hay, peats etc) to dry; dismantle and rebuild (a small stack etc of such) for drying
tussilago *n* Colt's-foot, a genus of broad-leaved plants
vinca *n* periwinkle
victual *n* ***1*** victual **2** corn, grain, *occas also* leguminous crops; a crop before or after harvesting
vulgo *adv* commonly, popularly
wan, wane *adj* lacking, deficient, misshapen
water furr, fur *see* FURR
wedder *n* wether, castrated ram
wile, wyle, wale *vt* ***1*** choose, select, pick out **2** arrange, separate into lots
win *vti* ***1*** *of cut corn, peats, hay etc* dry and make or become ready for storage by exposure to air and sun **2** dry out, season (wood, cheese etc)
windlestraw *n* a tall, thin, withered stalk of grass
winrae wind-raw a row or line into which mown hay is raked or in which small piles of cut peats are set to dry
yeld, yield *adj* ***1*** *of animals* barren, not bearing young because of age or accident **2** *of cows etc* not yielding milk because of age or being in calf
yoking, yokin ***1*** yoking (horses etc) **2** the starting of a spell of work **3** (1) the period during which a team of horses or oxen is in harness at one stretch, *usu* half a day's work (2) *in gen* a spell of work, a stint, shift

CAPACITY

Dry measure according to the standard of LINLITHGOW

1. For wheat, peas, beans, meal etc.

1 LIPPIE (OR FORPET)	.499 gallons	2.268 litres
4 lippies = 1 PECK	1.996 gallons	9.072 litres
4 pecks = 1 FIRLOT	3 pecks 1.986 galls	36.286 litres

4 firlots = 1 BOLL	3 bushels 3 pecks 1.944 galls	145.14 litres
16 Bolls = 1 CHALDER	7 quarters 7 bushels 3 pecks 1.07 galls	2322.32 litres

2. For barley, oats, malt

1 lippie (or forpet)	.728 gallons	3.037 litres
4 lippies = 1 peck	1 peck .912 gallons	13.229 litres
4 pecks = 1 firlot	1 bushel 1 peck 1.650 galls	52.916 litres
4 firlots = 1 boll	5 bushels 3 pecks 600 galls	211.664 litres
16 bolls = 1 chalder	11 quarters 5 bushels 1.615 galls	3386.624 litres

FIELD NAMES

and cultivated areas at Whiteriggs

Auld Toun Park *see* Old Town Park

Back Waters, Back Yards, Bank, Black Lands, Bridge Shade or Bridge of Leppie Shade, Brownies Haugh and leys, Burns (Long, Nether, Over or Upper).

Cottar Bank, Cottar Land, Shade, or Yards, Cowfolds or Cow Wards, Craig Braes or Craig Shade.

Floors

Gallownere

Gibson's Park, Gold's or Gould's Hillock

Haugh (Award, Dickies, East, Fallow, Ley).

Kavels

Loup Yards

North Park

Old Town Park or Old Town Hillock Park or Old Park

Red Hall Park, Rob's Hillock Park

Shepherd's Haugh, Smithy Park

Thornylee (Part leased to another farmer)

Waterlair (Part leased to another farmer)

SUGGESTED FURTHER READING

W. Cramond, *The Annals of Fordoun*, Montrose 1894.

A. Fenton, *Scottish Country Life*, Edinburgh 1976, reprinted 1977. A full bibliography is included.

A. Fenton, *Country Life in Scotland – Our Rural Past*, Edinburgh 1987

J. E. Handley, *Scottish Farming in the Eighteenth Century*, London 1953. A full bibliography is included.

Rev. J. Headrick, *A General View of the Agriculture of the County of Angus or Forfarshire*, 1813.

G. Robertson, *A General View of the Agriculture of Kincardineshire*, 1813.

Crabb Watt, Q.C., *The Mearns of Old*, 1914.

FORFAR SHIRE
PARISH of
Part of
EDZELL PARISH
FETTERCAIRN
Fasque
Balnakeite
Balbegno
Gannachy Bridge
Arnhall
Mill of Arnhall
Kings Yord
North Water Bridge
Shank of Finggey
Rough Bank
Ruins of Auchmull Castle
Tongha
Chattering Bridge
No. of INHABITANTS & Contents of the different Parishes
Strachan
Banchory Ternan
Part of Drumoak
Durris
Maryculter
Part of Banchory Devinich
Nigg
Fetteresso
Dunnottar
Kinneff & Caterline united
Arbuthnot
Glenbervy
Fordun
Fettercairn
Part of Edzell
Marykirk
Conveth
Garvock
Bervy
Benholm
St Cyrus
Sum
Altitude of the most remarkable Hills
from the Sea Yards
Kermancairn 340
Korleack 630
Klochnaben 790
Mountbattock 1150
Note.
In the Parish of St Cyrus the Lette
W.M. denote Lauriston Woodston M
& the Fields in which these Letters
serted belong respectively to these Esta
In the Parish of Marykirk the Let
M.D.B.G. denote Middleton Arnhall
sides Balbegno & Gallery for Thor
& the Fields so marked are the de
Tacts of Luthermoor that fell to the
of these Estates in the Division.